Revolutionizing Energy Conversion - Photoelectrochemical Technologies and Their Role in Sustainability

Edited by Mahmoud Zendehdel,
Narges Yaghoobi Nia and Mohamed Samer

Published in London, United Kingdom

Revolutionizing Energy Conversion - Photoelectrochemical Technologies and Their Role in Sustainability
http://dx.doi.org/10.5772/intechopen.1001665
Edited by Mahmoud Zendehdel, Narges Yaghoobi Nia and Mohamed Samer

Contributors
Xiaodan Wang, Beibei Wang, Hao Shen, Krishanu Bandyopadhyay, Abhineet Verma, Tamal Ghosh, Ravi Kumar Kanaparthi, Sudeena Nadendla, Satyen Saha, Mina Ahmadi-Kashani, Mahmoud Zendehdel, Mohammad Mahdi Abolhasani, Narges Yaghoobi Nia, Ife Elegbeleye, Eric Maluta, Regina Maphanga, Michael Walter, Oliver Brügner, Yasuhisa Maeda, Canhan Sen, Berk Alkan, Omid Mohammadmoradi, Alpay Taralp, Amit Kumar Verma, Prerna Tripathi, Akhoury Sudhir Kumar Sinha, Shikha Singh, Farai Dziike, Ryan Yow Zhong Yeo, Krishan Balachandran, Irwan Ibrahim, Mimi Hani Abu Bakar, Manal Ismail, Wei Lun Ang, Eileen Hao Yu, Swee Su Lim, Hajar Rajaei Litkohi, Hosein Yazdi Dehnavi, Ali Shirpay, M. Azizul Moqsud

First published in London, United Kingdom, 2024 by IntechOpen
IntechOpen is the global imprint of INTECHOPEN LIMITED, registered in England and Wales, registration number: 11086078, 167-169 Great Portland Street, London, W1W 5PF, United Kingdom

British Library Cataloguing-in-Publication Data
A catalogue record for this book is available from the British Library

Additional hard and PDF copies can be obtained from orders@intechopen.com

Revolutionizing Energy Conversion - Photoelectrochemical Technologies and Their Role in Sustainability
Edited by Mahmoud Zendehdel, Narges Yaghoobi Nia and Mohamed Samer
p. cm.
Print ISBN 978-0-85466-881-6
Online ISBN 978-0-85466-880-9
eBook (PDF) ISBN 978-0-85466-882-3

For EU product safety concerns:
IN TECH d.o.o., Prolaz Marije Krucifikse Kozulić 3, 51000 Rijeka, Croatia,
info@intechopen.com or visit our website at intechopen.com.

Meet the editors

Dr. Mahmoud Zendehdel is an Assistant Professor of Chemistry and Nanotechnology with a Ph.D. in Inorganic Chemistry. He has more than 18 years of research and industrial experience in electrochemistry, photochemistry, materials science, and nanotechnology. Dr. Zendehdel has coordinated several high-impact projects in areas such as natural gas desulfurization, methane conversion, and smart energy systems, leading to fourteen patents in advanced energy technologies and materials. He is currently the RDI Manager at Iritaly Trading Company S.r.l., where he is managing the "NEXTCCUS" EU project (ERA-NET ACT 2021) and CHOH-D (semi-industrial Italian project), a scientist at the University of Rome Tor Vergata, and CEO of Kimia Solar. His work focuses on innovative solutions in perovskite solar modules and photonic devices.

Dr. Narges Yaghoobi Nia holds a Ph.D. in Electronics Engineering and is a two-time winner of the prestigious Marie Skłodowska-Curie Fellowship, awarded in 2014 and 2022 for her work on perovskite technologies for space. She is the founder of the PVSPACE Organisation and currently serves as Principal Investigator of the P4SPACE project at the École Polytechnique Fédérale de Lausanne (EPFL), Switzerland, and the Aerospace Engineering School, Sapienza University of Rome, Italy. With more than 12 years of experience in emerging thin-film photovoltaics and electrochemistry, she has published more than fifty scientific papers and holds three patents. Dr. Yaghoobi Nia is also Editorial Board Manager of *Communications Engineering* in the Nature Portfolio and actively contributes to several EU-funded projects.

Prof. Dr. Mohamed Samer is laureate of the 2019 Cairo University Scientific Excellence Award and recipient of the 2017 Privilege Medal of First Class, issued by the Egyptian President. He is also laureate of the 2016 State Encouragement Award, conferred by the Egyptian Government. Dr. Samer obtained his Ph.D. from the University of Hohenheim, Germany. He has led nineteen research projects and established five research laboratories. He has 140 journal publications and 20 patents to his credit. Dr. Samer is a member of thirteen scientific societies and a peer reviewer for thirty high-impact journals. He is also a reviewer for seven research funding agencies.

Contents

Preface **IX**

Section 1
Advanced Photoelectrochemical Technologies **1**

Chapter 1 **3**
Advanced Hydrogenation Technologies for Photoelectrochemical Hydrogen Production
by Xiaodan Wang, Beibei Wang and Hao Shen

Chapter 2 **19**
Illuminating Advances: Photochemistry and Photophysics of N-Heterocyclic Carbenes (NHCs) and Its Structural Correlation
by Krishanu Bandyopadhyay, Abhineet Verma, Tamal Ghosh, Ravi Kumar Kanaparthi, Sudeena Nadendla and Satyen Saha

Chapter 3 **55**
Perovskite Paradigm: Unraveling Photoelectrochemical Synergies for Sustainable Transformations
by Mina Ahmadi-Kashani, Mahmoud Zendehdel, Mohammad Mahdi Abolhasani and Narges Yaghoobi Nia

Chapter 4 **87**
Photoanode Applications of Polyene-Diphenylaniline Dyes Molecules Adsorbed on TiO_2 Brookite Cluster
by Ife Elegbeleye, Eric Maluta, Regina Maphanga, Michael Walter and Oliver Brügner

Chapter 5 **105**
Photoelectrochemical Process on Semiconductor Electrodes
by Yasuhisa Maeda

Chapter 6 **125**
Polymer Dispersed Liquid Crystal Smart Film Technologies: Overview
by Canhan Sen, Berk Alkan, Omid Mohammadmoradi and Alpay Taralp

Chapter 7 167
Sustaining Sub-Bandgap Photons *via* Upconversion for Solar Splitting Cells
by Amit Kumar Verma, Prerna Tripathi, Akhoury Sudhir Kumar Sinha and Shikha Singh

Chapter 8 185
The Two-Step Innovative Smart Energy Storage: Photoelectrochemical Materials for the Fabrication of High-End, High-Efficiency Smart Energy Storage Devices
by Farai Dziike

Section 2
Microbial Fuel Cells (MFCs) 203

Chapter 9 205
3D Printing in Microbial Fuel Cell
by Ryan Yow Zhong Yeo, Krishan Balachandran, Irwan Ibrahim, Mimi Hani Abu Bakar, Manal Ismail, Wei Lun Ang, Eileen Hao Yu and Swee Su Lim

Chapter 10 219
Microbial Community
by Hajar Rajaei Litkohi and Hosein Yazdi Dehnavi

Chapter 11 239
Operational Principles of MFCs
by Ali Shirpay

Chapter 12 257
Plant Microbial Fuel Cell
by M. Azizul Moqsud

Preface

The global energy landscape is at a critical juncture, where the quest for sustainable and efficient energy conversion technologies is paramount. This book, *Revolutionizing Energy Conversion – Photoelectrochemical Technologies and Their Role in Sustainability*, addresses this imperative by exploring the cutting-edge developments in photoelectrochemical (PEC) technologies, which hold the potential to transform the way we generate and utilize energy.

PEC systems have emerged as a promising solution in the ongoing efforts to mitigate climate change and reduce our dependence on fossil fuels. This book delves into the latest research and innovations in PEC technologies, providing a comprehensive overview of their potential applications in sustainable energy conversion. Our aim is to present a well-rounded collection of chapters that not only highlight the scientific advancements in this field but also emphasize the practical implications of these technologies for a sustainable future.

The first section of this volume, "Advanced Photoelectrochemical Technologies", comprises eight chapters that explore a wide range of PEC systems and their applications. We begin with Chapter 1, "Advanced Hydrogenation Technologies for Photoelectrochemical Hydrogen Production", which provides an in-depth exploration of the latest advancements in hydrogenation processes. This chapter focuses on enhancing the efficiency of PEC hydrogen production, a crucial step toward realizing a hydrogen-based energy economy.

Following this, Chapter 2, "Illuminating Advances: Photochemistry and Photophysics of N-Heterocyclic Carbenes (NHCs) and Its Structural Correlation", delves into the photochemical and photophysical properties of N-Heterocyclic Carbenes (NHCs). This chapter highlights the structural Chapter 3, "Perovskite Paradigm: Unraveling Photoelectrochemical Synergies for Sustainable Transformations", presents a detailed examination of perovskite materials, which have gained significant attention due to their remarkable PEC properties. This chapter discusses the potential of perovskites in driving sustainable energy transformations, offering new perspectives on their role in the future of energy conversion.

In Chapter 4, "Photoanode Applications of Polyene-Diphenylaniline Dyes Molecules Adsorbed on TiO_2 Brookite Cluster", the focus shifts to the application of polyene-diphenylaniline dyes on TiO_2 brookite clusters. This chapter explores the potential of these materials in enhancing the performance of photoanodes, a key component of PEC cells.

Chapter 5, "Photoelectrochemical Process on Semiconductor Electrodes", provides a comprehensive overview of the processes occurring on semiconductor electrodes. This chapter discusses the critical role of semiconductor materials in PEC systems and their implications for the efficiency of energy conversion technologies.

Chapter 6, "Polymer Dispersed Liquid Crystal Smart Film Technologies: Overview", offers an in-depth review of polymer-dispersed liquid crystal smart films. These materials represent a novel approach to energy conversion and smart technology applications, and this chapter highlights their potential in driving innovation in PEC systems.

In Chapter 7, "Sustaining Sub-Bandgap Photons via Upconversion for Solar Splitting Cells", the authors explore the concept of upconversion for sub-bandgap photons. This chapter examines how this process can be harnessed to enhance the efficiency of solar splitting cells, a crucial development in the pursuit of efficient solar energy conversion.

The first section concludes with Chapter 8, "The Two-Step Innovative Smart Energy Storage: Photoelectrochemical Materials for the Fabrication of High-End, High-Efficiency Smart Energy Storage Devices". This chapter discusses the development of high-efficiency smart energy storage devices, emphasizing the innovative use of PEC materials in these technologies.

The second section of this volume, "Microbial Fuel Cells (MFCs)", presents a series of chapters dedicated to microbial fuel cells, an emerging technology with significant potential for sustainable energy production. The section begins with Chapter 9, "3D Printing in Microbial Fuel Cells", which examines the role of 3D printing in optimizing MFCs for practical applications.

Chapter 10, "Microbial Community", follows with a discussion on the composition and function of microbial communities within MFCs. This chapter provides insights into how these communities influence the performance and efficiency of MFCs.

Chapter 11, "Operational Principles of MFCs", offers a thorough review of the fundamental principles governing the operation of MFCs. This chapter serves as a foundation for understanding the technical aspects of these systems and their potential applications.

Finally, Chapter 12, "Plant Microbial Fuel Cell", explores the innovative integration of plant systems with MFCs. This chapter highlights the potential of this hybrid technology to contribute to sustainable energy production, offering a unique perspective on the future of MFCs.

The completion of this book would not have been possible without the dedication and hard work of many individuals. I would like to express my deepest gratitude to my co-editors, Dr. Narges Yaghoobi Nia and Prof. Dr. Mohamed Samer, for their invaluable contributions and insights throughout the editorial process. I would also like to thank the authors of the individual chapters for their exceptional work and dedication to advancing the field of photoelectrochemical technologies.

I am also grateful to the reviewers for their meticulous evaluations and constructive feedback, which have greatly enhanced the quality of this volume. Special thanks are due to our editorial assistants, whose behind-the-scenes efforts have been instrumental in bringing this book to fruition.

As we present this volume to the scientific community, we hope that it serves as a valuable resource for researchers, engineers, and policymakers alike. The advancements in PEC technologies and microbial fuel cells discussed in this book represent a significant step forward in our collective efforts to achieve a sustainable energy future.

Mahmoud Zendehdel
Head of Research and Innovation Office,
Iritaly Trading Company S.r.l.,
Rome, Italy

Narges Yaghoobi Nia
Laboratory of Photonics and Interfaces,
Ecole Polytechnique Fédérale de Lausanne,
Lausanne, Switzerland

Mohamed Samer
Cairo University,
Cairo, Egypt

Section 1

Advanced Photoelectrochemical Technologies

Chapter 1

Advanced Hydrogenation Technologies for Photoelectrochemical Hydrogen Production

Xiaodan Wang, Beibei Wang and Hao Shen

Abstract

Since it was first reported at Science in 2011, the hydrogenation technique to prepare black TiO_2 has received great attention worldwide. However, most of the reported hydrogenation techniques require harsh conditions and/or high costs that seriously limit their practical applications. In response to overcome the above technical bottleneck, two advanced hydrogenation technologies, low temperature hot wire H hydrogenation and room temperature H^+ hydrogenation, are developed. The chapter focuses on the two technologies to hydrogenate TiO_2 nanorods achieving the highest photocurrent density of black TiO_2 reported to date under photoelectrochemical (PEC) hydrogen production conditions, while simultaneously addressing issues like high temperature, high pressure and prolonged treatment as well as chemical residues with traditional hydrogenation approaches. Comparisons of the proposed technologies with conventional methods are conducted: Our advanced hydrogenations demonstrate more efficient and low-cost technologies beyond those of state-of-art hydrogenations, enabling them to move from basic research to large-scale practical application.

Keywords: black TiO_2, hydrogenation technologies, low temperature hot wire H hydrogenation, room temperature H^+ hydrogenation, photoelectrochemical hydrogen production

1. Introduction

The "green hydrogen" prepared by photoelectrochemical (PEC) water splitting absorbs sunlight through semiconductor photoanode, generates photogenerated electrons and holes to split water into hydrogen and oxygen, and efficiently converts solar energy into hydrogen energy for storage, which has great scientific significance and broad application prospects [1, 2].

Metal oxide semiconductors have been widely studied due to their excellent PEC stability, low cost, favorable band-edge position and bandgap, among which the most common oxide semiconductors such as Fe_2O_3 and $BiVO_4$ have suitable bandgaps, but due to their short hole diffusion length, the solar energy conversion efficiency is low, which seriously limits their practical applications. Among many semiconductors photocatalytic materials, TiO_2 has become one of the most promising semiconductor photocatalysts because of its stable chemical properties, strong oxidation-reduction and low cost, but it is a wide bandgap semiconductor material that can only absorb ultraviolet light in the solar spectrum, so improving the light response of TiO_2 to visible light is a research hotspot in this field. In recent years, in order to improve the response of TiO_2 to visible light, the surface doping of various elements has been reported, but the photocurrent is less than 1 mA cm^{-2} [1]. Therefore, new doping technology to improve the response of TiO_2 to visible light for enhanced solar-to-hydrogen conversion efficiency is highly demanded [1–10].

2. Experimentation with different hydrogenation technologies

The acquisition of black TiO_2 by high-pressure hydrogenated TiO_2 has attracted great interest [11]. In 2011, American scientists Xiaobo Chen and Samuel Mao published a study in the journal Science detailing their novel method for synthesizing black TiO_2 through high-pressure hydrogen treatment. **Figure 1** illustrates the distinctive core-shell structure of black TiO_2 and the corresponding energy diagram depicting a reduced band gap. The researchers confirmed that black TiO_2 possesses a broader band for light absorption, enabling it to absorb not only visible light but also ultraviolet and infrared light. Furthermore, black TiO_2 exhibits exceptional photocatalytic properties. Since its discovery, black TiO_2, with its combination of high sunlight absorption and ultra-high photocatalytic capabilities, has garnered significant attention in the scientific community.

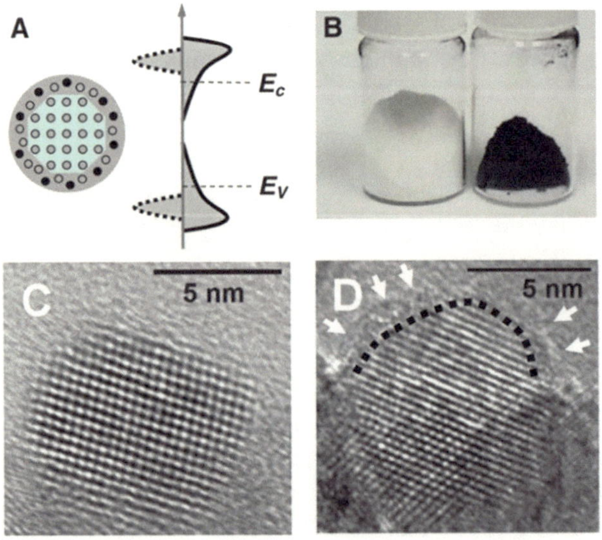

Figure 1.
(A) Core-shell crystal structure and narrowing band gap model; (B) photo: Hydrogenation of white pristine TiO_2 to black TiO_2; (C) and (D) high-resolution transmission electron microscope (HR-TEM) images of pristine TiO_2 and black TiO_2, respectively, confirming the core-shell structure of black TiO_2 by comparison.

Compared with the traditional elemental doped TiO_2, the hydrogenation method of hydrogen doping has a unique advantage: Because it causes the original crystal structure to be distorted without destroying the original structure, resulting in a new microstructure and electronic structure, which can change its band gap, conductivity and electrochemical properties, and has a wide range of application prospects [12]. Then, most of the hydrogenation methods reported in the existing literature, such as high-temperature hydrogen annealing and hydrogen plasma annealing, require high temperature or high pressure, long treatment time, high dose of H_2 or H^+ and other harsh conditions. In order to easily and economically prepare black TiO_2 with disordered shells, the researchers also developed a chemical reduction method. However, due to surface defects and chemical residues, the PEC activity of black TiO_2 is relatively low, and the typical photocurrent density is less than 2 mA cm^{-2} [13]. In the face of this bottleneck, we have invented two advanced hydrogenation methods, low temperature hot wire H hydrogenation and room temperature H^+ hydrogenation [14, 15], which successfully overcome not only the bottlenecks of the traditional hydrogenation process such as high temperature, high pressure and high energy consumption but also promoted the development of the world's hydrogenation field, so that the hydrogenation technology can truly move from basic research to large-scale practical application.

3. Attempts with advanced H and H^+ hydrogenation approaches

3.1 H hydrogenation

3.1.1 Work principle, experimental set-up and H treatment

Molecular hydrogen (H_2) has been experimentally shown to undergo catalytic dissociation into atomic hydrogen (H) when in contact with tungsten wires heated to a temperature exceeding 1000°C. Langmuir conducted research on the efficiency of tungsten wires in facilitating the dissociation of H_2 into H atoms in 1912 [16]. The ability of hot wires to dissociate hydrogen is effectively utilized in hot wire activated chemical vapor deposition (HWCVD) processes. The process of HWCVD is fundamentally distinct from plasma enhanced chemical vapor deposition (PECVD) due to the absence of hydrogen ions or energetic particles in the generation of atomic hydrogen H in HWCVD [14].

Inspired by the above previous work, we developed a low-temperature H hydrogenation method for the first time and studied the effect of hydrogenation parameters on enhanced PEC performance of hydrogenated TiO_2 photoanode materials [14].

Figure 2 illustrates the experimental set-up of low-temperature H hydrogenation method. A parallel array consisting of 10 tungsten wires, each 600 mm in length and 0.53 mm in diameter, was installed within a vacuum chamber. The spacing between the wires was set at 50 mm, resulting in a total activated area of 600 × 450 mm^2. The samples underwent treatment at varying wire temperatures (1600, 1700 and 1800°C), with a separation of 75 mm between the wires and the sample. Throughout the 20 mins H treatments, the sample temperature rises to 245, 265 and 290°C at T_{wire} = 1600, 1700 and 1800°C, respectively.

The flux rate of H hydrogen can be accurately controlled and adjusted by the wire temperature, as it is dependent on both the wire temperature and pressure, which the

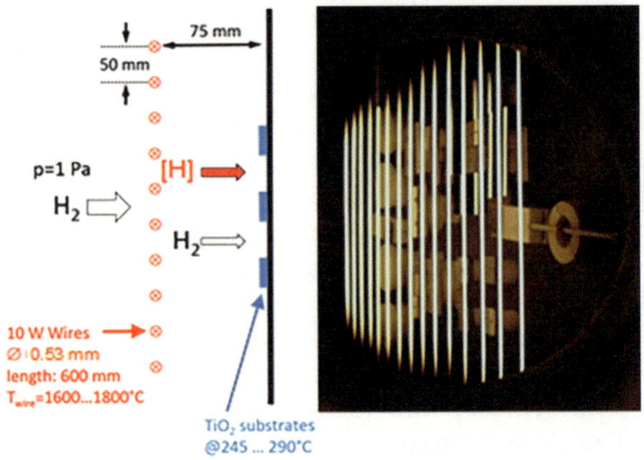

Figure 2.
Experimental set-up of low-temperature H hydrogenation.

pressure is usually preset at 1 Pa controlled by a gate valve. The hydrogen treatments, encompassing all processing parameters such as wire temperature, pressure, hydrogen flux and time, were meticulously regulated by a programmable recipe, enabling reproducible hydrogenations on TiO_2 nanorods.

In the subsequent sections, we have conducted a systematic investigation on the impact of wire temperatures of 1600, 1700 and 1800°C on the resulting structural, optical and PEC properties of H-treated TiO_2 nanorods.

3.1.2 Influence of wire temperature on the structural, optical, electrical and PEC properties

The investigation of the microstructure of H-TiO_2 nanorods was conducted using scanning electron microscopy (SEM) and high-resolution transmission electron microscopy (HR-TEM). The morphology of H-TiO_2 treated at a wire temperature of 1700°C closely resembles that of pristine TiO_2. The TEM and HR-TEM images indicate that the pristine TiO_2 exhibits a single-crystalline structure, whereas H-TiO_2 treated at a wire temperature of 1700°C composed of a core-shell structure which the thickness of the disordered shell is around 2 nm (**Figure 3a–f**). To investigate compositional changes of resulting disordered shell, the [O]/[Ti] ratio was determined through EELS analysis of Ti-$L_{2,3}$ and O-K spectra. It indicates that the [O]/[Ti] ratio of pristine TiO_2 remains constant at approximately 2 throughout the core and surface of the TiO_2 nanorod, whereas the [O]/[Ti] ratio of H-TiO_2 at a wire temperature of 1700°C is ca. 2 in the core region and decreases gradually in the disordered region (**Figure 3g-i**).

The optical absorption spectra of pristine TiO_2 and H-TiO_2 were analyzed in order to investigate the improved optical absorption properties of H-TiO_2. The results presented in **Figure 4a** indicate a red-shift in the band edge of H-TiO_2 as wire temperatures are elevated. Additionally, within the visible spectrum ranging from 420 to 850 nm, the absorption of H-TiO_2 demonstrates a gradual increase with rising wire temperatures. Further, Tauc plots depicted in **Figure 4b** illustrate that the band edge of H-TiO_2 is decreased to 2.9 eV relative to pristine TiO_2, which has a band edge of 3.0 eV, confirming the band gap narrowing of H-TiO_2.

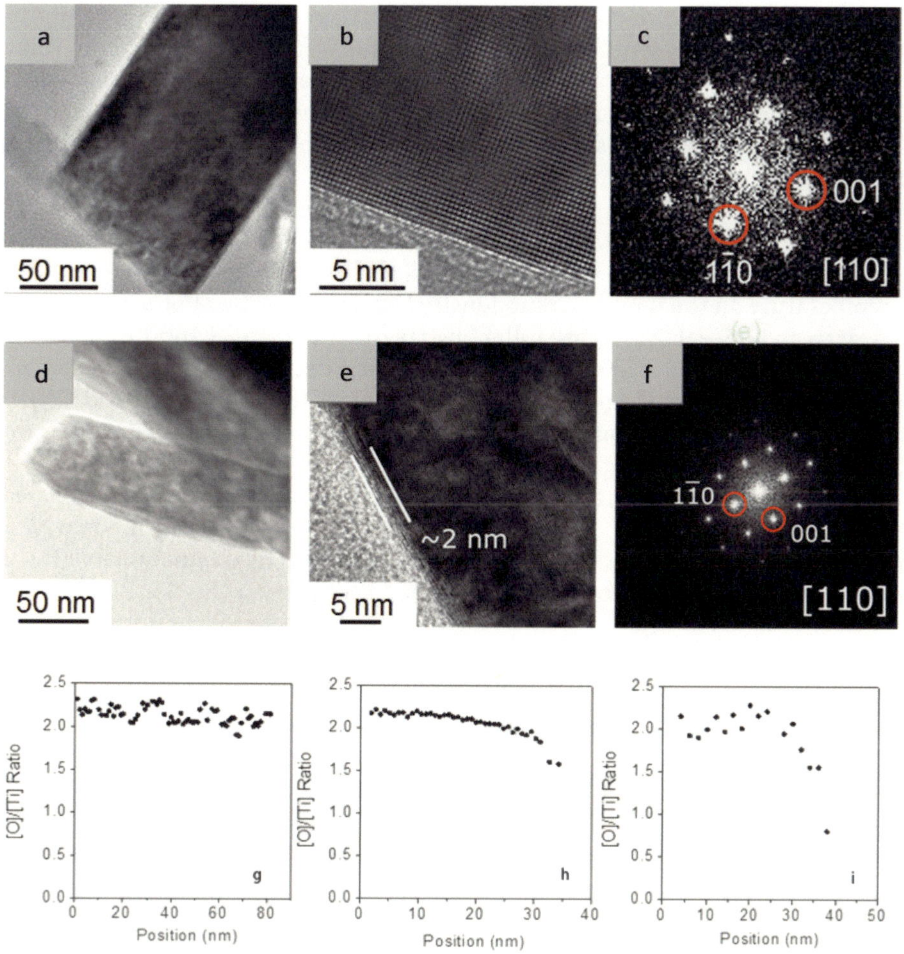

Figure 3.
TEM, HR-TEM images and fast Fourier transform (FFT) patterns of (a–c) pristine TiO_2 and (d–f) H-TiO_2 nanorods treated at T_{wire} = 1700°C. (g–i) the corresponding [O]/[Ti] ratios obtained from the quantification of the acquired electron energy loss spectroscopy (EELS) spectra.

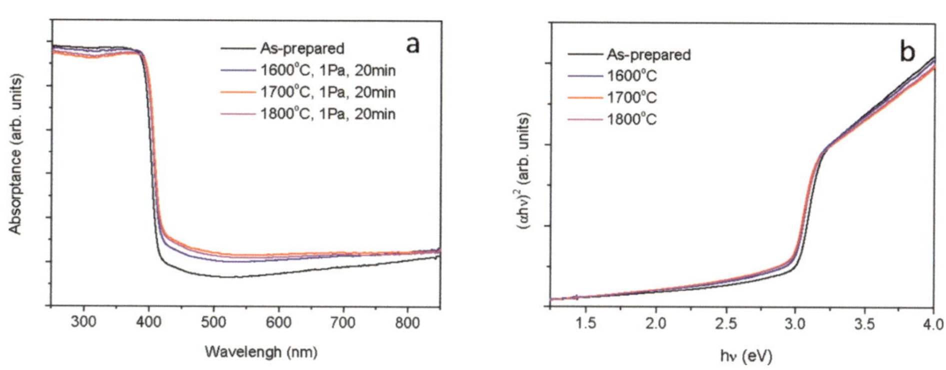

Figure 4.
(a) Optical absorption and (b) Tauc plots of optical absorption curves for pristine TiO_2 and H-TiO_2 nanorods treated at T_{wire} = 1600, 1700 and 1800°C, respectively.

In order to investigate the improved PEC properties of H-TiO_2 nanorods, photocurrent density versus potential (J-V) measurements were conducted using a three-electrode electrochemical set-up (Ag/AgCl in 3 M KCl as reference electrode and platinum wire as counter electrode). A Newport solar simulator 150 W Xe lamp with AM 1.5G filter acted as a light source for the PEC test. The lamp power was adjusted using a reference silicon solar cell to obtain 100 mW cm^{-2} (1 sun). All J-V curves of pristine TiO_2 and H-TiO_2 with an active area of 1 cm^2 were recorded in 1 M KOH using a Princeton 2273 potentiostat with a scan rate of 10 mV s^{-1}. A comparison of the J-V curves of pristine TiO_2 and H-TiO_2 samples was performed to examine the impact of wire temperature on the photocurrent density (**Figure 5a**). The photocurrent density demonstrates a noticeable increase following treatment at T_{wire} = 1600°C, reaching its peak value at T_{wire} = 1700°C, and subsequently decreasing at T_{wire} = 1800°C (**Figure 5a**). Specifically, the photocurrent density of H-TiO_2 at T_{wire} = 1700°C attains approximately 2.5 mA cm^{-2} (at 1.23 V vs. RHE), surpassing that of pristine TiO_2 by more than threefold.

The impact of wire temperature on the conductivity of H-TiO_2 was investigated through Mott-Schottky (M-S) measurements (dark, 1 M KOH) performed on the electrochemical workstation (Gamry, USA) (**Figure 5b**). The donor density of pristine TiO_2 and H-TiO_2 can be obtained by the M-S equation (Eq. 1):

$$N_d = \left(\frac{2}{e_0 \varepsilon_0 \varepsilon_r}\right)\left[\frac{d\left(1/C^2\right)}{dV}\right]^{-1} \qquad (1)$$

Here N_d is the donor density, e_0 is the electron charge, ε_0 is the permittivity of vacuum, ε_r is the dielectric constant of TiO_2 nanorods, C is capacitance and V is the applied bias voltage. It is anticipated that the slope of H-TiO_2 will decrease as the wire temperature increases, suggesting an increase in donor density and conductivity in accordance with the M-S equation. The donor densities of H-TiO_2 were calculated to be $2.5 \pm 0.1 \times 10^{17}$ cm^{-3} (1600°C), $4.6 \pm 0.1 \times 10^{17}$ cm^{-3} (1700°C) and $2.1 \pm 0.2 \times 10^{18}$ cm^{-3} (1800°C), surpassing those of pristine TiO_2 ($1.5 \pm 0.1 \times 10^{17}$ cm^{-3}), indicating a trend of increased conductivity in H-TiO_2 samples with higher wire temperatures.

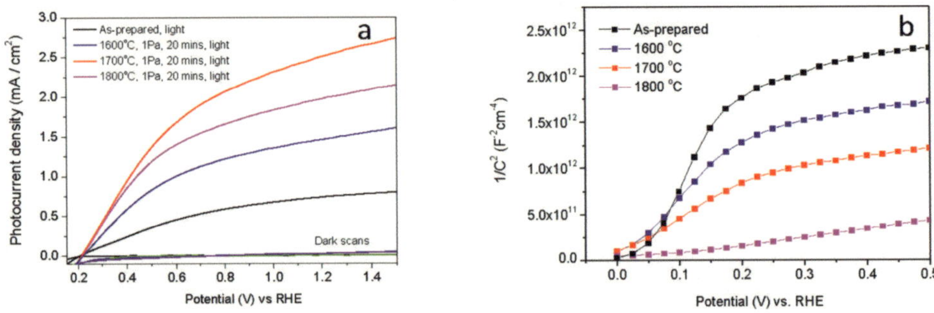

Figure 5.
(a) J-V curves of pristine TiO_2 and H-TiO_2 nanorods in 1 M KOH solution in the dark and under solar illumination. (b) Mott-Schottky plots of pristine TiO_2 and H-TiO_2 nanorods treated at T_{wire} = 1600, 1700 and 1800°C, respectively.

The optical absorption spectra indicate that the H hydrogenation significantly influences the reduction of the band gap from 3.0 to 2.9 eV. The J-V curves demonstrate that samples treated at a wire temperature of 1700°C exhibit the highest PEC activity, reaching 2.5 mA cm^{-2} at 1.23 V vs. RHE. The IPCE curves (not shown) further support H-TiO_2 at a wire temperature of 1700°C exhibit the most pronounced red-shift of the band edge, resulting in band gap narrowing.

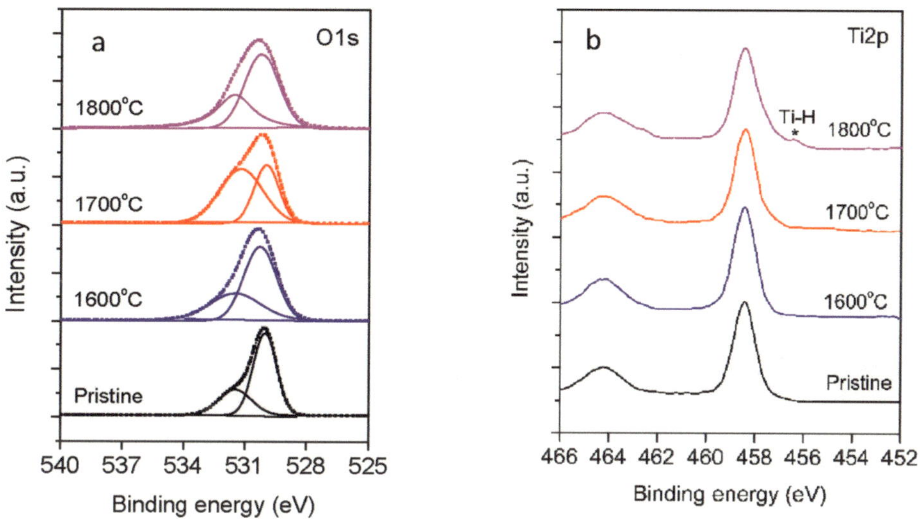

Figure 6.
(a) O1s and (b) Ti2p XPS spectra of pristine and H-TiO_2 samples treated at T_{wire} = 1600, 1700 and 1800°C, respectively.

In order to understand why H-TiO_2 treated at T_{wire} = 1700°C exhibits the highest photocurrent density, XPS analysis was conducted on pristine and H-TiO_2 samples at various wire temperatures. The XPS O1s spectra of the hydrogen-treated TiO_2 samples revealed two distinct peaks at binding energies of 530 and 531.5 eV, corresponding to TiO_2 and Ti-OH species, respectively (**Figure 6a**). The intensity of the Ti-OH peak was observed to increase as the wire temperature was raised from 1600 to 1700°C, but decreased as the temperature was further increased to 1800°C. The diminished Ti-OH peak observed in H-TiO_2 at a temperature of 1800°C suggests a reduced presence of Ti^{3+}, aligning with the findings of the EELS analysis. The XPS Ti2p spectra of both pristine and H-TiO_2 samples exhibit similar characteristics (**Figure 6b**). Specifically, the Ti2p XPS spectra display binding energies of 458.5 eV for Ti $2p_{3/2}$ and 464.3 eV for Ti $2p_{1/2}$, which align with the standard values for TiO_2. Notably, the presence of a Ti-H peak at 456.5 eV, rather than metallic Ti at 453.8 eV, is observed in the sample treated at Twire = 1800°C. The creation of surface Ti–H bonds in H-TiO_2 treated at a temperature of 1800°C is accompanied by a further reduction in surface Ti-OH bonds, leading to a decrease in the concentration of Ti^{3+} in the H-TiO_2 sample (**Figure 6b**). It is confirmed that H-TiO_2 treated at T_{wire} = 1700°C with the highest photocurrent density is attributed to the highest concentration of Ti^{3+}.

The findings indicate that low-temperature H hydrogenation results in the efficient and straightforward development of a disordered shell on the H-TiO_2 nanorods, leading to a reduction in band gap, enhancement in donor density, conductivity and concentration of Ti^{3+} and consequently, a notable enhancement in PEC activity.

3.2 H^+ hydrogenation

3.2.1 Work principle, experimental set-up and H^+ treatment

H^+ hydrogenation utilizing kinetic hydrogen ion species has garnered recent interest due to its independence from thermal activation. However, traditional H^+ hydrogenation methods are hindered by the requirement of high temperatures or the need for high power output from plasma equipment to generate high-energy hydrogen ions, resulting in inefficient and uncontrollable hydrogenation processes [17, 18].

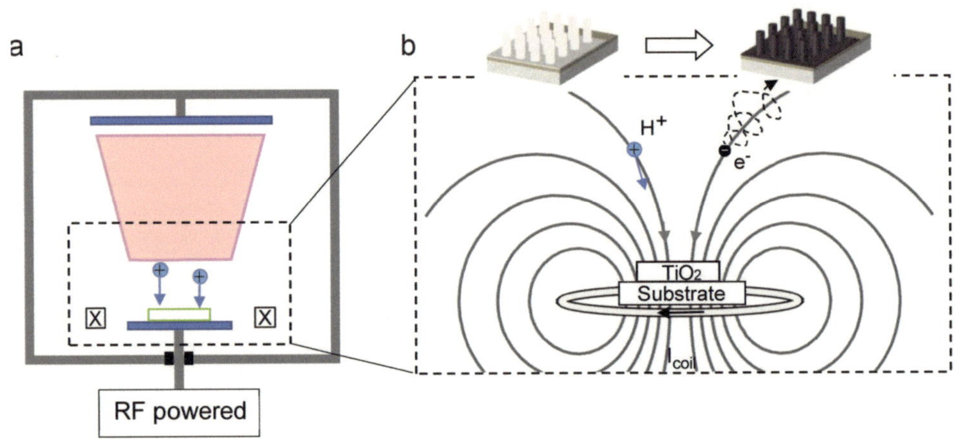

Figure 7.
(a) Experimental set-up of room temperature H^+ hydrogenation. (b) the role of a circular electromagnetic coil in the H^+ hydrogenation process.

The magnetic field generated by an electromagnetic coil in hydrogen plasma exerts distinct effects on hydrogen ions and free electrons. While the magnetic field can alter the trajectory of electrons, it is unable to significantly impact the motion of hydrogen ions due to their significantly greater mass. As free electrons orbit the magnetic field lines in a circular manner, their path is elongated, resulting in a heightened likelihood of hydrogen ionization and consequently an augmented density of hydrogen ions.

In contrast to traditional plasma hydrogenation methods, based on the above working principle, we incorporated a circular electromagnetic coil into the substrate electrode of a radio frequency (RF) plasma physical vapor deposition (PVD) system to generate a magnetic field (**Figure 7**) [15]. This magnetic field serves to enhance the concentration of hydrogen ions, and reduce RF power and self-bias voltage, thereby facilitating the production of controllable kinetic low-energy hydrogen ions. In our experimental set-up, a low self-bias voltage (V_{bias}) of −250V was achieved by applying 20 W of RF power to the substrate during the H^+ hydrogenation process.

A programmable recipe was utilized to meticulously regulate the processing parameters of RF power, magnetic field strength, hydrogen flow rate, processing pressure and treatment duration. In the context of the H^+ hydrogenation process, an RF power of 20 W was administered at the substrate electrode to attain a low self-bias voltage of −250V, with a magnetic field strength of 30 mT, hydrogen gas flow rate of 50 sccm and processing pressure of 1.5 Pa. To study the influence of treatment time on the structural, optical, electrical and PEC properties of H-TiO_2 nanorods, H-TiO_2 samples were synthesized at time intervals of 2.5, 5, 10, 20 and 40 mins, respectively.

3.2.2 Influence of treatment time on the structural, optical, electrical and PEC properties

SEM was used to analyze morphological changes in TiO_2 before and after H^+ treatment. No significant morphological changes were observed in SEM images of pristine TiO_2 and H-TiO_2 at 5 and 40 mins. X-ray diffraction (XRD) patterns showed only rutile phase with (101) and (002) peaks, indicating no phase change in either pristine TiO_2 or H-TiO_2 samples.

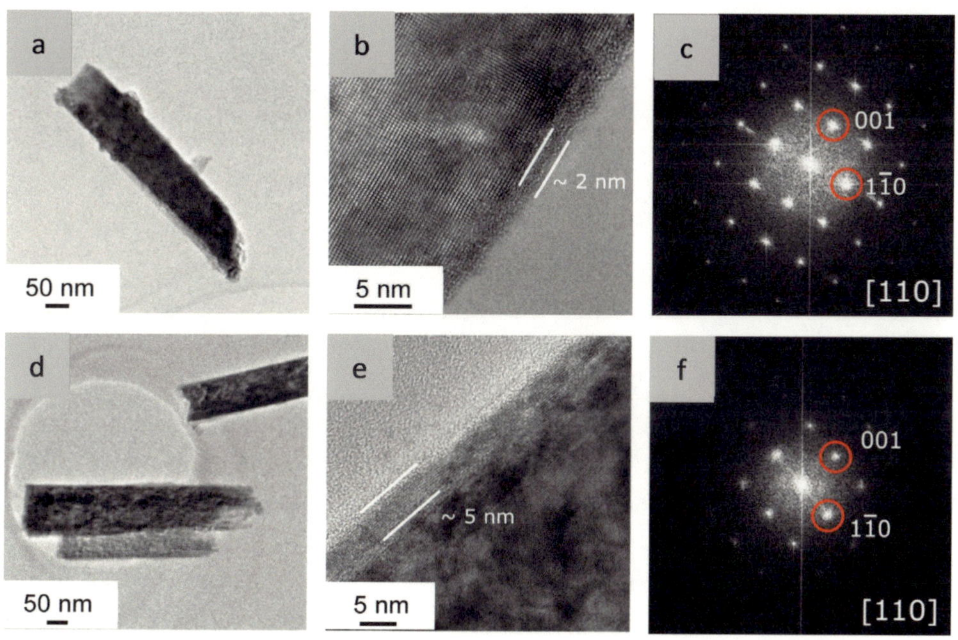

Figure 8.
TEM, HR-TEM and FFT images of H-TiO_2 at (a–c) 5 and (d–f) 40 mins treatment time, respectively.

TEM images show that in contrast to the unaltered TiO_2, H-TiO_2 samples exhibit a composite structure consisting of a crystalline core and a disordered shell (**Figure 8**). Analysis of the FFT image reveals that the core structure of H-TiO_2 remains single-crystalline, while the shell thickness measures approximately 2 nm for H-TiO_2 treated at 5 mins and increases to 5 nm for H-TiO_2 treated at 40 mins.

Figure 9 displays line-scan EELS data showing changes in [O]/[Ti] ratio in pristine TiO_2 and H-TiO_2 samples. The [O]/[Ti] ratio in the shell of H-TiO_2 decreases over time, indicating the formation of oxygen vacancies. In the core of H-TiO_2 treated at 5 mins, no oxygen vacancies are observed.

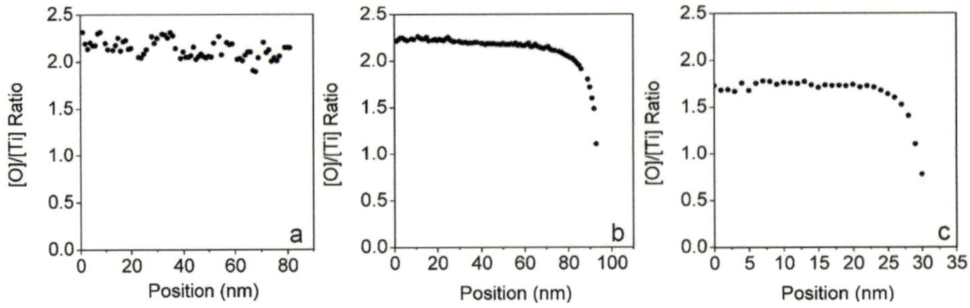

Figure 9.
[O]/[Ti] composition ratios of the (a) pristine TiO_2 and H-TiO_2 at (b) 5 and (c) 40 mins treatment time, respectively.

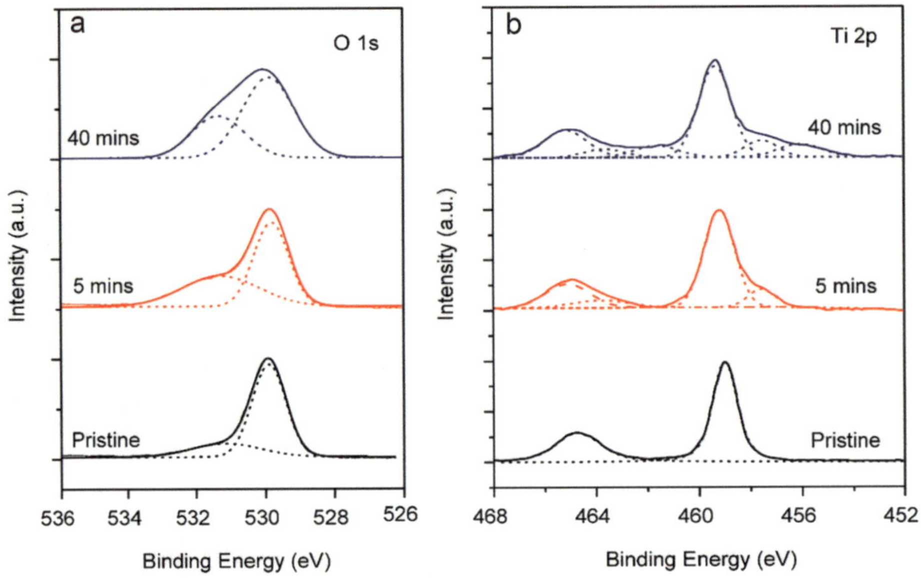

Figure 10.
(a) O1s and (b) Ti2p XPS spectra of pristine TiO_2 and H-TiO_2 at 5 and 40 mins treatment time, respectively.

The composition of the disordered shell was analyzed using XPS. **Figure 10** displays the O1s spectra of pristine TiO_2 and H-TiO_2 treated at 5 and 40 mins, respectively. The O1s peak at 530 eV corresponds to Ti-O, while the peak at 531.5 eV is indicative of Ti-OH, suggesting the presence of Ti-OH in the shell of H-TiO_2. Additionally, **Figure 10** illustrates the Ti2p spectra of pristine TiO_2 and H-TiO_2 at 5 and 40 mins, respectively. It is observed that Ti^{3+} is present in H-TiO_2 treated at 5 mins, and significant reduction ($Ti^{4+} \rightarrow Ti^{3+} \rightarrow Ti^{2+}$) is evident in H-$TiO_2$ treated at 40 mins.

We measured optical absorption spectra to compare the optical properties of pristine TiO_2 and H-TiO_2. The absorption band edges of H-TiO_2 are red-shifted and the absorption in the visible to infrared range increases with longer treatment time (**Figure 11a**). The band gap narrows from 3.0 eV in pristine TiO_2 to 2.91 eV in H-TiO_2 after 5 mins of treatment (**Figure 11b**).

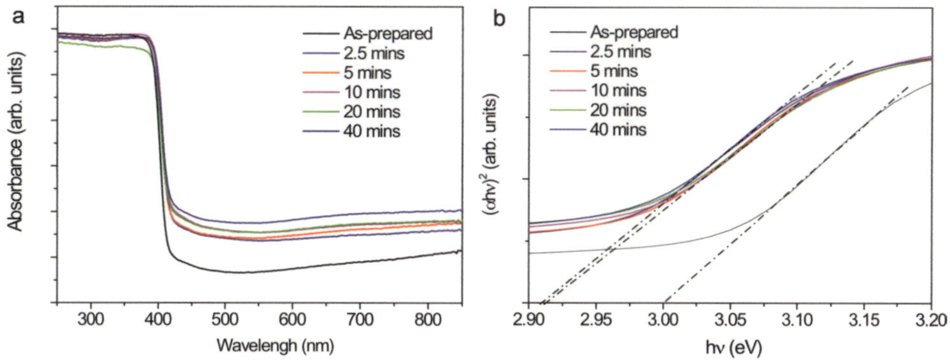

Figure 11.
(a) Optical absorption spectra and (b) Tauc plot of pristine TiO_2 and H-TiO_2 at different treatment times.

We calculated the photocurrent density of H-TiO_2 treated at 5 mins by integrating the measured IPCE spectrum (**Figure 12a**) with measured standard AM 1.5G solar spectrum [19]. **Figure 12b** shows that the calculated photocurrent density of H-TiO_2 at 5 mins is ca. 2.5 mA cm^{-2}, which is consistent with the measured photocurrent density, indicating the reliability of our PEC data. We note that a tiny contribution to the photocurrent density originates from absorbed photons with wavelengths larger than 427.5 nm. The formation of occupied Ti^{3+} midgap in the band gap promotes the visible-light absorption to enhance the visible-light-driven PEC activity (Inset, **Figure 12a**) [1, 14, 20].

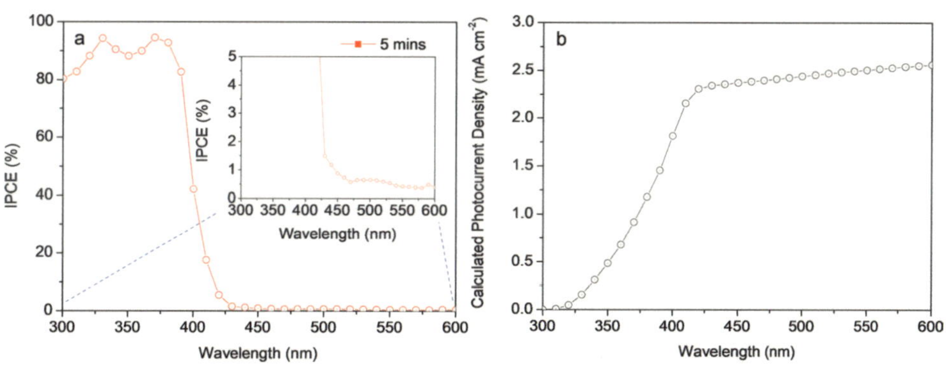

Figure 12.
(a) The IPCE spectrum of H-TiO_2 treated at 5 mins. (b) Calculated photocurrent density of H-TiO_2 treated at 5 mins by integrating the measured IPCE spectrum with measured standard AM 1.5G solar spectrum.

Figure 13a demonstrates an increase in photocurrent density of H-TiO_2 with increasing treatment time up to 5 mins, followed by a decrease in treatment time exceeding 5 mins. Specifically, at the 5-mins mark, a photocurrent density of 2.55 mA cm^{-2} at 1.23 V_{RHE} is attained, representing the highest value reported in studies of H-TiO_2.

Figure 13b shows that an increase in treatment time results in a rise in donor density but a decline in depletion region width. In comparison to pristine TiO_2, the negative shift in flat band potential of H-TiO_2 at 2.5 and 5 mins is attributed to a significant increase in donor density, subsequently causing a shift in the Fermi level toward the conduction band. We determined the band bending by flat band potential from M-S plots, the largest band bending of 1.1 V is achieved for H-TiO_2 at 5 mins.

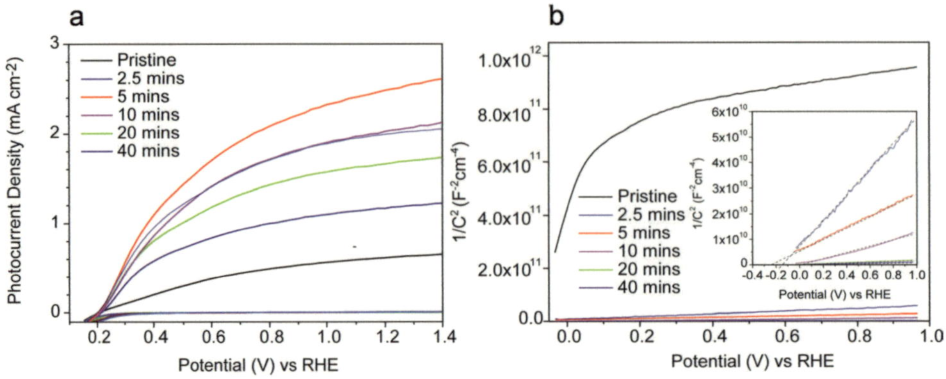

Figure 13.
(a) J-V curves of pristine TiO_2 and H-TiO_2 at different treatment times in 1 M KOH solution in the dark and under solar illumination. (b) M-S plots of pristine TiO_2 and H-TiO_2 at different treatment times.

The experiments on the hydrogenation-properties relation indicate that the increased photocurrent density of H-TiO_2 at 5 mins can be attributed to the elevated levels of Ti^{3+} within the disordered shell, which subsequently narrows the band gap and facilitates the generation of electron-hole pairs. EELS observes that oxygen vacancies are exclusively generated in the disordered shell, suggesting that the suppression of bulk defects enhances charge transport and facilitates the larger band bending to further promote charge transport and injection. Our conclusion is that the increased PEC performance can be primarily attributed to the narrower band gap, greater band bending and bulk defect suppression.

3.3 Comparison of H and H^+ methods with traditional hydrogenation methods

Compared with the conventional physical and chemical hydrogenation methods [11, 17–19, 21], we utilized low-temperature H and room-temperature H^+ approaches to achieve the highest photocurrent density of H-TiO_2 (**Figure 12**). Although H_2 hydrogenation can achieve high photocurrent densities of H-TiO_2, but high temperature and prolonged treatment time are needed (**Table 1**). Chemical reduction can prepare H-TiO_2 at room temperature, but the photocurrent density of H-TiO_2 is limited due to chemical residues and resulting charge carrier recombination centers. Although the conventional plasma hydrogenation can efficiently fabricate H-TiO_2, but high temperature or high power is required (**Table 1**). It is evident that we made a breakthrough in the development of efficient hydrogenation technology (**Figure 14**, **Table 1**).

Hydrogenation method	Temperature	RF power	Treatment time, pressure	Photocurrent density of H-TiO_2 at 1.23 V_{RHE}	Features and references
High pressure H_2	200°C	N/A	5 days, 20 bar	N/A	xHigh pressure; prolonged treatment [11]
High temperature H_2	400°C	N/A	1 h, 1 bar	2.4 mA cm^{-2}	xHigh temperature; prolonged treatment [19]

Hydrogenation method	Temperature	RF power	Treatment time, pressure	Photocurrent density of H-TiO_2 at 1.23 V_{RHE}	Features and references
Low temperature H	265 °C	N/A	20 mins, 1 Pa	2.5 mA cm^{-2}	√ Low temperature; short treatment [14]
Atomic H via spillover	250 °C	N/A	10 mins, 1 bar	1.5 mA cm^{-2}	xHigh cost: Noble metals; chemical residues [21]
High temperature H^+	500°C	15 W	3 h	0.2 mA cm^{-2}	xHigh temperature; prolonged treatment [17]
High temperature H^+	425°C	200 W	1 h	0.9 mA cm^{-2}	xHigh temperature and power; prolonged treatment [18]
Room temperature H^+	25°C	20 W	5 mins, 1.5 Pa	2.55 mA cm^{-2}	√ Room temperature; short treatment [15]

Table 1.
A comparison of the low temperature hot wire H hydrogenation and room temperature H^+ hydrogenation with state-of-the-art hydrogenations.

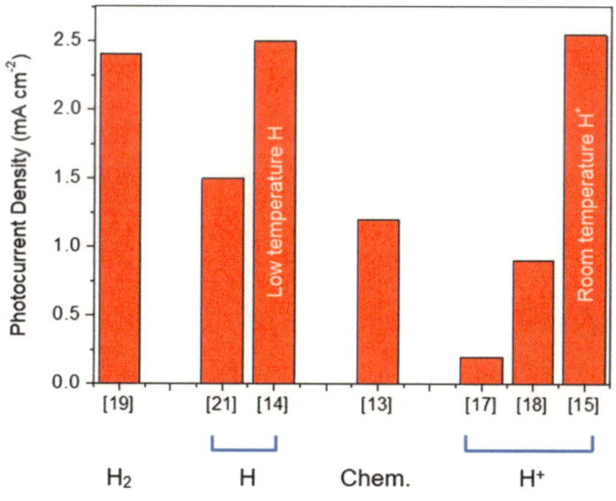

Figure 14.
Comparison of photocurrent density of black TiO_2 by our advanced hydrogenation approaches with state-of-the-art hydrogenations.

4. Conclusions

From the above discussion of hydrogenation efficiency, we experimentally confirmed that the low-temperature H and room-temperature H^+ approaches fundamentally break through the intrinsic limitations of poor conductivity and large band gap of TiO_2, and greatly improve its PEC properties. Compared to conventional hydrogenation methods, the two advanced hydrogenation technologies we have invented have the advantages of low cost, uniform hydrogenation, low temperature and low energy consumption, which can realize the optimal control of the physical and chemical properties of oxide photoanode materials, and have broad application prospects. Our findings indicated that two advanced hydrogenation technologies are future directions of development of practical hydrogenation technology and the influence of hydrogenation processing parameters on the structural, optical and PEC properties provides experimental guidance for the rapid synthesis of black TiO_2 with desirable PEC activity, which will advance the significant development of the hydrogenation technology to promote black TiO_2 for PEC hydrogen production from research to practical application.

Acknowledgements

X. Wang acknowledges the support of the research fund by the Institute of Carbon Neutrality, ShanghaiTech University. H. Shen and B. Wang thank the support of the research fund by Yiwu Research Institute, Fudan University.

Conflict of interest

The authors declare no conflict of interest.

Author details

Xiaodan Wang[1*], Beibei Wang[2] and Hao Shen[2,3*]

1 Institute of Carbon Neutrality, ShanghaiTech University, Shanghai, China

2 Yiwu Research Institute, Fudan University, Zhejiang, China

3 Institute of Optoelectronics, Fudan University, Shanghai, China

*Address all correspondence to: wangxd2@shanghaitech.edu.cn
and shenhao@ywfudan.cn

References

[1] Zhang W, He H, Li H, Duan L, Zu L, Zhai Y, et al. Visible-light responsive TiO_2-based materials for efficient solar energy utilization. Advanced Energy Materials. 2021;**11**(15):2003303. DOI: 10.1002/aenm.202003303

[2] El Ouardi M, El Idrissi A, Arab M, Zbair M, Haspel H, Saadi M, et al. Review of photoelectrochemical water splitting: From quantitative approaches to effect of sacrificial agents, oxygen vacancies, thermal and magnetic field on (photo) electrolysis. International Journal of Hydrogen Energy. 2024;**51**:1044-1067. DOI: 10.1016/j.ijhydene.2023.09.111

[3] Cui H, Zhao W, Yang C, Yin H, Lin T, Shan Y, et al. Black TiO_2 nanotube arrays for high-efficiency photoelectrochemical water-splitting. Journal of Materials Chemistry A. 2014;**2**(23):8612-8616. DOI: 10.1039/C4TA00176A

[4] Zhu L, Ma H, Han H, Fu Y, Ma C, Yu Z, et al. Black TiO_2 nanotube arrays fabricated by electrochemical self-doping and their photoelectrochemical performance. RSC Advances. 2018;**8**(34):18992-19000. DOI: 10.1039/C8RA02983K

[5] Li Z, Bian H, Xiao X, Shen J, Zhao C, Lu J, et al. Defective black TiO_2 nanotube arrays for enhanced photocatalytic and photoelectrochemical applications. ACS Applied Nano Materials. 2019;**2**(11):7372-7378. DOI: 10.1021/acsanm.9b01878

[6] Wu P, Liu C, Luo Y, Wu K, Wu J, Guo X, et al. A novel black TiO_2/ZnO nanocone arrays heterojunction on carbon cloth for highly efficient photoelectrochemical performance. Frontiers of Materials Science. 2019;**13**(1):43-53. DOI: 10.1007/s11706-019-0447-2

[7] Yoon D-H, Rokon BM, Dowla U, Sakthisabarimoorthi A. Composite nanostructures of black TiO_2/WO_3 on rGO nanosheets for photoelectrochemical water splitting. Diamond and Related Materials. 2022;**129**:109363. DOI: 10.1016/j.diamond.2022.109363

[8] Yoon D-H, Biswas UD, Rokon M, Sakthisabarimoorthi A. Enhancement of photoelectrochemical activity by Ag coating on black TiO_2 nanoparticles. Materials Chemistry and Physics. 2022;**291**:126675. DOI: 10.1016/j.matchemphys.2022.126675

[9] Meng M, Feng Y, Li C, Gan Z, Yuan H, Zhang H. Black 3D-TiO_2 nanotube arrays on Ti meshes for boosted photoelectrochemical water splitting. Nanomaterials. 2022;**12**(9):1447. DOI: 10.3390/nano12091447

[10] Yoon D-H, Rokon BM, Dowla U, Sakthisabarimoorthi A. Impact of crystalline core/amorphous shell structured black TiO_2 nanoparticles on photoelectrochemical water splitting. Optical Materials. 2022;**133**:113030. DOI: 10.1016/j.optmat.2022.113030

[11] Chen XB, Liu L, Yu PY, Mao SS. Increasing solar absorption for photocatalysis with black hydrogenated titanium dioxide nanocrystals. Science. 2011;**331**(6018):746-750. DOI: 10.1126/science.1200448

[12] Sun X, Guo Y, Wu C, Xie Y. The hydric effect in inorganic nanomaterials for nanoelectronics and energy applications. Advanced Materials. 2015;**27**(26):3850-3867. DOI: 10.1002/adma.201500546

[13] Yan P, Liu G, Ding C, Han H, Shi J, Gan Y, et al. Photoelectrochemical water splitting promoted with a disordered surface layer created by electrochemical reduction. ACS Applied Materials & Interfaces. 2015;**7**(6):3791-3796. DOI: 10.1021/am508738d

[14] Wang X, Mayrhofer L, Hoefer M, Estrade S, Lopez-Conesa L, Zhou H, et al. Facile and efficient atomic hydrogenation enabled black TiO_2 with enhanced photo-electrochemical activity via a favorably low-energy-barrier pathway. Advanced Energy Materials. 2019;**9**(33):1900725. DOI: 10.1002/aenm.201900725

[15] Wang X, Mayrhofer L, Keunecke M, Estrade S, Lopez-Conesa L, Moseler M, et al. Low-energy hydrogen ions enable efficient room-temperature and rapid plasma hydrogenation of TiO_2 nanorods for enhanced photoelectrochemical activity. Small. 2022;**18**(46):2204136. DOI: 10.1002/smll.202204136

[16] Langmuir I. A chemically active modification of hydrogen. 5. Journal of the American Chemical Society. 1912;**34**(10):1310-1325. DOI: 10.1021/ja02211a004

[17] Lepcha A, Maccato C, Mettenbörger A, Andreu T, Mayrhofer L, Walter M, et al. Electrospun black titania nanofibers: Influence of hydrogen plasma-induced disorder on the electronic structure and photoelectrochemical performance. Journal of Physical Chemistry C. 2015;**119**(33):18835-18842. DOI: 10.1021/acs.jpcc.5b02767

[18] Tian Z, Cui H, Zhu G, Zhao W, Xu J, Shao F, et al. Hydrogen plasma reduced black TiO_2-B nanowires for enhanced photoelectrochemical water-splitting. Journal of Power Sources. 2016;**325**:697-705. DOI: 10.1016/j.jpowsour.2016.06.074

[19] Wang GM, Wang HY, Ling YC, Tang YC, Yang XY, Fitzmorris RC, et al. Hydrogen-treated TiO_2 nanowire arrays for photoelectrochemical water splitting. Nano Letters. 2011;**11**(7):3026-3033. DOI: 10.1021/nl201766h

[20] Liu X, Zhu G, Wang X, Yuan X, Lin T, Huang F. Progress in black titania: A new material for advanced photocatalysis. Advanced Energy Materials. 2016;**6**(17):1600452. DOI: 10.1002/aenm.201600452

[21] Xu Y, Zhang C, Zhang L, Zhang X, Yao H, Shi J. Pd-catalyzed instant hydrogenation of TiO_2 with enhanced photocatalytic performance. Energy & Environmental Science. 2016;**9**(7):2410-2417. DOI: 10.1039/C6EE00830E

Chapter 2

Illuminating Advances: Photochemistry and Photophysics of N-Heterocyclic Carbenes (NHCs) and Its Structural Correlation

Krishanu Bandyopadhyay, Abhineet Verma, Tamal Ghosh, Ravi Kumar Kanaparthi, Sudeena Nadendla and Satyen Saha

Abstract

This chapter discusses the significant impact of N-heterocyclic carbene (**NHC**) ligands on the optical properties of metal complexes. It highlights the key role of **NHC** ligands in shifting the d–d transitions of these complexes to shorter wavelength regions, rendering them luminescent. The study focuses on NHC-based metal complexes involving copper, silver, platinum, iridium, and coinage metals, showcasing their luminescent properties and potential applications in organic light-emitting diodes (OLEDs). However, it also underscores a challenge: the quantum yields of these metal complexes may not be sufficient for practical applications. To address this issue, this chapter proposes modifying the NHC ligand with electron-rich substituents, aiming to reduce the HOMO-LUMO gap and shift d-d transitions into the UV region. Additionally, it introduces cyclic (alkyl)(amino)carbene (CAACs) as promising ligands for tuning the photophysical properties of metal complexes. CAACs are noted for their ability to decrease the HOMO-LUMO gap compared to NHCs and their electrophilic nature. Luminescent CAAC metal complexes are found to exhibit sub-microsecond lifetimes, making them suitable candidates for OLED applications. Ultimately, the research challenges researchers to better understand the structure, bonding, and properties of these complexes to create stable and highly luminescent materials using abundant and cost-effective metals like iron, copper, zinc, and silver.

Keywords: N-heterocyclic carbene (NHC) ligands, optical properties of metal complexes, luminescent metal complexes, cyclic (alkyl)(amino)carbene (CAACs) ligands, structure-property of NH-Carbenes

1. Introduction

Carbenes are an intriguing class of carbon-containing compounds with a neutral divalent carbon atom with six electrons on its valence shell (**Figure 1**). However,

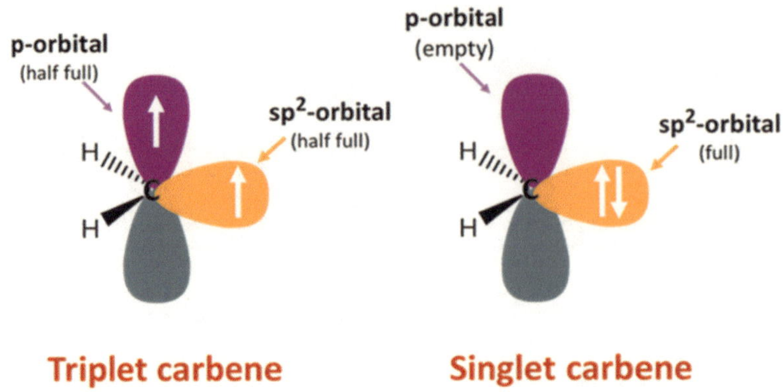

Figure 1.
Schematic representation of triplet and singlet carbene.

unbound carbenes are extremely unstable due to their incomplete electron octet and coordinative unsaturation. Therefore, they have traditionally been considered extremely reactive, transient intermediates in chemical transformations like cyclopropanation [1].

The exploration of carbenes has remained a significant and enduring focus of interest for researchers over the years, sparking numerous theoretical and experimental inquiries [2–10]. Among the most prominent questions that have arisen are those concerning the electronic configuration of carbenes – whether they exist in a singlet or triplet state – and the precise geometry around the carbenic carbon atom. Ongoing research endeavors, spanning from the realms of physical chemistry to organic chemistry, have provided a more systematic understanding of the nature of carbenes and their applications [3].

Initially conceived as intermediates in the context of organic transformations, these intriguing chemical entities later emerged as a central subject of study within the field of coordination chemistry [4, 11, 12]. The stability of carbenes became a crucial consideration for their utilization in coordination chemistry [5]. Consequently, the quest for isolating and definitively characterizing free, uncoordinated carbenes became a notable challenge in the late 1990s. In 1991, a milestone achievement was made when an isolable and "bottleable" carbene was successfully incorporated into a nitrogen heterocycle, giving rise to what is now known as N-heterocyclic carbene (NHC) [6].

1.1 Definition and a brief history of NHCs

NHCs are electron-rich nucleophilic heterocyclic species containing a carbene carbon and at least one directly bonded nitrogen atom within the ring structure NHCs. Compared with most carbenic systems, which are inherently unstable and exist only as transient intermediates in organic processes (e.g., cyclopropanations and C—H insertions) without any additional stabilizing features. NHCs are stable compounds that can often be isolated and stored. NHCs, nowadays, find many applications across the life sciences, ranging from catalysis to materials science and biomedical applications [7].

The preparation of carbon having six valance electrons was a synthetic challenge for a long time, as there were many testified failed attempts to prepare them [8]. In

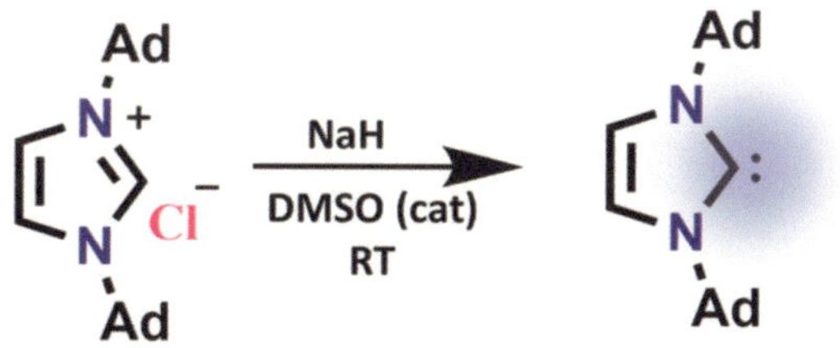

Figure 2.
Preparation of N-heterocyclic carbene by the deprotonation of bis-(1-adamantyl) imidazolium chloride.

the 1960s, Wanzlick tried to isolate carbenes but was unsuccessful. Later he proposed that a carbene center at the imidazole ring's 2-position would be stable due to the electron-donating effects of nearby nitrogen atoms provided the conceptual foundation for the evolution of these species' chemistry [9]. Bertrand et al. in 1988 proposed for the first time that the presence of heteroatoms, such as silicon or phosphorus, close to the carbon center increased the stability of carbenic species [10]. In 1991, Arduengo established the integration of the carbenic carbon atom into a nitrogen heterocycle decorated with bulky substituents [6, 13]. The inaugural achievement in isolating stable crystalline N-heterocyclic carbene (NHC) involved a meticulously orchestrated process. This milestone was realized through the deprotonation of bis-(1-adamantyl) imidazolium chloride using sodium hydride within the medium of tetrahydrofuran. The success of this reaction was facilitated by the presence of a catalytic quantity of dimethyl sulfoxide, as depicted in **Figure 2** [6].

With the discovery of the first NHCs in 1991, this field has expanded rapidly (as shown in **Figure 3**) in understanding the subtle behavior of NHCs and their fundamental applications in natural sciences. It is fascinating to observe that new applications for NHCs are continuously being discovered at a remarkable speed [1, 15]. An example of advancements in catalytic activity involves using NHCs as ligands in transition metal species. This approach goes beyond merely replacing phosphines, aiming to achieve increased stability and turnover frequencies [16]. In **Figure 3**, the studies carried out on NHCs from 1991 to 2021 are summarized very well.

1.2 Structure and properties of NHCs

NHCs have proven cutting-edge classes of compounds with a significant influence on chemical science [17–35]. NHCs are known for their intrinsic reactivity; NHCs are not different from them. However, the presence of a cyclic scaffold and the hetero-atom increased the level of stability in contrast to other congeners [14]. This chapter briefly describes their primary characteristics, structural variation, metal complexation, and photophysical behavior. Although heteroatoms, particularly nitrogen atoms, can enhance the stability of carbenes, NHCs possess various structural characteristics that contribute to their stability in terms of both kinetics and thermodynamics, as well as their unique reactivity binding patterns [36]. Roughly some of the core features of the NHC structure can contribute: (1) presence and location of heteroatoms (nitrogen and others); (2) substituents at proximal positions; (3) backbone structure (e.g. unsaturation) and substituents; (4) ring size. A representation of the general structures of NHCs, firstly reported compound 1,3-di(adamantyl)imidazol-2-ylidene (IAd) presented in **Figure 2** [37].

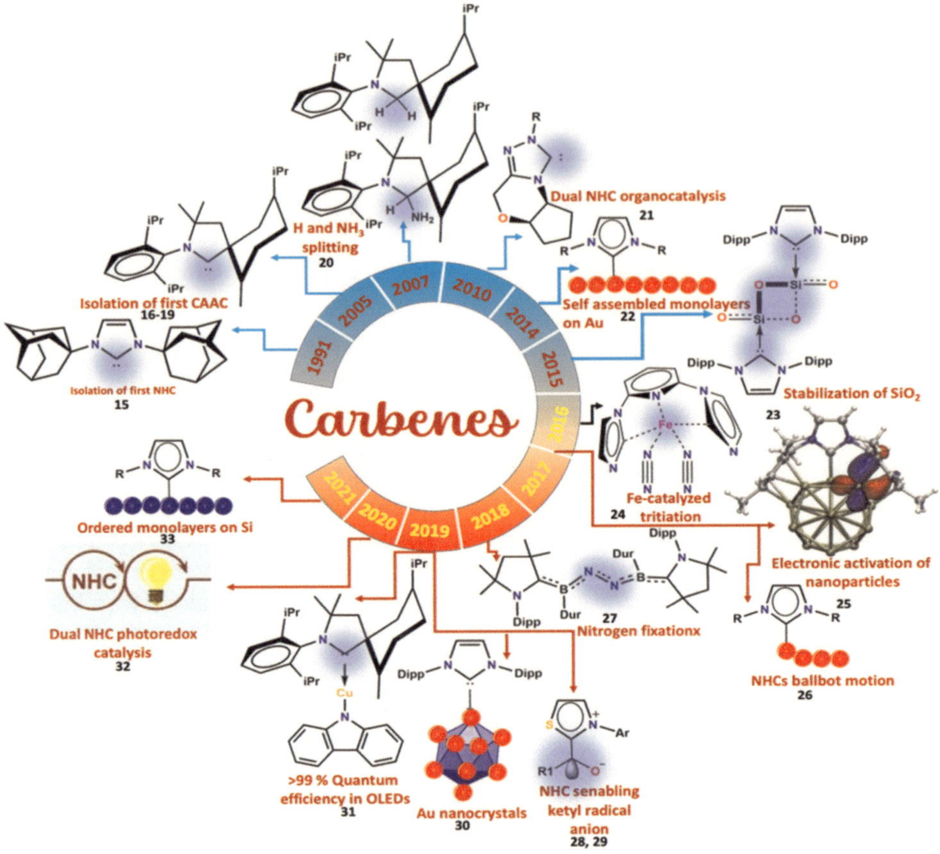

Figure 3.
Selected high impact contribution in the field of carbenes during last three decades. Ar = aryl; OLEDs = organic light-emitting diodes; NHC = N-heterocyclic carbene; Dur = 2,3,5,6-tetramethylphenyl; Dipp = 2,6-diisopropylphenyl; CAAC = cyclic (alkyl)(amino)carbene. Part of the figure was adapted with permission from ref. [14].

The accessible lone pairs can overlap with the vacant p-orbital (LUMO, "electron donating") while the neighboring nitrogen atoms can "withdraw electron density" from the non-bonding lone pair at the carbenic carbon (HOMO, "-electron withdrawing") due to their high electronegativity (**Figure 4**) [38]. The reactive center is crucially protected from possible dimerization or other breakdown routes by the steric bulk on the N-substituents (such as adamantyl in Arduengo *et al.* first reported NHC) [6], yet their identity can modify the reactivity of the ring system. Regarding the cyclic backbone, unsaturation might contribute to the heterocyclic system's aromatic nature and provide greater thermodynamic stability (about 25 kcalmol-1 for imidazol-2-ylidenes compared to the comparable imidazolin-2-ylidenes) [39, 40].

1.3 Generation of carbenes and N-heterocyclic carbenes (NHCs)

Elimination and fragmentation processes are typically used to create carbene and NHC compounds. When groups linked to the carbon atom are broken as a result of photolysis, thermolysis, or a reaction with metals, carbenes are produced as intermediate product.

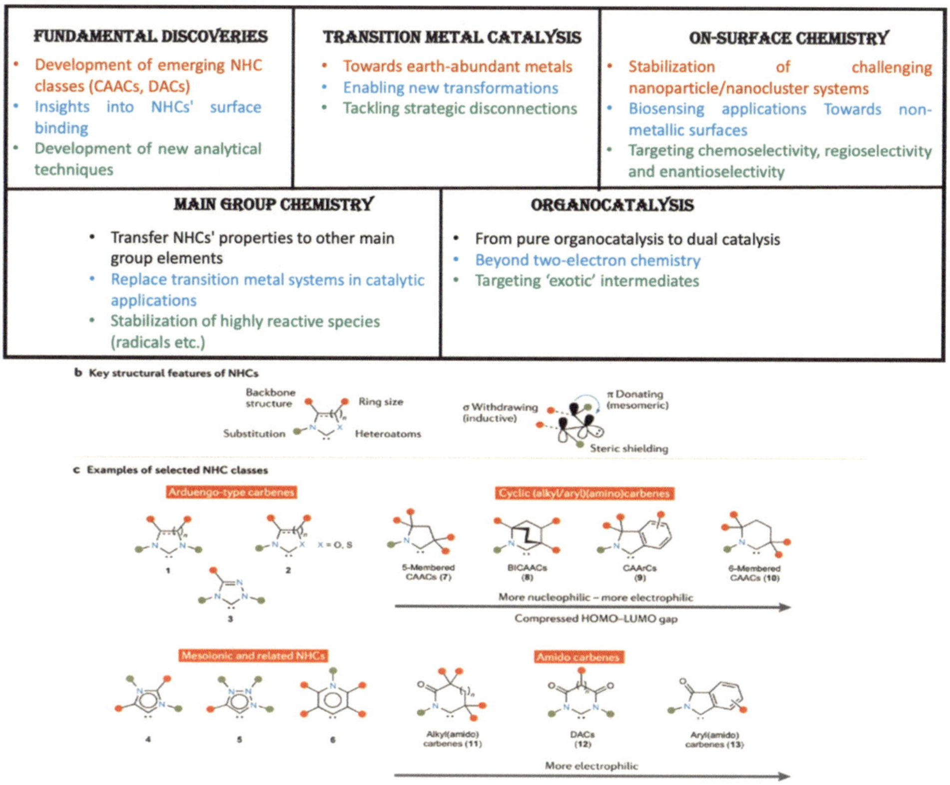

Figure 4.
Recent key advances, structural features, and selected N-heterocyclic carbenes classes. (a) Summary of the most significant N-heterocyclic carbene (NHC)-related discoveries during the past decades. (b) Crucial NHC structural characteristics. (c) Overview of a few established and new classes of NHCs. ***Note:*** *Green dots = heteroatom substituents and red dots = backbone substituents. Part of the figure was adapted with permission from ref. [14].*

1.3.1 Generation of carbenes

1.3.1.1 α-elimination

The hydrolysis of chloroform in basic medium was probably the first reaction in which intermediate NHC formation was suggested [41]. In the early 1950s, Hine and coworkers investigated the mechanism of this reaction, ushering in the current age of carbene chemistry [42–44]. The acidic proton of chloroform separates in the basic environment, forming the trichloromethyl anion. When the chloride anion is taken out of the carbanion, dichlorocarbene is created. Carbon monoxide is produced *via* the hydrolysis of dichlorocarbene in aqueous environments (**Figure 5**). Because the hydrogen and chlorine released during the formation of carbenes are connected to the same carbon atom, these reactions are referred to as α-elimination reactions.

Cyclopropane derivative can be prepared by the addition of NHC to the double bond electron system using aprotic solvents and strong bases, as presented in **Figure 6**. Before the reaction, the solvent used must be dried very well. Because the water in the environment can easily react with the NHC formed, and it also reacts with the base in the environment and completely prevents the formation of NHC.

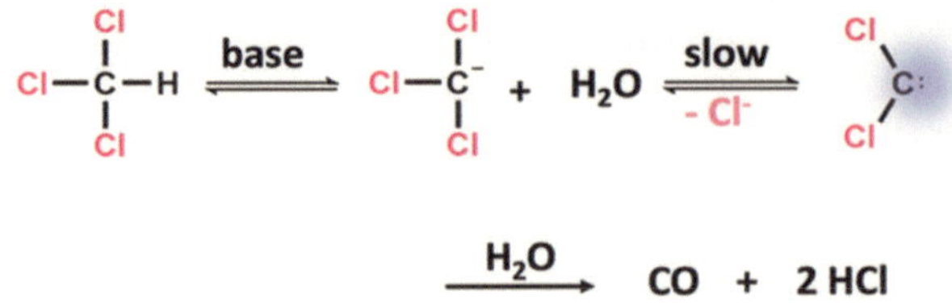

Figure 5.
The generation of intermediate carbenes through α-elimination reactions in chloroform.

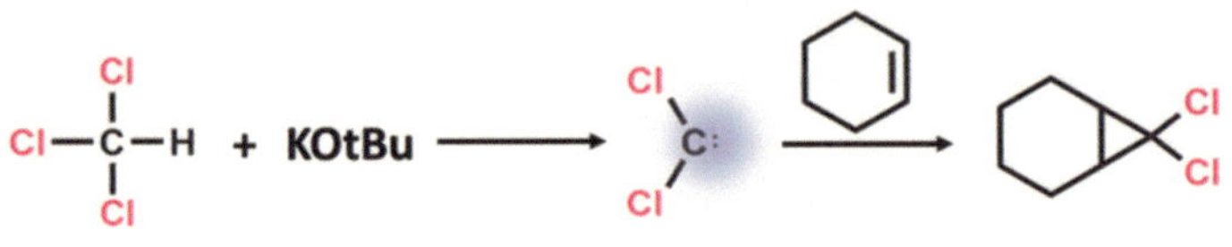

Figure 6.
Addition of carbenes to a double bond system to yield cyclopropane derivatives.

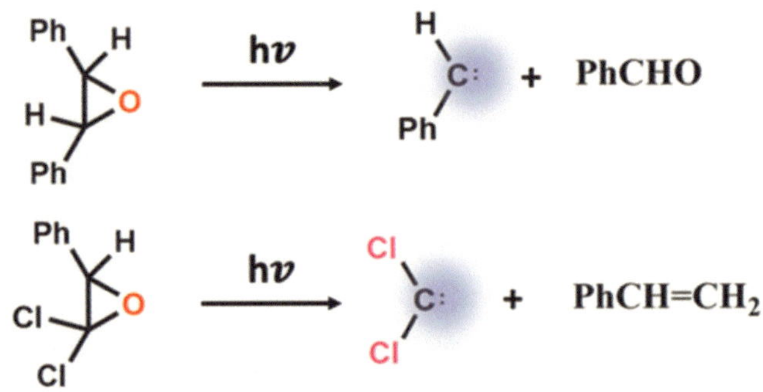

Figure 7.
Generation of carbenes from cyclopropanes and epoxides.

1.3.1.2 Three-membered carbon ring for carbenes

Due to high steric strain, three-membered rings frequently breakdown into carbene intermediates when heated or exposed to radiation. The photolysis of 1,1-dichloro-2-phenylcyclopropane produces CCl_2, while oxirane photolytic decomposition gives arylcarbenes, as illustrated in **Figure 7**. For substituted arylcarbenes, such as diphenylcarbene [45], phenylmethylcarbene [46], cyanophenylcarbene [47, 48], and methoxycarbonylphenylcarbene [49], substituted phenyloxiranes serve as handy precursors. With unsymmetrical precursors, selective cleavage was seen; it appears that the thermodynamically more stable isomer is favorable.

1.3.1.3 Simmon-Smith reaction

Methylene carbene cannot be produced using the α-elimination method. There are techniques for producing methylene carbene adducts, though. The reaction of alkenes with diiodomethane in the presence of zinc to yield cyclopropane compounds is the most frequently employed method (**Figure 8**) these reactions are popularly known as Simmon-Smith reaction [50–52].

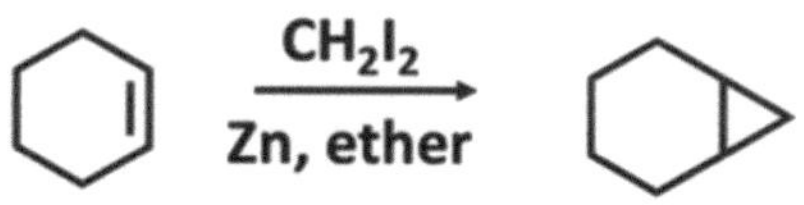

Figure 8.
Simmon-Smith reaction.

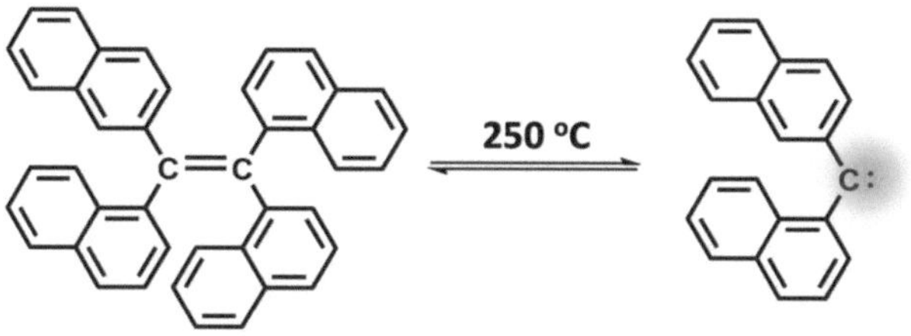

Figure 9.
Thermal dissociation of tetranaphth-1-ylethene via *breakdown of π-bond for the generation of carbenes.*

1.3.1.4 Strained alkenes

The π-bond is weakened by a greatly diminished p-p overlap and distortion from planarity when the alkene is very sterically hindered. As a result, the energy of the ground state is increased, making it feasible for heating to cause dissociation to NHCs. Tetranaphth-1-ylethene's reversible dissociation into bis(naphth-1-yl) carbene at 250°C [53] is a well-known example of this mechanism (**Figure 9**).

1.3.2 Generation of N-heterocyclic carbenes

1.3.2.1 Cyclization by introduction of the precarbenic atom moiety

The incorporation of the C1 precarbenic unit stands out as the most widely employed strategy for NHC precursors. Renowned for its high efficiency and flexibility, this method allows for the integration of diverse substituents. Depending on the desired core of the N-heterocyclic carbene precursor, various reagents can be utilized to introduce the carbenic carbon moiety. This section provides an overview of the different possibilities, accompanied by representative examples [14].

1.3.2.1.1 Trialkyl orthoformate (HC(OR)3) as the precarbenic unit

In 1991, Saba and Kaloustian reported that the treatment of *N,N'*-disubstituted α to ω -alkanediamines with trialkyl orthoformate, in the presence of a source of protons and anions, results in the formation of the corresponding cyclic formamidinium salt. This method consistently yields high product yields (>90%), as demonstrated in **Figure 10**. Additionally, the efficiency of this approach extends to the synthesis of triazolium salts [54].

This reaction serves as the final step in a significant number of NHC-precursor syntheses due to its typically high yield and ease of execution. The resulting salts often precipitate, facilitating recovery through simple filtration. Essentially, any tethered secondary diamine can be transformed into the corresponding cyclic formamidinium salt, with the main challenge lying in finding a suitable synthesis for the starting

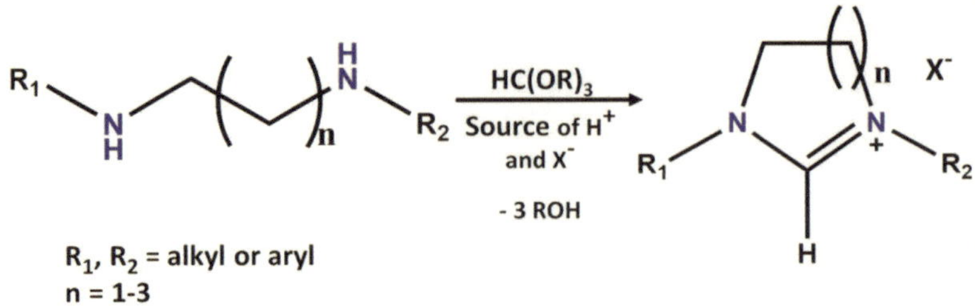

Figure 10.
General route to the cyclic formamidinium salt by condensation of a diamine with trialkyl orthoformate.

diamine (**Figure 10**). Depending on the characteristics of the target NHC precursor (such as the nature of the linking backbone, nitrogen substituents, presence of chirality, or annelated cycles), several synthetic pathways to substituted diamines have been developed. These can be categorized into six major strategies: (a) formation of a bisimine and reduction; (b) bisacylation of two amines and reduction; (c) monoalkylation and monoacylation of two amines [55–59].

1.3.2.1.1.1 Condensation-reduction route

In a typical synthetic procedure, a diimine is initially formed through the condensation of an aryl or alkylamine with glyoxal. This diimine is then reduced to the corresponding diamine, which is isolated either as the free base or as the dihydrochloride salt at this stage. The cyclization step involves the use of triethyl orthoformate as the C1-building block, along with one equivalent of acid if starting from the free diamine, leading to the formation of the corresponding imidazolinium salt (**Figure 11**). This synthetic route remains the most widely adopted standard protocol for obtaining common imidazolinium salts, proving applicable to a broad range of primary amines, as demonstrated by selected examples in **Figure 11** [60].

Hoveyda and co-workers successfully employed the method for synthesizing the unsymmetrical (and chiral) imidazolinium salt 4 from (S)-NOBIN in a two-step sequence, using Boc-protected amino aldehyde as the key coupling partner (**Figure 12**). The deprotection of the Boc group took place during the formation of the dihydrochloride salt of the diamine. However, it is worth noting that compound had to be synthesized first from mesitylamine in a three-step route, contributing to an overall lengthy synthetic procedure [61].

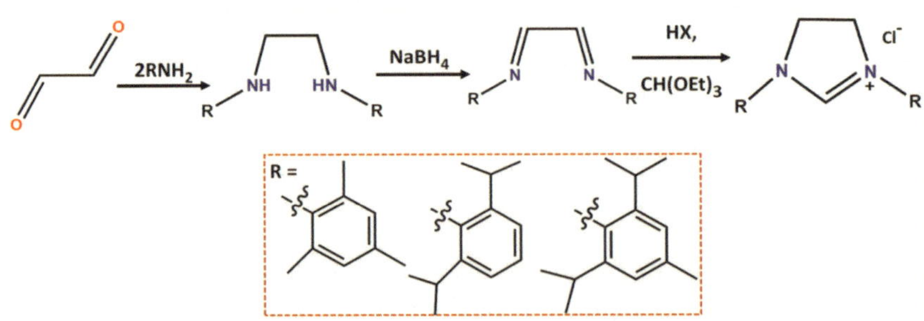

Figure 11.
Synthesis of the symmetrical imidazolinium chlorides 1a-q in a three-step sequence: Glyoxal condensation with an arylamine, reduction of the resulting diimine, and cyclization with triethyl orthoformate.

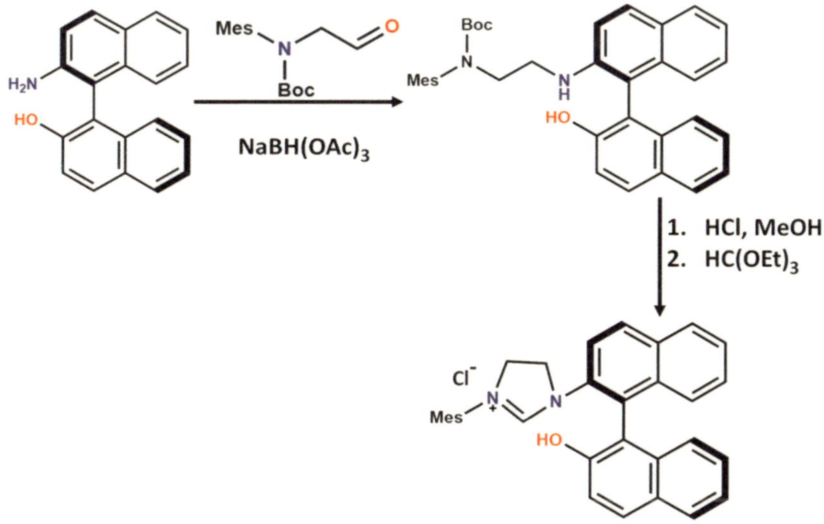

Figure 12.
Synthesis of the unsymmetrical, chiral imidazolinium chloride 4 according to Hoveyda.

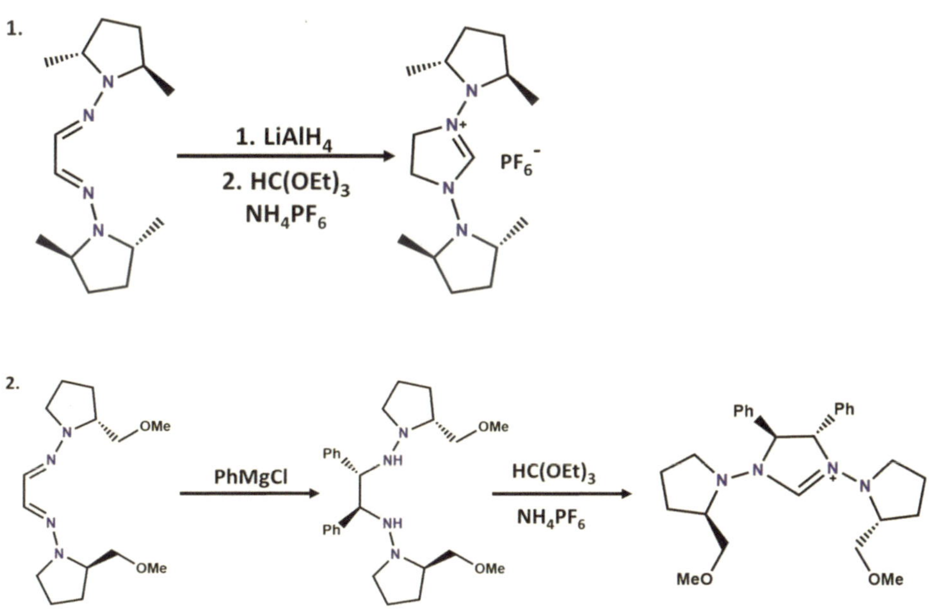

Figure 13.
Synthesis of the 1,3-Bis(N,N-dialkylamino)- imidazolinium salts 6a and 6b derived from chiral N,N-Dialkylhydrazines.

Lassaletta and co-workers extended this method to the cyclization of bis-hydrazine, resulting in the formation of 1,3-bis(*N*,*N*-dialkylamino)-imidazolinium salt, introducing a new family of chiral N-heterocyclic carbene precursors (**Figure 13**, Eq. 1). Interestingly, the incorporation of a chiral backbone into the target structure can also be achieved using proline derivatives. For instance, the stereoselective addition of PhMgCl to glyoxal RAMP bis-hydrazone, followed by

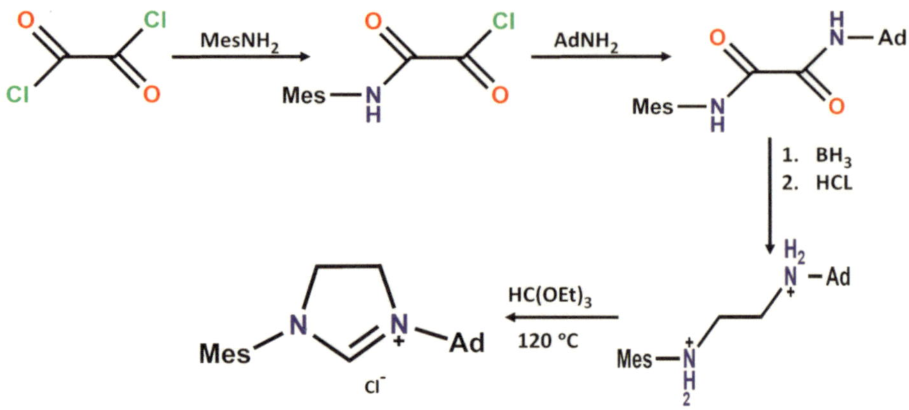

Figure 14.
Synthesis of 1-Adamantyl-3-mesitylimidazolinium chloride, according to Mol and co-workers.

the direct treatment of the air-sensitive intermediate bis-hydrazine with $HC(OEt)_3$ and NH_4PF_6, yielded [product] in 44% yield [62, 63].

1.3.2.1.1.2 Bisacylation/reduction route

The use of oxalyl chloride or derivatives, such as monoethyl oxalyl chloride, for synthesizing *N,N′*-disubstituted-1,2-diamines retains the advantages of Arduengo's methodology. This approach requires only inexpensive reagents, making it practical for large-scale syntheses. Additionally, it offers increased modularity and flexibility, allowing for the synthesis of unsymmetrical imidazoliniums. Mol and co-workers initially reported the synthesis of the mixed adamantyl/mesityl N-heterocyclic precursor 1-(1-adamantyl)-3-mesityl-imidazolinium chloride. This was achieved through the reaction of an amine with oxalyl chloride, followed by the reaction of a second amine with the intermediate acyl chloride (**Figure 14**). The resulting oxalamide was subsequently reduced and cyclized using triethyl orthoformate. In cases where controlling mono- vs. bisamidation with oxalyl chloride posed challenges, ethyl oxalyl chloride was alternatively employed to access unsymmetrical imidazolinium salts [64].

2. s, p, d and f block metal complexes

2.1 Complexes with s-block elements

Complexes of s-block metals in their zero oxidation state offer chemists the opportunity to handle these elements in soluble and distinct molecular forms, typically in their electron-rich electronic state. Beryllium (Be) is one of them which very rarely investigated chemical element of the first octal row of the periodic table, except inert neon. The reason behind this is the high toxicity of Be, which makes the experiment **risky and hazardous**.

Beryllium chemistry is a productive field that utilizes the superior lewis acidity of beryllium compounds compared to other main-group elements. Beryllium oxide

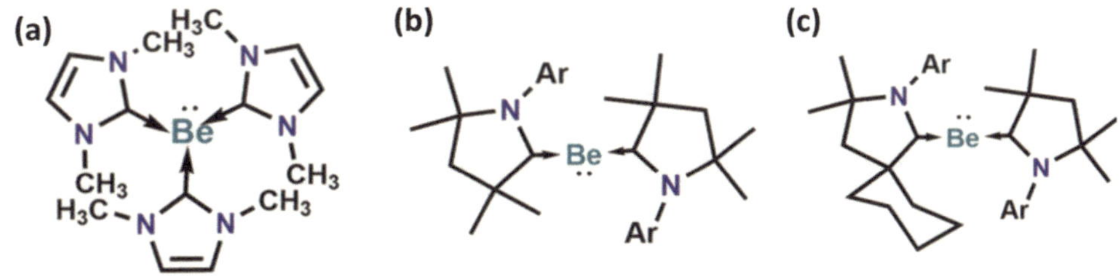

Figure 15.
Rare Be complexes reported a decade ago (a) tri-coordinated beryllium complexes, (b) di-coordinated beryllium complexes, [Be(cAACMe)2], and (c) di-coordinated beryllium complexes with zero oxidation state, [Be (cAACMe)(cAAC)cy]. Ar = 2,6 diisopropylphenyl.

(BeO) is the strongest neutral main group Lewis acid known to date. Dehnicke et al. conducted a systematic study on the stable complexes that Beryllium dihalides BeX2 (X = F, Cl) form with lewis bases. In its zero oxidation state, beryllium atom showcase a significant lewis acidity. In 2013, studies predicted tri-coordinated beryllium complexes with CO, PR3, or NHC that would lead to an electron octet at beryllium, which could be potentially synthesized and isolated, as a free molecule (**Figure 15a**) [65].

Braunschweig and co-workers pointed out one unexpected result, where they synthesized di-coordinated beryllium complexes, found to be less predicted, as shown in **Figure 15b**, where two cyclic (alkyl)(amino)carbene (cAAC) ligands stabilize a beryllium atom to form [Be(cAACMe)2] and [Be(cAACMe)(cAAC)Cy], in the formal oxidation state zero (**Figure 15b,c**) [66].

Lithium, another versatile alkaline metal, can form numerous complexes with NHC that have the ability to deprotonate. Alkaline metal bases are commonly used as deprotonating agents for bis(imidazolium) and related salts. Similarly, bis (trimethylsilyl) amides (HMDS) of sodium, potassium, and lithium are regularly used for deprotonation due to their bulkiness and low nucleophilicity. Fehlhammer's et al. synthesized the lithium complexes that involve monoanionic BH2-connected bis (NHC)s, having terminal groups of methyl, ethyl, and isopropyl. However, they have solely served as intermediates to produce corresponding gold, palladium, and platinum complexes. Later, Smith et al. synthesized an air and temperature-sensitive lithium complex (**Figure 16**) having tert-butyl analogue. Despite their high sensitivity, the complex was successfully characterized through NMR and single-crystal X-ray diffraction analysis. The complex forms a dimer consisting of two inequivalent lithium ions with coordination numbers of two and three, respectively [66].

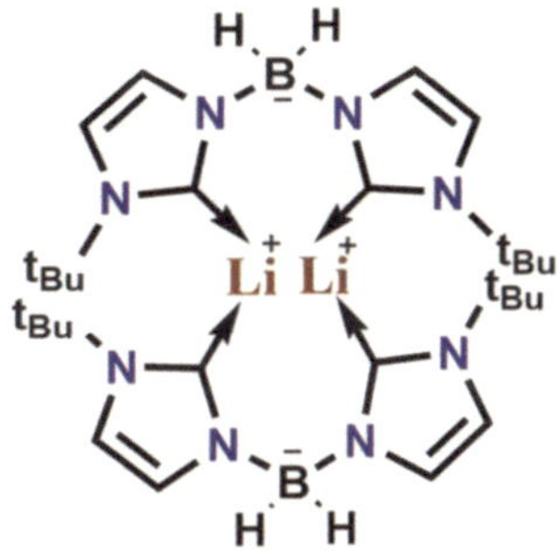

Figure 16.
Lithium complex with imidazolium-based NHC having tert-butyl analog.

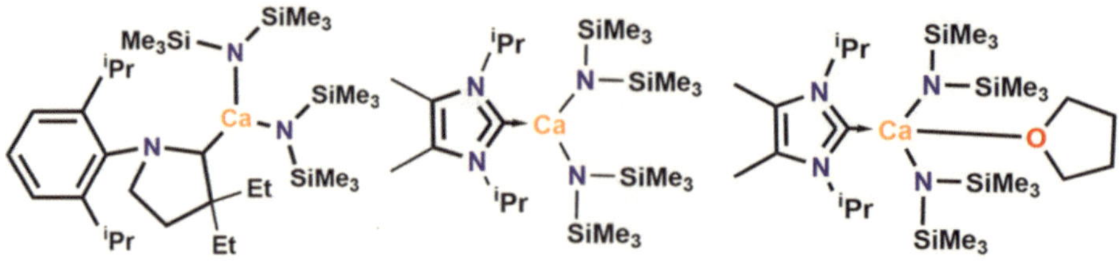

Figure 17.
Typical examples of calcium NHC complexes with imidazolium derivatives based NHC.

Calcium on the other hand, multifaceted alkaline earth metal which shows great potential in organometallic chemistry due to its high lewis acidity and large ionic size. It was investigated that when it comes to stabilizing calcium amides, smaller NHCs are a superior choice in comparison to larger NHCs. **Here, the point of interest was the formation of calcium-based reducing agents such as hydrides and amino boranes.** The utilization of sterically unencumbered NHCs enables the extraction of Me_2N = BH_2 from $Ca(NMe_2BH_3)$ or Ca-$(NMe_2BH_2NMe_2BH_3)$ units by reacting with calcium amido borane complexes. This procedure yields a thermally-stable calcium complex that is highly enriched with amido borane [67]. Some of the examples of typical calcium NHCs complexes is presented in **Figure 17**.

Recent years have seen significant improvements in the production of molecular alkaline earth metal catalysts and reagents. Beryllium and magnesium molecular aryloxide and alkoxide complexes as prospective starting points for the synthesis of light, alkaline earth metals. Bis-NHC beryllium/magnesium dichloride complex, (sIPr)2$BeCl_2$/(sIPr)2$BeCl_2$ is created when beryllium chloride or magnesium chloride reacts with two equivalents of 1,3-diisopropyl-4,5-dimethylimidizol-2-ylidine (presented in **Figure 18**). The terminal aryloxide (sIPr)Be$(ODipp)_2$ and alkoxide dimer $[(sIPr)Mg(OEt)Me]_2$, $[(sIPr)Mg(OEt)Brl]_2$ and $[(sIPr)Be(OEt)Cl]_2$ are produced with lithium diisopropylphenoxide (LiODipp) or sodium ethoxide (NaOEt), respectively depicted in **Figure 18** [68].

2.2 Coordination with p-block element

Synthetic targets for stable heavy main group element radicals are difficult. Despite several methods for stabilizing such odd-electron entities, there are not many heavier pnictogen-centered radicals. Radical cations with two coordinated pnictogen centers [(MeCAAC)EGa(Cl)L]$[B(C_6F_5)_4]$ (MeCAAC = $[H_2C(CMe_2)_2NDipp]C$; Dipp = 2,6-I-$Pr_2C_6H_3$; E = Sb, Bi, As; L = $HC[C(Me)NDipp]_2$) synthesized by one-electron oxidation of L(Cl)Ga-substituted pnictinidenes (MecAAC)EGa(Cl)L (E = Bi^{III}, Sb^{II}, As^{I}) presented in **Figure 19** [69].

Synthetic chemists have used NHCs, cyclic alkyl amino carbenes, and aberrant NHCs as neutral ligands for stabilizing odd compounds containing main group elements as a result of the isolation of stable NHC. Below are a few of the instances showing the silicon NHC complex presented in **Figure 20** [70].

2.3 Coordination with transition metal

The two most common techniques for producing transition metal-NHC complexes are carbene transfer procedures or the utilization of free carbenes (isolated or synthesized *in situ*). Due to their excellent ligand characteristics, stable NHCs exhibit a rich

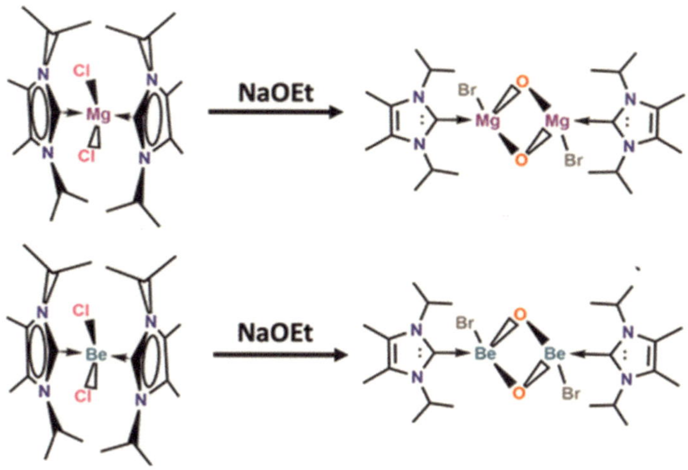

Figure 18.
The synthesis of Be and Mg NHCs [(sIPr)Mg(OEt)me]$_2$ and [(sIPr)Be(OEt)Cl]$_2$.

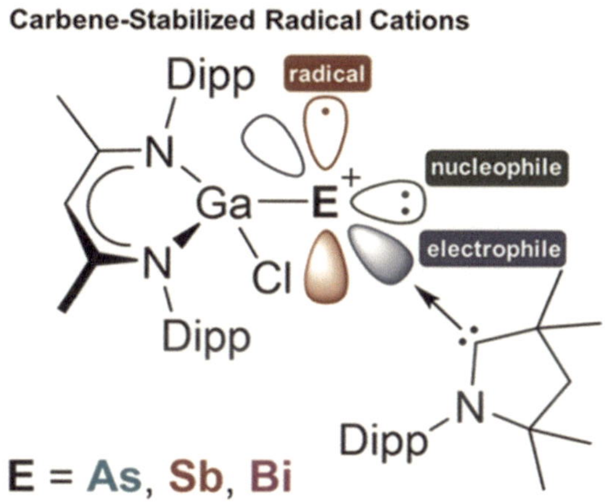

Figure 19.
A rare examples of main group NHCs complex [(MeCAAC)EGa(Cl)L][B(C_6F_5)$_4$] (MecAAC = [H_2C (CMe_2)$_2$NDipp]C; Dipp = 2,6-I-$Pr_2C_6H_3$; E = Sb, Bi, As; L = HC[C(me)NDipp]$_2$). The figure is reprinted with permission from ref. [69].

coordination chemistry and have a wide range of application such as catalyst components [71]. Five-membered rings are by far the most prevalent derivatives, which can have different substitution patterns and ring sizes ranging from four to six. Numerous inorganic counterparts have also been created, in which isolobal fragments have been used in place of the N_2C: moiety.

The potential of carbenes with an inorganic backbone as ligands in transition metal chemistry is illustrated, given examples where it acts as a better-donor than classical carbenes. This also highlights the significant impact of the inorganic backbone on the carbon ligand's coordination capacity. Currently, being researched is the synthesis of analogues with less bulky substituents on the nitrogen atoms next to the carbene carbon. An examples of W NHCs is illustrated where imidazolium based OTf- is complexes with W by the exposer with radiation presented in **Figure 21** [71].

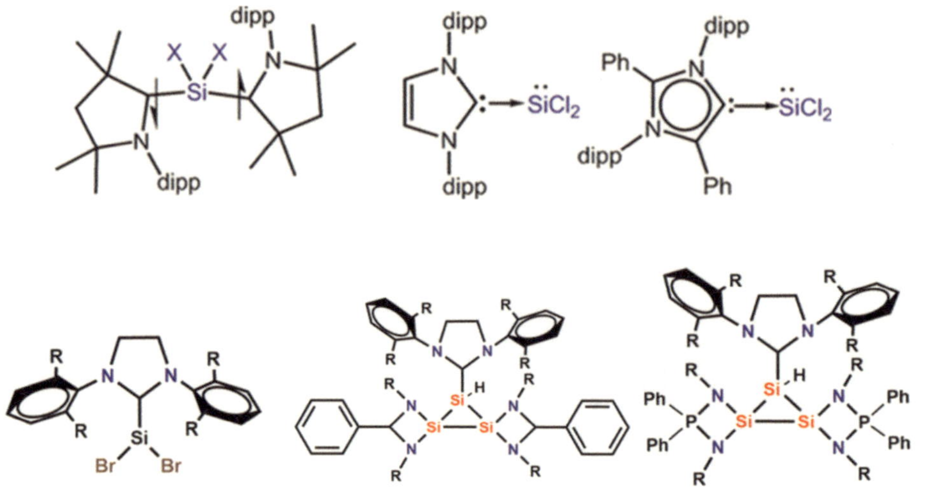

Figure 20.
Few examples of SI NHC complexes.

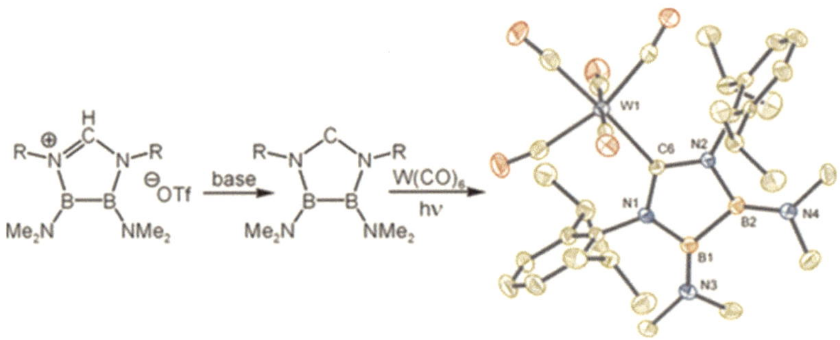

Figure 21.
A W NHCs complex with imidazolium based OTf-. Reprinted with the permission from ref [71].

The chemotherapy medication cisplatin $PtCl_2(NH_3)_2$ has gained significant importance. The development of new platinum-based medications with lower toxicity is a vibrant and highly active area of research because despite having a number of serious side effects that significantly restrict the drug's efficacy. The usage of NHCs ligands has shown to be a potential method for creating cisplatin alternatives a such example is depicted in **Figure 22**. For the creation of prospective metallodrugs, not only platinum NHC complexes but also Au, Ag, Pd, Cu, Ru, or Ir NHC complexes have been studied. The antiproliferative effect of platinum (II) NHCs complexes carrying dithiocarbamate ligands was tested against several cancer cell lines. These complexes exhibit remarkable stability in solution [72].

Because of their powerful propensity to donate electrons, NHCs make excellent alternatives for phosphine ligands because they display more permanent metal–ligand interaction, which increases the robustness of the complex. In addition to complexes made of noble metals, Fe-NHC complexes have lately drawn a lot of interest because they might be used as affordable, ecologically friendly, and potent catalyst precursors. In homogeneous catalysis, well-defined and *in situ* generated FeNHC complexes have been used; the majority of these include monodentate NHC ligands.

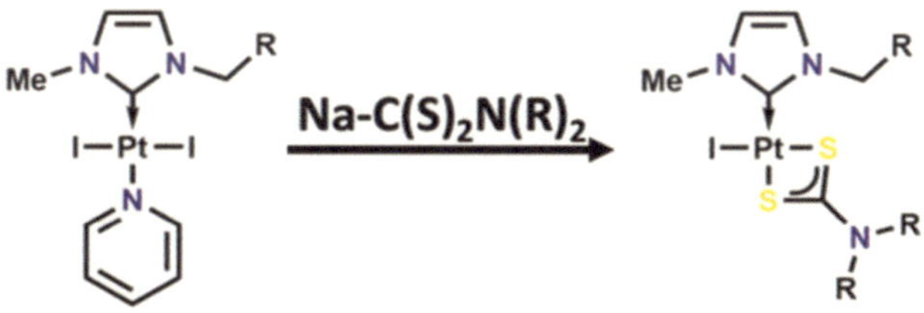

Figure 22.
Dithiocarbamate ligand is used with platinum (II) NHCs complexes as an alternative for cisplatin $PtCl_2(NH_3)_2$.

Figure 23.
Bis(imidazo[1,5-a]pyridine-3-ylidene is used with the iron to form Fe-NHC complex.

Nonheme iron(II) catalysts with tetradentate nitrogen ligands have cis-coordination geometry, resulting in two cis-oriented binding sites at the iron center. Instances of this include BPMEN and BQEN (**Figure 23**), which each use two pyridine/quinoline and amine donors. It has been demonstrated that these instances provide iron complexes useful for difficult C—H oxidation processes. Initial results on the hydrosilylation of acetophenone catalyzed by the novel Fe-NHC complex [73] using the bis(imidazo[1,5-a]pyridine-3-ylidene) ligand, which has two amine moieties in addition to two NHC groups.

Using the transmetalation reaction between the equivalent silver(I) complex $[Ag_2L_2](PF_6)_2$ and $KAuBr_4$ in the presence of $AgPF_6$, one may create gold(III) complexes ($[AuL_2](PF_6)_3$ (L = diNHC)). t-butyl and other bulky wingtip substituents hinder the production of the bis(diNHC) gold(III) complex, which results in the synthesis of the dinuclear gold(I) complex $[Au_2L_2](PF_6)_2$ following a redox rearrangement presented in **Figure 24** [74].

The sodium salt $Na[Ph_2P(O)eC(H)eSO_2Ph]$ has been synthesized and isolated as a methanide ligand system for transition metal complexes. The structural elements of this ligand have been shown to facilitate the separation of nucleophilic late transition metal NHCs complexes. By using a straightforward salt metathesis procedure, the appropriate Ru(cymene) chlorido complex was quickly made accessible. However, dehydrohalogenation of the chlorido complex resulted in the development of a cyclometallated Ru complex as opposed to the NHCs complex, as was the case with previously reported thio- and iminophosphoryl-tethered ligand systems (**Figure 25**) [75].

Iron(0) species with coordination unsaturation have drawn a lot of attention in recent years due to the resurgence of iron catalysis [76]. These species are thought to be reactive intermediates in processes like alkyne cyclotrimerization, the Diels-Alder reaction between dienes and alkynes, hydrogenation of alkenes and alkynes, and

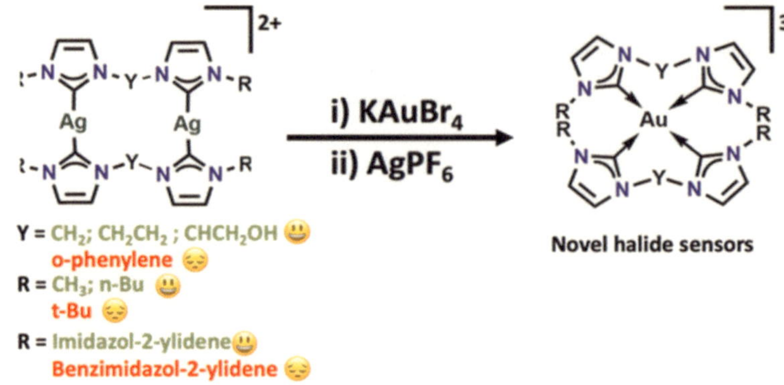

Figure 24.
Preparation of Au-NHCs using reinstallation reaction where $[Ag_2L_2](PF_6)_2$ and $KAuBr_4$ reacts with $AgPF_6$, yields gold(III) complexes ($[AuL_2](PF_6)_3$ (L = diNHC)).

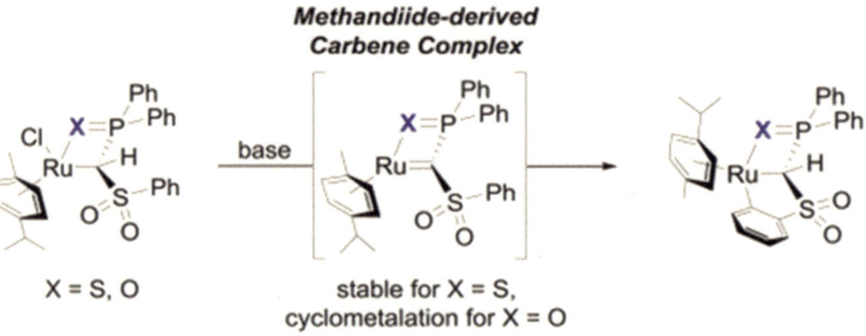

Figure 25.
Dehydrohalogenation of the chlorido complex resulted in the development of a cyclometallated Ru complex. Reprinted with permission from ref [75].

others. Coordinatively unsaturated iron(0) complexes are known to be responsive to C—H bond activation processes. For instance, it is known that in an intramolecular C (sp3)H bond oxidative addition reaction, four coordinate iron(0) complex [Fe(PMe3) 4] in solution phase to reversibly transform into the iron(II) complex [Fe-$(PMe_3)_3(CH_2PMe_2)H$] (**Figure 26**) [76].

2.4 Complexes with inner transition metal

Due to their distinct electrical characteristics, lanthanide organometallic reagents and catalysts have gained increasing interest over the past two decades in the activation of chemical bonds and polymerization. Due to their excellent stability, lanthanide amides are one of the most useful groups among them. In order to investigate their reactivity, numerous lanthanide amides have been synthesized using various kinds of anionic spectator ligands. Despite the fact that certain of the homoleptic Ln[N $(SiMe_3)_2]_n$ complexes have already been employed as organometallic reagents and catalysts, there are currently no practical methods to control their reactivity and selectivity.

NHC ligands can stabilize the heteroleptic amido-ytterbium(II) alkynyl complex and anilido complex. This is most likely due to the NHC's relativistically strong

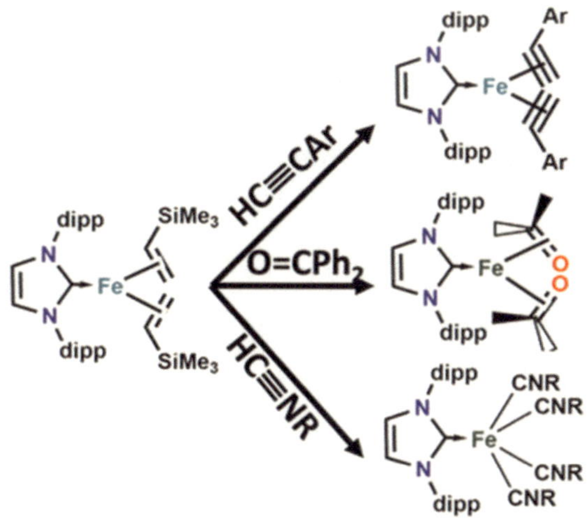

Figure 26.
Addition reaction to yield different derivatives on the Fe-NHCs complexes.

interaction with the divalent ytterbium ion and appropriate steric effect. Furthermore, rac-LA's ROP has strong activity for NHC-stabilized samarium-(II) bis(amido) complexes even in the absence of an alcohol initiator (**Figure 27**). Ongoing research examines the reactivity of heteroleptic ytterbium(II) alkynyl and anilido complexes stabilized by NHC as well as the catalysis of samarium(II) bis(amido) complexes [78].

3. Use of NHCs in organic synthesis

Divalent carbon compounds called carbenes are a common building component in chemical synthesis. They are specifically employed in the synthesis of small rings. Diazo compounds and ketene are traditional sources of carbenes and easily undergo

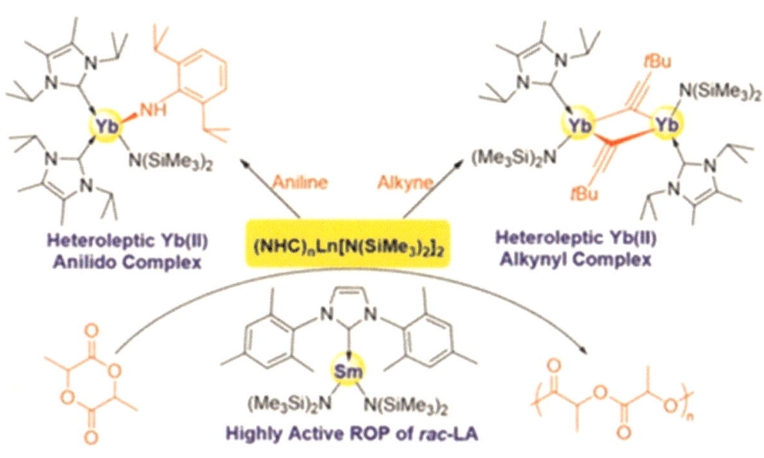

Figure 27.
NHC ligands can stabilize the heteroleptic amido-ytterbium(II) alkynyl complex and anilido complex. Rac-LA's ROP has strong activity for NHC-stabilized samarium-(II) bis(amido) complexes even in the absence of an alcohol initiator reprinted with permission from ref. [77].

thermal, photochemical, or metal-induced cleavage. Additionally, a-haloalkyl lithium compounds [79] and trihalomethylmercury compounds have been shown to be promising organometallic precursors. The choice of precursor and the manner of formation both affect the reactivity of the carbene species.

NHCs complexes contain metal-stabilized carbenes. These compounds can be divided into two classes: the "Fischer-type" and the "Schrock type", named after their discoverers. Subdivision into "carbene complexes" and "alkylidene complexes", which is encountered now and then, makes little sense didactically and has at most only historical significance [80].

4. Photophysics of NHCs

The NHC-based metal complexes exhibit very interesting photophysical properties and many of them showed applications in OLEDs. Therefore, the origin of the electronic transitions in the NHC metal complexes and fluorescence emission bands has been discussed. A brief note on the design principles and the structure of NHC metal complex and their photophysical properties are highlighted in the book chapter.

Fluorescent/luminescent complexes that consist of N-heterocyclic carbine (NHC) ligands have been studied quite extensively over the last two decades [81–83]. Unlike many of the other ligands, the NHC ligands are strong σ-donors and stabilize most of the coordination complexes which in turn have shown very interesting applications in material chemistry [83–89]. The NHC ligands tend to push the d−d transitions to a shorter wavelength (high-energy) side thereby the designed metal−NHC complexes becomes fluorescent/luminescent (**Figure 28**). Photophysical properties such as UV-absorption, fluorescence, phosphorescence, and singlet-state lifetimes of NHC-metal complexes are important indicators to assess the utility of the compounds in opto-electronic applications. Hence, the photophysical properties of different NHC-metal complexes were studied and documented in the literature [81–83].

For example, photophysical properties of the coordination polymers supported by a pyridyl-substituted NHCs ligand were reported along with the monomers [90]. Electronic absorption spectra of monomers NHC1 and NHC2 exhibit absorption bands between 260 and 270 nm which are attributed to the π–π* transition and these are

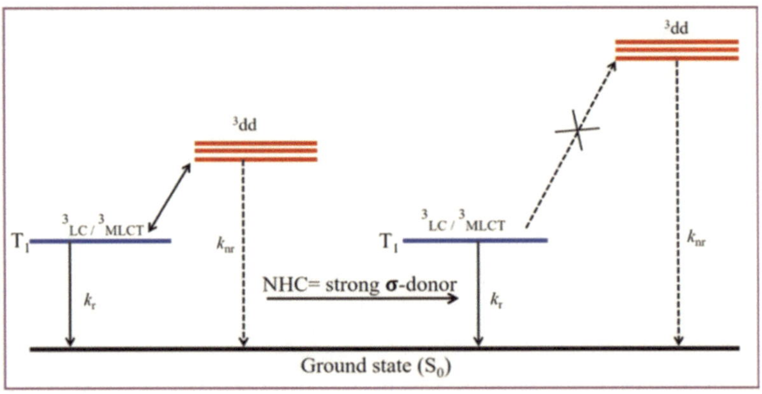

Figure 28.
Schematic representation of excited states of metal-NHC complexes emphasizing the NHCs as strong-donors ligands and destabilization of 3dd metal cantered states [83].

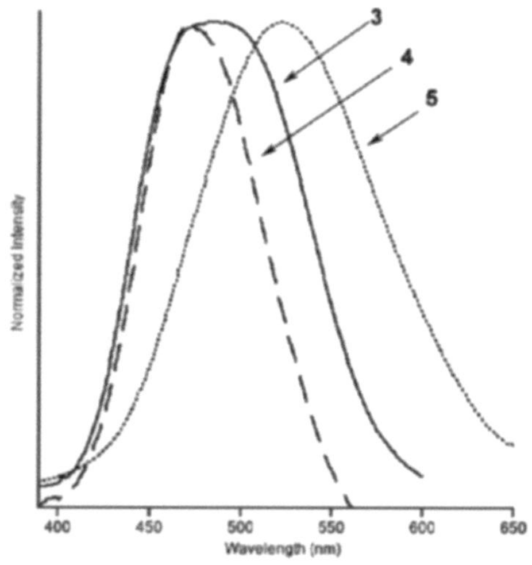

Figure 29.
Room temperature emission spectra of solid samples of the nitrile-containing polymers NHC3 (3), NHC4 (4), and NHC5 (5). The nitrate-containing polymer has a spectrum nearly identical to that of NHC4.

similar to the precursor of [HCH_3impy]PF_6 molecule. Because of the dissociation, the polymeric NHC-complexes show similar absorption respective to their monomers in solution which is evident from the electronic absorption spectra of NHC3, NHC4, NHC5 & NHC6 and its similarity with that of the monomer, NHC2. It was found that that all the coordination polymers as well as NHC monomers are photo-luminescent upon UV excitation in both solution and solid-state phases. While carbine precursors exhibit an emission band at 435 nm, the monometallic Au-containing monomer species, NHC2, shows an emission band at 541 nm in solid state. It was observed that upon cooling the solid samples to 77 K, the emission band shifts toward blue side (506 nm). Similarly, the nitrile-containing coordination polymers show intense bands at 480, 474, 522, and 469 nm at room temperature for the acetonitrile, benzonitrile, benzylnitrile, and nitrite containing polymers NHC3, NHC4, NHC5 & NHC6, respectively. Upon cooling these solid samples to 77 K, the emission bands sharpen and blue-shifted (**Figures 29** and **30**).

The same research group has also reported four NHC coordination polymers by modifying pyridyl-substituted N-methylimidazole backbones, {[AuAg(CH_3impy)$_2$(L)](BF_4)$_2$}$_n$ and {[AuAg(CH_3impy)$_2$(NO_3)]NO_3}$_n$ where L is CH_3CN, C_6H_5CN, or $C_6H_5CH_2CN$ [91]. The electronic absorption spectra and fluorescence emission properties are found to be similar to the other polymers NHC3, NHC4, NHC5 & NHC6. Further, the data also suggest that the optical properties of the polymers can be tailored by making appropriate changes to the ligands and composition of the polymers.

In another study, the photophysical properties of picolyl-substituted NHC-bridged triangular complexes of Ag(I) and Cu(I) were reported in different solvents [92]. In acetonitrile, the electronic absorption spectra of Cu-containing complexes NHC10a, NHC10b, and NHC10c exhibit two distinct absorption bands: a single absorption band between 253 and 265 nm and additional peaks at 374, 366, and 373 nm, respectively. Based on the ligand precursor absorption study, these absorption bands are attributed to the pyridyl π-π^* transition and MLCT, respectively. Upon UV-excitation, the Cu-complexes NHC10a, NHC10b, and NHC10c emit at 459, 427, and 429 nm in acetonitrile, whereas in the solid state, these bands shift to 433, 429, and 432 nm,

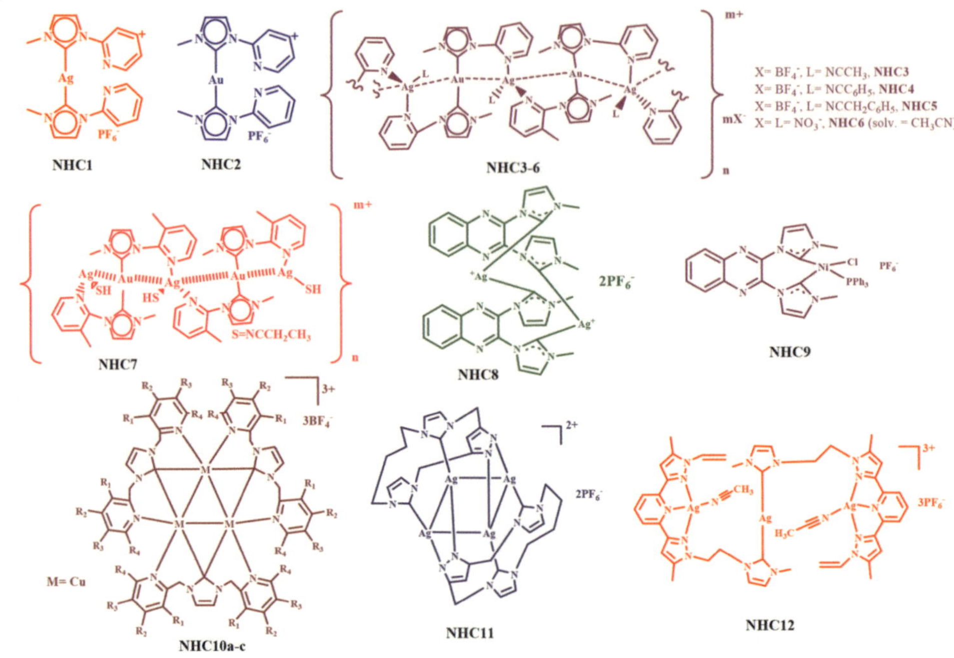

Figure 30.
NHC-based metal complexes.

respectively. The emission maxima of the complexes are roughly following their substitution pattern on the pyridine ring. The excitation spectra suggest that NHC10a, NHC10b, and NHC10c show that these emission bands are effectively populated by two states. The emission maxima hardly change with a change in the solvent. On the other hand, the silver-containing NHC-bridged triangular complexes are found to be less emissive partly because of the dissociation of the complexes in acetonitrile (**Figure 31**).

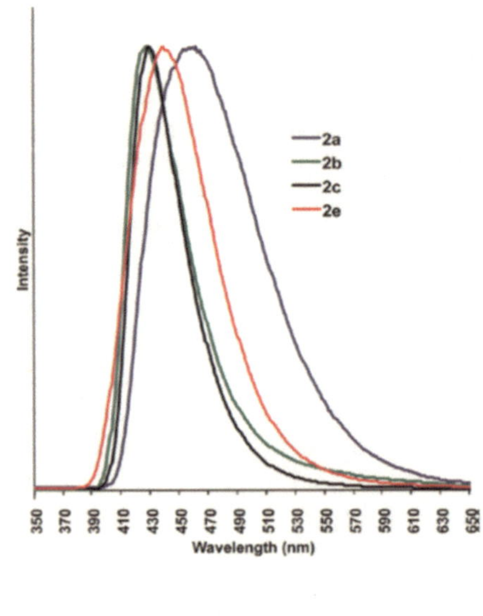

Figure 31.
Normalized room-temperature emission spectra of NHC10a, NHC10b, and NHC10c in acetonitrile.

Photophysical properties of pyrazole-linked bis(imidazolium) based multidentate mixed N,C-ligands, and corresponding tetranuclear and trinuclear silver–NHC complexes were reported [93]. UV-visible absorption spectra of precursor $H_2L_2PF_6$, silver–NHC complex NHC8, and nickel–NHC complex NHC9 in acetonitrile show maximum absorption of 248, 245, and 251 nm, respectively. The small spectral difference is attributed to the absorption of the quinoxaline moiety in these three compounds and the dominance of the HOMO electrons localized on quinoxaline π-orbitals instead of metal-ligand bonding in the silver(I) and nickel(II) complexes. The solid-state luminescent properties reveal that the emission spectrum of precursor ligand shows a single emission band at 416 nm and nickel–NHC complex NHC9 exhibits a single broad emission at 471 nm. However, the silver–NHC complex NHC8 was found to be non-fluorescent due to the decomposition of the complex upon excitation. It was also found that by varying the N-alkyl groups of the silver cluster, photoluminescent properties can be tuned, considering the fact that luminescence originates from intraligand or metal-based excitation states. Luminescence properties of imidazolium salts $[H_3L_1](PF_6)_2$ and $[HL_4](PF_6)$, and multinuclear silver complexes that consist of pyrazole-functionalized NHC ligands were reported [94] and it was observed that the ligand $[H_3L_1](PF_6)_2$ shows a single broad band at 401 nm, whereas its tetranuclear complex NHC12 shows two broad emission bands centered at 362 and 463 nm. Similarly, the ligand $[HL_4](PF_6)$ exhibits emission bands at 371 and 461 nm, while its silver complex NHC11 emits a broad band at 442 nm. The intense emission behavior of the Ag(I)–NHC complexes is attributed to the intraligand excited states or metal-centered excited states (**Figures 32** and **33**).

Photophysical properties of cyclometalated alkynyl platinum(II) complexes (NHC13-18) baring tridentate pyridine-based NHC ligands were studied in detail [95]. The absorption spectrum of $[Pt^{II}(CNC)Cl][PF_6]$ consists of a strong absorption band at 272 nm and a moderately intense absorption band at 364 nm. Molar absorption coefficients (ε) are estimated to be of the order of 10^3–10^4 dm^3mol^{-1} cm^{-1} in acetonitrile. The shorter wavelength absorption band cantered 272 nm is assigned π-π* transitions of the C-N-C pincer ligands, whereas the longer wavelength absorption band (364 nm) is assigned to the metal-to-ligand charge-transfer (MLCT) [dπ(Pt) → π*(C^N^C)] transitions which are partly mixed with the IL π-π* transitions of the C^N^C pincer ligands. However, upon substitution of alkynyl ligands in place

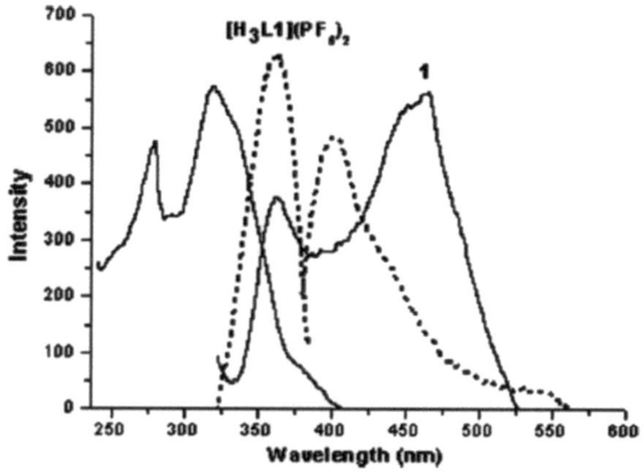

Figure 32.
The solid-state emission (right) and excitation (left) spectra of $[H_3L_1](PF_6)_2$ (dashed) and NHC8 (solid).

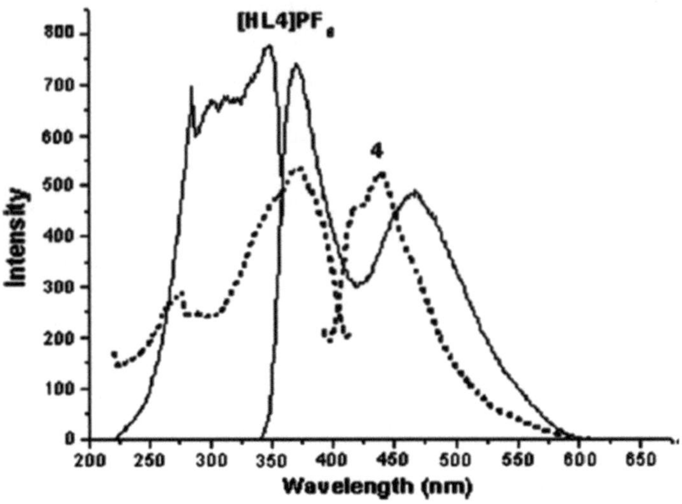

Figure 33.
The solid-state emission (right) and excitation (left) spectra of [HL_4](PF_6) (dashed) and NHC9 (solid).

of chloro ligand, the low-energy absorption bands are red-shifted and situated between 383 and 471 nm with ε values of the order of 103 dm^3mol^{-1} cm^{-1} whereas the shorter wavelength band remained unchanged. The high-energy absorption band (277–291 nm) is assigned to the IL π-π* transitions of the C≡CR and C^N^C pincer ligands and whereas the lower-energy band (383–471 nm) is attributed to as an admixture of the MLCT [dπ(Pt) → π*(C^N^C)] transition and alkynyl-to-C^N^C ligand-to-ligand charge-transfer (LLCT) [π(C≡CR)π* → (C^N^C)] transition. As compared to the alkylalkynyl complex NHC13, the low-energy absorption band of the arylakynyl complexes NHC14-18 is red-shifted which is attributed to the π-donating nature of the arylakynyl ligands. The longer wavelength absorption maxima of the NHC14-18 complexes are as follows: NHC14 (376) > NHC15 (405) > NHC16 (422) > NHC17 (439) > NHC18 (471 nm), which is consistent with the π-electron-donating ability of the arylalkynyl ligand. The red shift of the longer wavelength absorption band is assigned to the negative solvatochromism in which the dipole moments of the complexes decrease upon electronic excitation (**Figure 34**).

Fluorescence study reveal that the chloroplatium(II) precursor complex, [Pt^{II}(C^N^C)Cl][PF_6], is non-fluorescent in acetonitrile solvent due to the presence of low-energy d–d orbitals that promote the quenching of the 3MLCT/3 IL state through nonradiative processes. On the other hand, the alkynylplatinum(II) complexes NHC13-17 exhibit strong structureless fluorescence emission bands in acetonitrile whose emission maxima are found between 497 and 631 nm. (**Figure 35**). The fluorescence emission maxima of complexes NHC13-17 vary based on the substituent on the phenyl ring of the alkynyl ligands. For example, the complex NHC17 which contains the most electron-rich methoxy group show emission maxima at 631 nm whereas the complex NHC14, which consists of the most electron-poor trifluoromethyl substituent exhibits emission maxima at 500 nm. This shift is consistent with the assignment of 3MLCT/3LLCT in which the electron-donating substituents on the phenyl ring raises the energy level of the dπ(Pt) and π(C≡CR) orbitals which in turn the lowers CT energy. In other words, the fluorescence emission maxima of the complexes NHC13-17 can be modulated from green to yellow with the substitution of alkynyl/arylalkynyl ligands. More interestingly, unlike many of the

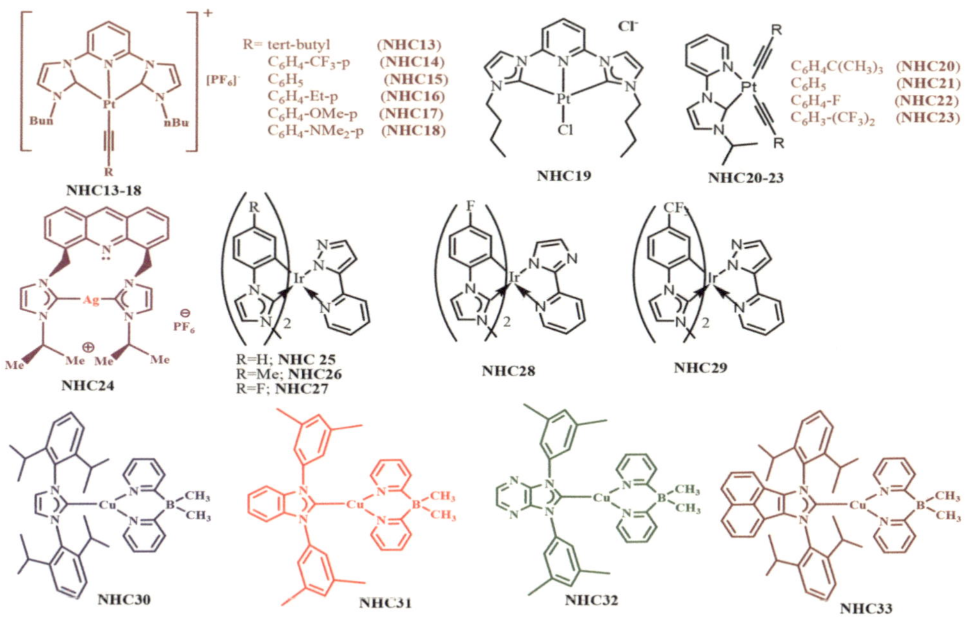

Figure 34.
NHC based metal complexes.

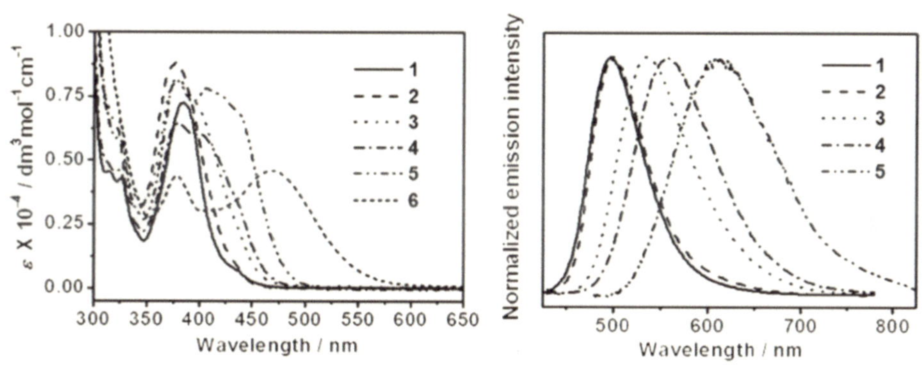

Figure 35.
Electronic absorption spectra of NHC13–18 in acetonitrile (left) and normalized emission spectra of NHC13–17 in degassed CH_3CN at 298 K.

NHC-metal complexes, the complexes NHC13-15 are found to be highly fluorescent, which is evident from the estimated fluorescent quantum yields. The observed large Stokes shifts and long fluorescence emission lifetimes (microseconds) further confirm the emitting state has a triplet character and is assigned to the 3MLCT [dπ(Pt)→π* (C^N^C)] state with the mixing of a 3LLCT [π(C≡CR)π*→(C^N^C)] state. In solution, the complex NHC18 is found to be non-fluorescent at room temperature, which is attributed to photoinduced electron transfer (PET) quenching of the emissive 3MLCT state by amine functional group. The non-fluorescent nature can also be interpreted in terms of the low-lying non-emissive 3LLCT excited state that originated from the higher-lying π (C≡CR) orbital of the aminophenyl alkynyl unit. Further, the complexes NHC13-18 are fluorescent in the solid state as well as in glass matrices at 77 K, in which vibronically structured emission bands were observed between 461 and 512 nm.

Lee and co-workers have reported photophysical properties of a square-planar Pt (II) complex (NHC19) baring a pyridine-based pincer-type NHC ligand, $[Pt^{II}(C^\wedge N^\wedge C)(Cl)]Cl$ where C^N^C is 2,6-bis(1-butylimidazol-2-ylidenyl)pyridine [96]. The hydrated form of the complex shows exhibits strong fluorescence emission at 614 nm (orange) with a lifetime of 34 μs (quantum yield, φ = 0.30), whereas the anhydrate form of the complex shows intense fluorescence emission at 550 nm with a lifetime of 22 μs (quantum yield, φ = 0.35). In other words, the photoluminescence of the complex in the solid state switches from orange to green upon hydration and dehydration, respectively. Based on TDDFT calculations, the lowest electronic transition is attributed to the metal-metal to ligand charge transfer (MMLCT) and the observed fluorescence emission band to the 3MMLCT (**Figure 36**).

In another study, the photophysical properties of a new class of cyclometalated pyridine NHC (NHC) Pt(II) complexes with electronically different alkyne derivatives (NHC20-23) have been investigated [97]. The electronic absorption spectra of NHC20-23 in dichloromethane solvent display a strong band below 300 nm and a moderately intense absorption band between 320 and 420 nm. The spectroscopic data combined with the DFT and TDDFT calculations demonstrate that shorter wavelength absorption bands mainly originate from the 1LLCT[πalk→πN^C*] and 1MLCT [d→πN^C*] transitions. The UV/vis absorption spectra of the four complexes were also studied in different solvents and a negative solvatochromic behavior was observed, which demonstrates the different dipole moments for both ground and excited states. It is interesting to note that this observation is similar to the bipyridine Pt(II) dialkyne complexes behavior which is reported in the literature [98–100].

The complexes NHC20–23 are non-fluorescent in solution at room temperature which is quite different than the previously reported bipyridine Pt(II) and bis-NHC Pt (II) complexes [98–101]. The solvent molecules promote the thermal quenching of the fluorescence emission in solution at room temperature [102]. However, the 1 wt % of complexes NHC20-23 in poly(methyl methacrylate) (PMMA) matrix gives a broad and structureless single fluorescence emission band whose maxima is found between 459 and 480 nm. Surprisingly fluorescence quantum yields are found to be very high, 51–59%. The fluorescence emitting state has been attributed to a mixed excited state of 1LLCT, 3LLCT, 1MLCT, and 3MLCT based on related complexes fluorescence behavior [103–105]. It was observed that with an increase in the weight percentage of the complex from 10 to 80 wt % in the PMMA films, a new fluorescence emission band emerged in the longer wavelength region (575–595 nm) at the expense of shorter

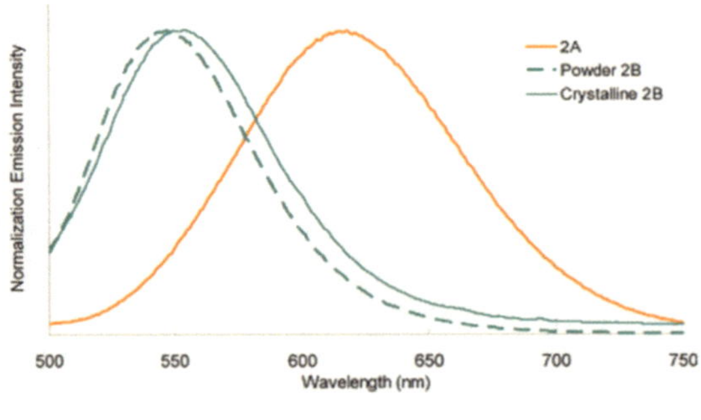

Figure 36.
Fluorescence emission spectra of hydrated and dehydrated NHC19 at 300 K in the solid state. 2A.

wavelength band w-energy emission was observed. However, the overall fluorescence quantum yield decreased to 20 wt % due to an increase in the concentration and its self-quenching [106]. Further, based on the excitation spectral analysis, the longer fluorescence emission band has been attributed to the excimer emission rather than a ground-state dimer which generally arises due to the strong metal-metal interactions. The ground-state dimer formation was ruled out from the UV-visible spectral measurements in which the MMLCT band is absent with an increase in the concentration of the molecules. The PMMA films show different fluorescence emission bands at 460 and 570 nm due to the presence of monomeric and excimeric species and cover the entire visible region which essentially produces the white light. The CIE coordinates of complex NHC23 at a concentration of 20 wt % in PMMA were found to be x = 0.31 and y = 0.33 and these are very close to the white light CIE coordinates (x = 0.33, y = 0.33) (**Figures 37** and **38**).

Bis-NHC silver complex baring acridine chromophoric derivative was synthesized and photophysical properties were studied [107]. The complex NHC24 is shown to be fluorescent and emits blue light in solution state with CIE coordinate x = 0.154 and y = 0.061 (**Figure 39**).

Photophysical and electrochemical properties of five iridium bis(carbene) complexes NHC25, NHC26, NHC27, NHC28, and NHC29 were reported by Cheng and co-workers [108]. Electroluminescent (EL) devices of NHC26, NHC27, and NHC29 were fabricated as dopant emitters. The EL devices emit deep-blue light with external quantum efficiencies of 15.2, 14.1, and 7.6%, and the CIE coordinates were estimated to be of (0.14, 0.27), (0.14, 0.18) and (0.14, 0.10), respectively.

Phosphorescent three-coordinated NHC copper complexes that consist of anionic nonconjugated di(2-pyridyl)-dimethyl borate ligand have been reported by Thompson and co-workers [109]. The complexes NHC30, NHC31, and NHC32 emit sky-blue, yellow, and orange light in solid-state with quantum yields of 0.8 to 0.7, and to 0.16, respectively. Further, the emission lifetimes are observed to be 11, 15, and 7.5 μs, respectively which shows a modest increase in the emission lifetimes: 36, 17, and 21 μs, respectively at 77 K. Therefore, the emission of the NHC30, NHC31, and NHC32 complexes at room temperature is attributed to the phosphorescent in nature.

Inspired by the promising photochemical properties of ruthenium polypyridine complexes, Chung and co-workers have synthesized Ruthenium complexes NHC33 and NHC34 of NHC [110]. These NHC complexes are designed to be structural

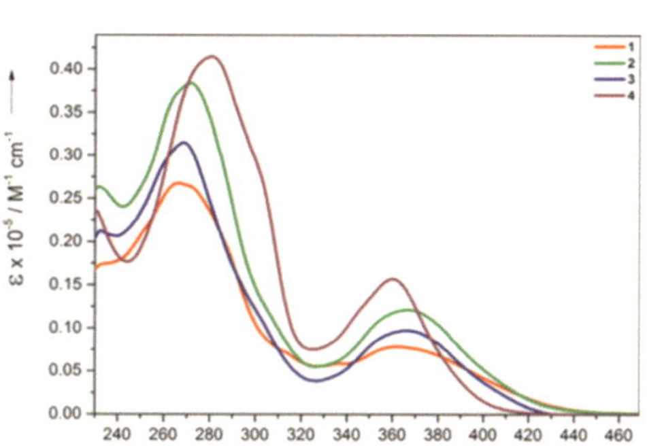

Figure 37.
UV/Vis spectra of NHC20 (c = 1.62 × 10–5 M), NHC21 (c = 4.58 × 10–5 M), NHC3 (c = 1.87 × 10–5 M), and NHC4 (c = 3.39 × 10–5 M) in CH_2Cl_2.

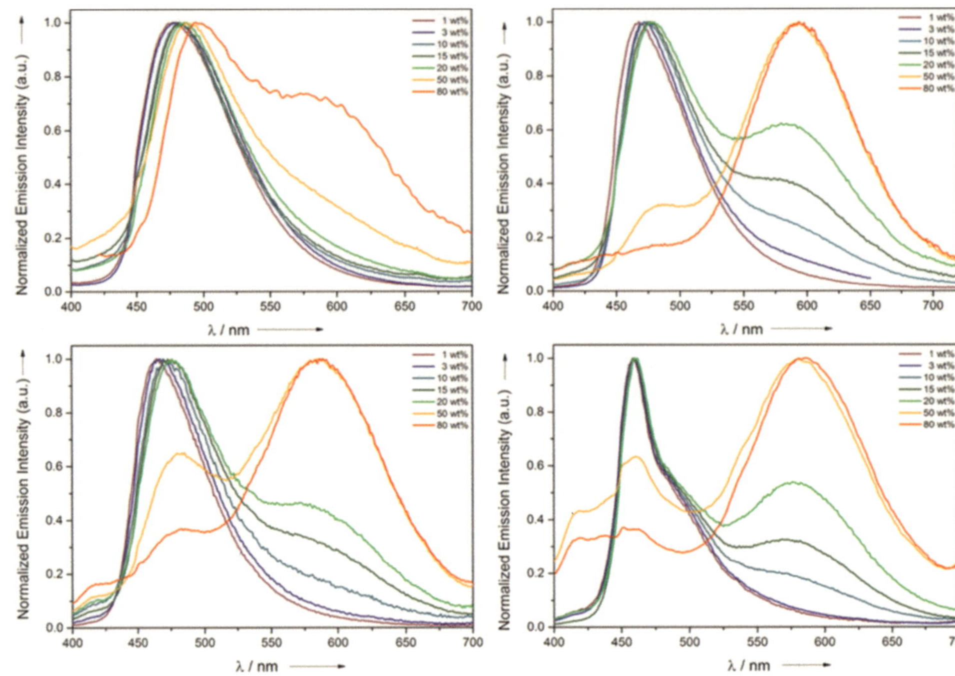

Figure 38.
Concentration-dependent emission spectra of complexes NHC20 (top left), NHC21 (top right), NHC22 (bottom left), NHC23 (bottom right) in PMMA films.

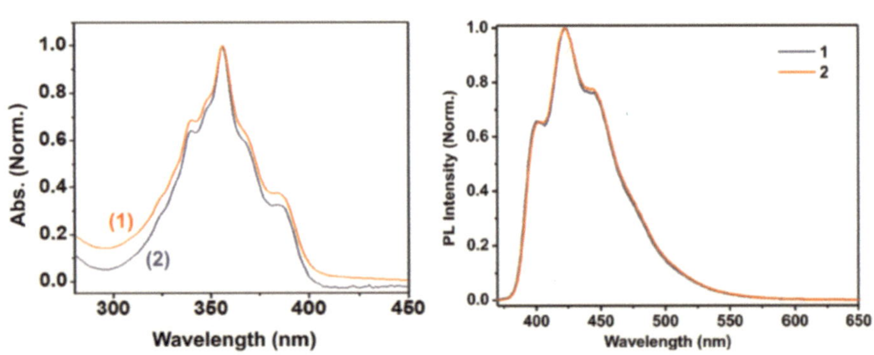

Figure 39.
Electronic absorption and photoluminescence of NHC24 complex (red line) and its precursor ligand (blue line).

analogs of $[Ru(bpy)_3]^{2+}$ and $[Ru(terpy)_2]^{2+}$, respectively. The electronic absorption spectrum of the Complex 14 is blue-shifted (368 nm) as compared to the $[Ru(bpy)_3]^{2+}$ complex. Similarly, complex $15(PF_6^-)$ shows two absorption bands with a maximum of 343 and 382 nm, and as compared to the $[Ru(terpy)_2]^{2+}$, these two band are highly blue-shifted. The blue-shifts of the MLCT absorption bands are assigned to the relative electron-richness of the NHC ligands over the bipyridine and terpyridine ligands. Complex NHC33 is nonemissive in acetonitrile at room temperature. In sharp contrast to $[Ru(terpy)_2]^{2+}$, the complex 15 (PF_6^-) exhibits a strong fluorescence emission with a maximum at 532 nm. Further, the terpy-like complex NHC34 displays a long fluorescence lifetime of 820 ns in MeCN and 3100 ns in H_2O (**Figure 40**).

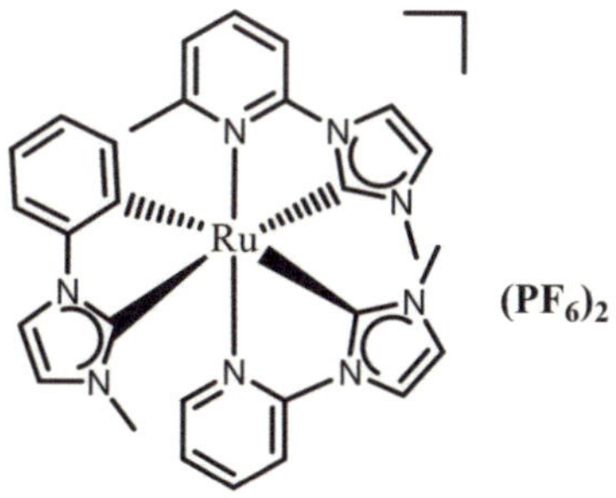

NHC33

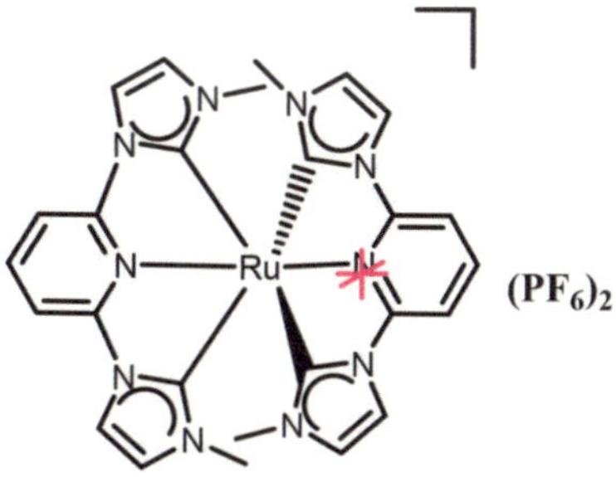

NHC34

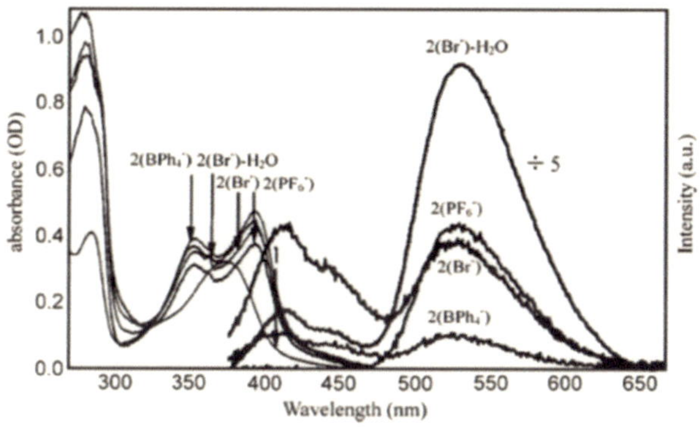

Figure 40.
Absorption and emission spectra, with excitation at 354 nm, of Ru2+ complexes in acetonitrile (if not specified) and water under argon.

5. Conclusion

The strong σ-donating nature of the NHC is the key factor in manipulating the optical properties of NHC complexes. The NHC ligands shift the d–d transitions to a shorter wavelength region and the complexes become luminescent. The studies on NHC-based metal complexes of copper, silver, platinum, iridium, and coinage metals reveal that several metal complexes are luminescent and some of them are even found in applications in organic light-emitting diodes (OLEDs). However, serval studies also reveal that the quantum yields of the metal complexes are not sufficient to be used in applications. This perhaps can be overcome by modifying the NHC ligand with electron-rich substituents which is expected to reduce the HOMO-LUMO gap and push the d-d transitions toward the UV-region. Apart from the NHC, the cyclic (alkyl)(amino)carbene (CAACs) were found to the promising in tuning the photophysical properties. The CAACs reduce the HOMO–LUMO gap as compared to NHCs and the CAAC is found to be more electrophilic in nature. It has been found that luminescent CAAC metal complexes exhibit sub-microsecond lifetimes which make them suitable for OLED applications. The challenge remains open in this research to master the structure, bonding, and property in order to tailor very stable and highly luminescent materials based on abundant and cheap metals such as iron, copper, zinc, and silver complexes.

Acknowledgements

SS thanks the Department of Chemistry, BHU, for infrastructure and IoE – BHU (faculty incentive grant) for financial assistance.

Conflict of interest

There is no conflict of interest.

Author details

Krishanu Bandyopadhyay[1], Abhineet Verma[1], Tamal Ghosh[2], Ravi Kumar Kanaparthi[3], Sudeena Nadendla[3] and Satyen Saha[1*]

1 Department of Chemistry, Institute of Science, Banaras Hindu University, Varanasi, India

2 Department of Chemistry, Motilal Nehru National Institute of Technology, Allahabad, India

3 Department of Chemistry, Central University of Kerala, Kasaragod, Kerala, India

*Address all correspondence to: satyen.saha@gmail.com

References

[1] Davies HML, Manning JR. Catalytic C–H functionalization by metal carbenoid and nitrenoid insertion. Nature. 2008;**451**:417-424

[2] Hopkinson MN, Richter C, Schedler M, Glorius F. An overview of N-heterocyclic carbenes. Nature. 2014;**510**: 485-496

[3] de Frémont P, Marion N, Nolan SP. Carbenes: Synthesis, properties, and organometallic chemistry. Coordination Chemistry Reviews. 2009;**253**(7–8): 862-892

[4] Chinoporos E. Carbenes. Reactive Intermediates Containing Divalent Carbon. Chemical Reviews. 1963;**63**:235

[5] Lappert MF, R.S. Rowe the role of group 14 element carbene analogues in transition metal chemistry. Coordination Chemistry Reviews. 1990;**100**:267

[6] Arduengo AJ, Harlow RL, Kline M. A stable crystalline carbene. Journal of the American Chemical Society. 1991;**113**: 361-363

[7] Sau SC, Pradip KH, Swadhin KM, Michele S, Guy B. Stable abnormal N-heterocyclic carbenes and their applications. Chemical Society Reviews. 2020;**49**(4):1233-1252

[8] Lin IJB, Vasam CS. Preparation and application of N-heterocyclic carbene complexes of Ag(I). Coordination Chemistry Reviews. 2007;**251**(5–6): 642-670

[9] Wanzlick H-W, Schikora R. Aspects of Nucleophilic Carbene Chemistry. Angewandte Chemie. 1960;**72**:494

[10] Igau A et al. 'Analogous.alpha.,. alpha.'-bis-carbenoid, triply bonded species: Synthesis of a stable. lambda. 3-phosphino carbene-.lambda.5-phosphaacetylene'. Journal of the American Chemical Society. 1988;**110** (19):6463-6466

[11] Schubert U. Structural consequences of bonding in transition metal carbene complexes. Coordination Chemistry Reviews. 1984;**55**:261

[12] Bertrand G, Reed R. λ^3-Phosphinocarbenes λ^5-phosphaacetylene. Coordination Chemistry Reviews. 1994;**137**:323

[13] Runbo P, Liancheng H, Yue Z, Xinping W. The dynamic Lewis acid–carbene hybrid: Pushing the electrophilicity of carbenes to the limit. Journal of the American Chemical Society. 2023;**145**(40):21733-21737

[14] Benhamou L, Chardon E, Lavigne G, Bellemin-Laponnaz S, César V. Synthetic routes to N-heterocyclic carbene precursors. Chemical Reviews. 2011;**111**: 2705-2733

[15] Emmanuel BP-R, Karen G-S, Daniel M-E. Mercaptopyridine-functionalized MIC-Pd(II)-PEPPSI complexes: Synthesis and hemilability effect in catalytic applications. Organometallics. 2023;**42**(19):2893-2901

[16] Jazzar R, Soleilhavoup M, Bertrand G. Cyclic (alkyl)- and (aryl)-(amino) carbene coinage metal complexes and their applications. Chemical Reviews. 2020;**120**:4141-4168

[17] Melaimi M, Jazzar R, Soleilhavoup M, Bertrand G. Cyclic (alkyl)(amino) carbenes (CAACs): Recent developments. Angewandte Chemie, International Edition. 2017;**56**: 10046-10068

[18] Soleilhavoup M, Bertrand G. Cyclic (alkyl)(amino) carbenes (CAACs): Stable carbenes on the rise. Accounts of Chemical Research. 2015;**48**:256-266

[19] Lavallo V, Canac Y, Präsang C, Donnadieu B, Bertrand G. Stable cyclic (alkyl)(amino)carbenes as rigid or flexible, bulky, electron-rich ligands for transition-metal catalysts: A quaternary carbon atom makes the difference. Angewandte Chemie, International Edition. 2005;**44**:5705-5709

[20] Frey GD, Lavallo V, Donnadieu B, Schoeller WW, Bertrand G. Facile splitting of hydrogen and ammonia by nucleophilic activation at a single carbon center. Science. 2007;**316**: 439-441

[21] Zhao L, Zeng X. Transition-metal-catalyzed reactions promoted by cyclic (alkyl or aryl)(amino)carbene ligands. Chemistry. 2022;**8**:2082-2113

[22] Wang Y, Tran PM, Lahm ME, Xie Y, Wei P, Adams ER, et al. Activation of ammonia by a carbene-stabilized dithiolene zwitterion. Journal of the American Chemical Society. 2022;**144**: 16325-16331

[23] Raup DEA, Cardinal-David B, Holte D, Scheidt KA. Cooperative catalysis by carbenes and Lewis acids in a highly stereoselective route to γ-lactams. Nature Chemistry. 2010;**2**:766-771

[24] Crudden CM et al. Ultra stable self-assembled monolayers of N-heterocyclic carbenes on gold. Nature Chemistry. 2014;**6**:409-414

[25] Mahoney JK et al. Air-persistent monomeric (amino) (carboxy) radicals derived from cyclic (alkyl)(amino) carbenes. Journal of the American Chemical Society. 2015;**137**:7519-7525

[26] Pony Yu R, Hesk D, Rivera N, Pelczer I, Chirik PJ. Iron-catalysed tritiation of pharmaceuticals. Nature. 2016;**529**:195-199

[27] Ernst JB et al. Molecular adsorbates switch on heterogeneous catalysis: Induction of reactivity by N-heterocyclic carbenes. Journal of the American Chemical Society. 2017;**139**:9144-9147

[28] Wang G et al. Ballbot-type motion of N-heterocyclic carbenes on gold surfaces. Nature Chemistry. 2017;**9**: 152-156

[29] Légaré MA et al. Nitrogen fixation and reduction at boron. Science. 2018; **359**:896-900

[30] Ishii T, Kakeno Y, Nagao K, Ohmiya H. N-heterocyclic carbene-catalyzed decarboxylative alkylation of aldehydes. Journal of the American Chemical Society. 2019;**141**:3854-3858

[31] Han Y-F, Huang Y, Liu H, Gao Z-H, Zhang C-L, Ye S. Photoredox cooperative N-heterocyclic carbene/ palladium-catalysed alkylacylation of alkenes. Nature Communications. 2022; **13**:5754

[32] Narouz MR et al. N-Heterocyclic carbenefunctionalized magic-number gold nanoclusters. Nature Chemistry. 2019;**11**:419-425

[33] Hamze R et al. Eliminating nonradiative decay in Cu(I) emitters: >99% quantum efficiency and microsecond lifetime. Science. 2019;**363**: 601-606

[34] Davies AV, Fitzpatrick KP, Betori RC, Scheidt KA. Combined photoredox and carbene catalysis for the synthesis of ketones from carboxylic acids. Angewandte Chemie, International Edition. 2020;**59**:9143-9148

[35] Franz M et al. Controlled growth of ordered monolayers of N-heterocyclic carbenes on silicon. Nature Chemistry. 2021;**13**(9):828-835

[36] Huynh HV. Electronic properties of N-heterocyclic carbenes and their experimental determination. Chemical Reviews. 2018;**118**:9457-9492

[37] Runyon JW et al. Carbene-based Lewis pairs for hydrogen activation. Australian Journal of Chemistry. 2011; **64**:1165-1172

[38] Dröge T, Glorius F. The measure of all rings — N-heterocyclic carbenes. Angewandte Chemie, International Edition. 2010;**49**:6940-6952

[39] Heinemann C, Müller T, Apeloig Y, Schwarz H. On the question of stability, conjugation, and "aromaticity" in imidazol-2-ylidenes and their silicon analogs. Journal of the American Chemical Society. 1996;**118**:2023-2038

[40] Vivancos Á, Segarra C, Albrecht M. Mesoionic and related less heteroatom-stabilized N-heterocyclic carbene complexes: Synthesis, catalysis, and other applications. Chemical Reviews. 2018;**118**:9493-9586

[41] Mąkosza M, Fedoryński M. Dichlorocarbene and analogs: Discovery, properties and reactions. Russian Chemical Bulletin. 2021;**70**(11): 2045-2050

[42] Hine J, Dowell AM. Carbon dihalides as intermediates in the basic hydrolysis of haloforms. III. Combination of carbon dichloride with halide ions. Journal of the American Chemical Society. 1954;**76**: 2688-2692

[43] Hine J, Langford PB. Methylene derivatives as intermediates in polar reactions. IX. The concerted mechanism for α-eliminations of haloforms. Journal of the American Chemical Society. 1957; **79**:5497-5500

[44] Hine J. Divalent Carbon. New York: Ronald Press Co.; 1964

[45] Green SP et al. Thermal stability and explosive hazard assessment of diazo compounds and diazo transfer reagents. Organic Process Research & Development. 2019;**24**(1):67-84

[46] Liu W et al. Copper-catalyzed oxidation of hydrazones to diazo compounds using oxygen as the terminal oxidant. ACS Catalysis. 2021;**11**(5): 2676-2683

[47] Wei F et al. Gold carbene chemistry from diazo compounds. Science Bulletin. 2015;**60**(17):1479-1492

[48] Hutchins RO, Maryanoff B, Milewski C. Selective reduction of aliphatic ketones and aldehydes to hydrocarbons with sodium cyanoborohydride and P-toluenesulfonyl hydrazide in dimethylformamide-sulfolane. Journal of the American Chemical Society. 1971;**93**(7):1793-1794

[49] Visser P et al. Reactivity of carbenes and ketenes in low-temperature matrices. Carbene co trapping, Wolff rearrangement, and ketene–pyridine ylide (zwitterion) observation. Journal of the American Chemical Society. 1996; **118**(50):12598-12602

[50] Simmons HE, Smith RD. A new synthesis of cyclopropanes from olefins. Journal of the American Chemical Society. 1958;**80**(19):5323-5324

[51] LeGoff E. Cyclopropanes from an easily prepared, highly active zinc-copper couple, dibromomethane, and olefins. The Journal of Organic Chemistry. 1964;**29**(7):2048-2050

[52] Simmons HE, Smith RD. A new synthesis of cyclopropanes. Journal of the American Chemical Society. 1959;**81** (16):4256-4264

[53] Petrellis PC, Griffin GW. Photofragmentation of oxirans. Precursors for phenylcyanocarbene and phenylmethoxycarbonylcarbene. Chemical Communications. 1967;(14): 691-692

[54] Saba S, Brescia A-M, Kaloustian MK. One-pot synthesis of cyclic amidinium tetrafluoroborates and hexafluorophosphates; the simplest models of N^5,N^{10}0 methenyltetrahydrofolate coenzyme. Tetrahedron Letters. 1991;**32**:5031

[55] Delaude L, Szypa M, Demonceau A, Noels AF. Synthesis and application of new N-heterocyclic carbene ruthenium. Advanced Synthesis and Catalysis. 2002; **344**:749

[56] Leuthausser S, Schwarz D, Plenio H. Tuning the electronic properties of N-heterocyclic carbenes. Chemistry - A European Journal. 2007;**13**:7195

[57] Kantchev EAB, Ying JY. Practical One-Pot, Three-Component Synthesis of N-Heterocyclic Carbene (NHC) Ligated Palladacycles Derived from *N,N*-Dimethylbenzylamine. Organometallics. 2009;**28**:289

[58] Leuthausser S, Schmidts V, Thiele CM, Plenio H. π-Face Donor Properties of N-Heterocyclic Carbenes in Grubbs II Complexes. Chemistry: A European Journal. 2008;**14**:5465

[59] Fleckenstein C, Roy S, Leuthausser S, Plenio H. Sulfonated N-heterocyclic carbenes for Suzuki coupling in water. Chemical Communications. 2007;(27): 2870

[60] Arduengo AJ III, Krafczyk R, Schmutzler R, Craig HA, Goerlich JR, Marshall WJ, et al. Imidazolylidenes, imidazolinylidenes and imidazolidines. Tetrahedron. 1999;**55**:14523

[61] Van Veldhuizen JJ, Garber SB, Kingsbury JS, Hoveyda AH. A Recyclable Chiral Ru Catalyst for Enantioselective Olefin Metathesis. Efficient Catalytic Asymmetric Ring-Opening/Cross Metathesis in Air. Journal of the American Chemical Society. 2002;**124**:4954

[62] Alcarazo M, Roseblade SJ, Alonso E, Fernandez R, Alvarez E, Lahoz FJ, et al. 1,3-Bis(*N,N*-dialkylamino)imidazolin-2-ylidenes: Synthesis and Reactivity of a New Family of Stable N-Heterocyclic Carbenes. Journal of the American Chemical Society. 2004;**126**:13242

[63] Enders D, Meiers M. Diastereo- and Enantioselective Synthesis of Differently *N,O*-Protected 1,3-Amino Alcohols with Three Neighbouring Stereogenic Centers. Synthesis. 2002:2542

[64] Dinger MB, Nieczypor P, Mol JC. Adamantyl-Substituted *N*-Heterocyclic Carbene Ligands in Second-Generation Grubbs-Type Metathesis Catalysts. Organometallics. 2003;**22**:5291

[65] Dutton JL, Frenking G. New avenues in S-block chemistry: Beryllium(0) complexes. Angewandte Chemie International Edition. 2016;**55**(43): 13380-13382

[66] Arrowsmith M et al. Neutral zero-valent S-block complexes with strong multiple bonding. Nature Chemistry. 2016;**8**(9):890-894

[67] Obi AD et al. Carbene–calcium silylamides and amidoboranes. Organometallics. 2022;**41**(21):3064-

3072. DOI: 10.1021/acs.organomet.2c00464

[68] Walley JE et al. N-heterocyclic carbene-supported aryl- and alk- oxides of beryllium and magnesium. Catalysts. 2019;**9**(11):934. DOI: 10.3390/catal9110934

[69] Krüger J et al. Single-electron oxidation of carbene-coordinated pnictinidenes–entry into heteroleptic radical cations and metalloid clusters. Inorganic Chemistry. 2022;**61**(15):5878-5884. DOI: 10.1021/acs.inorgchem.2c00249

[70] Ghosh, M.l Khan, S. N-heterocyclic silylenes in coinage metal chemistry: An account of recent advances. Dalton Transactions. 2021;**50**(31):10674-10688

[71] Pei R, He L, Zhao Y, Wang X. The dynamic Lewis acid–carbene hybrid: Pushing the electrophilicity of carbenes to the limit. Journal of the American Chemical Society. 2023;**145**:21733-21737

[72] Díez-González S, N-Heterocyclic Carbenes. From Laboratory Curiosities to Efficient Synthetic Tools. 2nd ed. The Royal Society of Chemistry; 2017

[73] Weskamp T et al. A novel class of ruthenium catalysts for olefin metathesis. Angewandte Chemie, International Edition. 1998;**37**(2490–2493):20

[74] Huang J et al. Olefin metathesis-active ruthenium complexes bearing a nucleophilic carbene ligand. Journal of the American Chemical Society. 1999;**121**(2674–2678):21

[75] Scholl M et al. Synthesis and activity of a new generation of ruthenium-based olefin metathesis catalysts coordinated with 1,3-dimesityl-4,5-dihydroimidazol-2-ylidene ligands. Organic Letters. 1999;**1**:953-956

[76] Viciano M et al. C—H oxidative addition of bisimidazolium salts to iridium and rhodium complexes, and N-heterocyclic carbene generation. A combined experimental and theoretical study. Organometallics. 2006;**25**:1120-1134

[77] Pan Z, Gao D, Zhang C, Guo L, Li J, Cui C. Synthesis and reactivity of N-heterocyclic carbene stabilized lanthanide(II) bis(amido) complexes. Organometallics. 2021;**40**(11):1728-1734

[78] Clement ND et al. Oxidative addition of imidazolium salts to Ni (0)and Pd (0): Synthesis and structural characterization of unusually stable metal–hydride complexes. Angewandte Chemie, International Edition. 2004;**43**:1277-1279

[79] Lappert MF, Pye PL. Electron-rich olefin-derived neutral mono- and bis-(carbene) complexes of low-oxidationstate manganese, iron, cobalt, nickel, and ruthenium. Journal of the Chemical Society Dalton Transactions. 1977;(21):2172-2180

[80] Wang HMJ, Lin IJB. Facile synthesis of silver (I)—carbene complexes. Useful carbene transfer agents. Organometallics. 1998;**17**(972–975):23

[81] Visbal R, Gimeno MC. N-heterocyclic carbene metal complexes: Photoluminescence and applications. Chemical Society Reviews. 2014;**43**:3551-3574

[82] Strassner T. Phosphorescent Platinum(II) Complexes with C∧C* Cyclometalated NHC Ligands. Accounts of Chemical Research. 2016;**49**:2680-2689

[83] Amouri H. Luminescent Complexes of Platinum, Iridium, and Coinage Metals Containing *N*-Heterocyclic Carbene Ligands: Design, Structural Diversity, and Photophysical Properties. Chemical Reviews. 2023;**123**:230-270

[84] Crabtree RH. Recent Developments in the Organometallic Chemistry of N-Heterocyclic Carbenes. Coordination Chemistry Reviews. 2007;**251**:595

[85] Herrmann WA, Kocher C. N-Heterocyclic Carbenes. Angewandte Chemie, International Edition in English. 1997;**36**:2162-2187

[86] Dušková T et al. N-heterocyclic carbenes with three and six fluorous ponytails and their highly fluorophilic Rh and Ir complexes. Journal of Organometallic Chemistry. 2023;**986**: 122605

[87] Bourissou D, Guerret O, Gabbai FP, Bertrand G. Stable Carbenes. Chemical Reviews. 2000;**100**:39-91

[88] Diez-Gonzalez S, Nolan SP. Stereoelectronic parameters associated with N-heterocyclic carbene (NHC) ligands: A quest for understanding. Coordination Chemistry Reviews. 2007; **251**:874-883

[89] Lin JCY, Huang RTW, Lee CS, Bhattacharyya A, Hwang WS, Lin IJB. Coinage metal-N-heterocyclic carbene complexes. Chemical Reviews. 2009; **109**:3561-3598

[90] Catalano VJ, Etogo AOJ. Luminescent coordination polymers with extended Au(I)-Ag(I) interactions supported by a pyridyl-substituted NHC ligand. Organometallic Chemistry. 2005; **690**:6041-6050

[91] Catalano VJ, Etogo AO. Preparation of Au(I), Ag(I), and Pd(II) N-Heterocyclic Carbene Complexes Utilizing a Methylpyridyl-Substituted NHC Ligand. Formation of a Luminescent Coordination Polymer. Inorganic Chemistry. 2007;**46**: 5608-5615

[92] Catalano VJ, Munro LB, Strasser CE, Samin AF. Modulation of metal-metal separations in a series of Ag(I) and intensely blue photoluminescent Cu(I) NHC-bridged triangular clusters. Inorganic Chemistry. 2011;**50**(17): 8465-8476

[93] Xia C, Yang K, Sun W, Lu X, Xie JJ. Silver(I), nickel(II) N-heterocyclic carbene complexes based on bidentate bis-imidazolium salt with a quinoxaline linker: Syntheses, structures, and characterization. Coordination Chemistry. 2017;**70**(4):615-625

[94] Zhou Y, Zhang X, Chen W, Huayu Qiu H. Synthesis, structural characterization, and luminescence properties of multinuclear silver complexes of pyrazole-functionalized NHC ligands containing Ag-Ag and Ag-π interaction. Journal of Organometallic Chemistry. 2008;**693**:205-215

[95] Leung SY-L, Lam ES-H, Lam WH, Wong KM-C, Wing-Tak Wong W-T, Yam VW-W. Luminescent Cyclometalated Alkynylplatinum(II) Complexes with a Tridentate Pyridine-Based N-Heterocyclic Carbene Ligand: Synthesis, Characterization, Electrochemistry, Photophysics, and Computational Studies. Chemistry - A European Journal. 2013;**19**: 10360-10369

[96] Lee C-S, Sabiah S, Wang J-C, Hwang W-S, Lin IJB. Water-Induced Changes of Photoluminescence of a Pincer-Type N-Heterocyclic Carbene Platinum(II) Complex. Organometallics. 2010;**29**: 286-289

[97] Bachmann M, Suter D, Blacque O, Venkatesan K. Tunable and Efficient White Light Phosphorescent Emission Based on Single Component N-Heterocyclic Carbene Platinum(II) Complexes. Inorganic Chemistry. 2016; **55**(10):4733-4745

[98] Connick WB, Geiger D, Eisenberg R. Excited-State Self-Quenching Reactions of Square Planar Platinum(II) Diimine Complexes in Room-Temperature Fluid Solution. Inorganic Chemistry. 1999;**38**: 3264-3265

[99] Hissler M, Connick WB, Geiger DK, McGarrah JE, Lipa D, Lachicotte RJ, et al. Platinum Diimine Bis(acetylide) Complexes: Synthesis, Characterization, and Luminescence Properties. Inorganic Chemistry. 2000;**39**:447-457

[100] Hissler M, McGarrah JE, Connick WB, Geiger DK, Cummings SD, Eisenberg R. Platinum diimine complexes: Towards a molecular photochemical device. Coordination Chemistry Reviews. 2000;**208**:115-137

[101] Zhang Y, Garg JA, Michelin C, Fox T, Blacque O, Venkatesan K. Synthesis and Luminescent Properties of *cis* Bis-*N*-Heterocyclic Carbene Platinum(II) Bis-Arylacetylide Complexes. Inorganic Chemistry. 2011;**50**:1220-1228

[102] Wang X, Chang YL, Lu JS, Zhang T, Lu ZH, Wang SN. Bright Blue and White Electrophosphorescent Triarylboryl-Functionalized C^N-Chelate Pt(II) Compounds: Impact of Intramolecular Hydrogen Bonds and Ancillary Ligands. Advanced Functional Materials. 2014;**24**: 1911-1927

[103] Hua F, Kinayyigit S, Cable JR, Castellano FN. Platinum(II) Diimine Diacetylides: Metallacyclization Enhances Photophysical Properties. Inorganic Chemistry. 2006;**45**: 4304-4306

[104] Pomestchenko IE, Castellano FN. Solvent switching between charge transfer and intraligand excited states in a multichromophoric platinum (II) complex. The Journal of Physical Chemistry. A. 2004;**108**:3485-3492

[105] Damrauer NH, Cerullo G, Yeh A, Boussie TR, Shank CV, McCusker JK. Femtosecond Dynamics of Excited-State Evolution in $[Ru(bpy)_3]^{2+}$. Science. 1997; **275**:54-57

[106] Penzkofer A, Lu Y. Fluorescence quenching of rhodamine 6G in methanol at high concentration. Chemical Physics. 1986;**103**:399-405

[107] Prabusankar G, Muthukumaran N, Vaddamanu M, Raju G, Velappan K, Sathyanarayana A, et al. Blue-emitting acridine-tagged silver(I)-bis-N-heterocyclic carbine. RSC Advances. 2019;**9**:7543

[108] Hsieh C-H, Wu F-I, Fan C-H, Huang M-J, Lu K-Y, Chou P-Y, et al. Design and Synthesis of Iridium Bis (carbene) Complexes for Efficient Blue Electrophosphorescence. Chemistry - A European Journal. 2011;**17**:9180-9187

[109] Krylova VA, Djurovich PI, Conley BL, Haiges R, Whited MT, Williams TJ, et al. Control of emission colour with N-heterocyclic carbene (NHC) ligands in phosphorescent three-coordinate Cu(i) complexes. Chemical Communications. 2014;**50**:7176-7179

[110] Son SU, Park KH, Lee Y-S, Kim BY, Choi CH, Lah MS, et al. Synthesis of Ru (II) Complexes of N-Heterocyclic Carbenes and Their Promising Photoluminescence Properties in Water. Inorganic Chemistry. 2004;**43**:6896-6898

Chapter 3

Perovskite Paradigm: Unraveling Photoelectrochemical Synergies for Sustainable Transformations

Mina Ahmadi-Kashani, Mahmoud Zendehdel, Mohammad Mahdi Abolhasani and Narges Yaghoobi Nia

Abstract

Owing to the tunable electronic properties, high carrier lifetimes, low recombination possibility, and long diffusion lengths, perovskites have gained attention for potential use in photoelectrocatalytic processes. Photoelectrochemical methods can convert sunlight into electricity or useful fuels, resulting in extensive research to develop PEC applications. This chapter embarks on a comprehensive exploration of the perovskite structure's transformative influence on diverse photoelectrochemical cell (PEC) and monolithic Photovoltaic-Electrocatalytic (PV-EC) devices including water splitting, H_2 evolution, CO_2 reduction, N_2 reduction, degradation of pollutants, (bio) sensing, and organic synthesis. By assessing the mechanisms and kinetics involved, we aim to disclose the potential of perovskite-based photoelectrochemical systems in shaping the landscape of green energy and environmental stewardship. Furthermore, the chapter addresses the progress and challenges in enhancing the stability, selectivity, and efficiency of perovskite-based PEC and monolithic PV-EC reactions to unravel the synergistic potential for sustainable transformations in the realm of photoelectrochemistry.

Keywords: perovskite, photoelectrochemical cell, monolithic PV-EC, water splitting, H_2 evolution, CO_2 reduction, N_2 reduction, (bio) sensing

1. Introduction

Photoelectrochemical (PEC) processes, also referred as artificial photosynthesis, serve as primary means to harness solar energy. These procedures transform sunlight, the most plentiful renewable energy source, into chemical fuels that can be stored, providing versatility in sequential energy transportation and storage. PEC techniques combine the advantages of both particulate photocatalytic and electrochemical systems, leading to higher energy conversion efficiencies by enhancing light absorption and improving charge separation and utilization [1]. Moreover, the PEC methods offer advantages of being cost-effective and easy to handle, enabling the production of the desired final products through thermodynamic and kinetics control [2]. By mimicking natural photosynthesis, PEC methods also provide fast reaction kinetics and selectivity for the products, all without requiring high input energy costs, which helps to minimize the rate of electron-hole recombination [3]. Unlike PEC devices, which rely solely on sunlight for power, monolithic photovoltaic-electrochemical (PV-EC) devices as a type of PEC systems, also known as integrated solar fuel generators or tandem devices, have the ability to utilize solar light during analysis and can also be powered by renewable electricity into a single device [4]. The objective of such systems is to surpass the constraints of individual solar cells and boost the overall efficiency of energy conversion. The photovoltaic cell captures sunlight and converts it into electricity. However, the electrochemical cell utilizes the generated electricity to power chemical reactions that generate fuels or other valuable products. By combining high-efficiency photovoltaic and electrochemical cells, these systems are able to capture a wide range of solar radiation and utilize surplus energy that would otherwise be wasted in individual PEC cells. PV-EC systems have the potential to enhance the durability and stability of devices. Furthermore, PV and EC components can be easily adjusted and optimized, allowing greater operational flexibility and better overall performance [5].

These techniques demonstrate significant potential in various fields, including in the case of water splitting, solar fuel production, environmental remediation, and chemical sensing [6–8].

At the core of these PEC systems lies a photoelectrode, typically crafted from nanostructured semiconductors with suitable bandgaps, enabling them to absorb light and produce electron-hole pairs for involvement in reduction/oxidation reactions [9]. Optimal semiconducting materials for photoelectrodes should exhibit favorable characteristics in light absorption, charge separation, and transmission. Yet, challenges persist due to crystal defects, unsuitable energy band structures, and electron-hole recombination, hindering the development of superior photoelectrode materials capable of efficient PEC conversions for solar energy utilization. Over the past decade, researchers have unveiled promising inherent power utilization capabilities in perovskites [10–12]. As an example, perovskite solar cells have attained a remarkable power conversion efficiency of 25.5% [13]. This remarkable improvement in photovoltaic performance has positioned perovskite solar cells (PSCs) at the forefront of developing cost-effective next-generation solar power and integrated technologies [14, 15]. Besides their ability to convert solar energy to electricity, perovskites have several other desirable characteristics including suitable band gap, high light-harvesting capability, good electron mobility, well-adjusted charge transport, redox behavior, multicolor emission, compatibility with localized surface plasmon effect, as well as cost-effective scalable fabrication processes [16–18]. These make perovskites highly valuable for various applications in the photoelectrochemical field [19].

Thanks to its compositional and structural flexibility, approximately 90% of the elements in the periodic table can be partially replaced at anionic or cationic sites in the perovskite structure [20, 21]. Furthermore, by various synthesis methods and operating factors, the multifaceted shapes, and different particle sizes of perovskite can be easily obtained. Considering the exceptional properties of perovskites, incorporating them into photoelectrode materials for PEC conversions could be deemed a prudent decision. Although they have promising applications in various fields, there are concerns about toxicity of certain perovskite materials, particularly lead-based perovskites. Also, perovskite materials are strongly criticized for their lack of stability in practical applications. For instance, halide perovskites exhibit high sensitivity to water vapor and oxygen, and they can readily dissolve or decompose in polar solvents, posing significant challenges in device operations [22]. Therefore, the development of stable perovskites through compositional engineering may be a practical effective strategy. This can be further enhanced through approaches such as crystal engineering [23–25], doping strategy [26], polaron arrangement [27], barrier layers [28], and encapsulation methods which not only can enhance the device stability but also can prevent from toxic exposure and environmental impacts.

This chapter aims to comprehensively explore the transformative influence of perovskite structures on various photoelectrochemical cell (PEC) and monolithic Photovoltaic-Electrochemical (PV-EC) devices. These devices include water splitting, H_2 evolution, CO_2 reduction, N_2 reduction, degradation of pollutants, (bio) sensing, and organic synthesis. By examining the mechanisms and kinetics involved, our goal is to uncover the potential of perovskite-based photoelectrochemical systems in promoting green energy and environmental stewardship. Additionally, the chapter addresses the progress and challenges in improving the stability, selectivity, and efficiency of perovskite-based PEC and monolithic PV-EC systems. Through these efforts, we aim to reveal the synergistic potential for sustainable transformations in the field of photoelectrochemistry and solar fuel production.

2. Perovskite structure

In 1839, Gustav Rose first discovered $CaTiO_3$ in the Ural Mountains. It was later named by Russian mineralogist Count Lev Alekseyevich von Perovski [29]. The term "perovskite" has come to encompass a group of organic and inorganic materials with crystal structures similar to ABX_3, now identified as the perovskite structure.

2.1 Perovskite oxide

In perovskite oxide with a typical ABO_3 structure, A-site is frequently occupied by a metal ion including alkaline, alkaline-earth, or lanthanide rare-earth metals, while B-site cations can be occupied by metal ions such as the transition metals (TMs). As shown in **Figure 1(a)**, the ideal perovskite oxide has a cubic crystal with space group Pm-3 m, which has a high degree of symmetry. The B-sites are coordinated by six oxygens, forming the corner-sharing octahedrons of BO_6 and the cations in the A-site are surrounded by eight BO_6 octahedral arrangement [30]. Due to the lattice distortions caused by variations in ionic radii of the constituent cations, the symmetry of perovskite can be reduced to orthogonal, tetragonal, rhombohedral, monoclinic, and triclinic structures. These crystal lattice distortions impact on dipole moment and the electronic structure, leading to the emergence of new properties and applications in various fields [31].

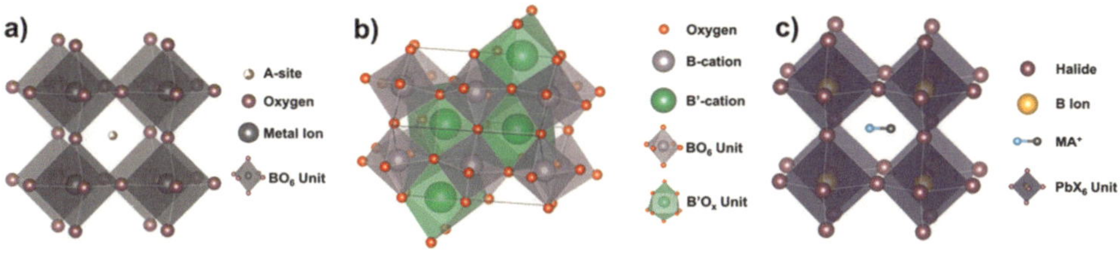

Figure 1.
Schematic of the crystal unit cell of an ideal perovskite oxide with a cubic structure (a), Ruddlesden–Popper perovskite oxide with a layered structure (b) and halide perovskite with methylammonium as A-site cation (c).

Perovskite oxides can serve as proton and oxygen conductors, redox catalysts or multi-functional catalysts, depending on the various metal ions located in the A and B-sites [32]. Besides, the composition of perovskite oxides can be customized to produce specific perovskite structures such as double-perovskite structures, layered perovskite structures, and Ruddlesden–Popper (RP) perovskite structures with some exceptional physicochemical properties. Double perovskite oxides have the general formula $A_2BB'O_6$, where A is a monovalent or divalent cation, B and B′ are different trivalent cations that occupy equivalent positions in the crystal lattice. Layered perovskite oxides are another class of perovskite oxides with general formula $A_2B_{n-1}B_nO_{3n+1}$, where A is a monovalent or divalent cation and, B and B′ are different trivalent cations that form distinct layers separated by layers of A cations. Similar to layered perovskite oxide, Ruddlesden–Popper perovskite oxide has a layered structure, with alternative layers of perovskite-like and rock salt-like structures (see **Figure 1(b)**). However, in A-site was partially replaced by a larger cation, resulting in a layered structure with an extra oxygen layer [33]. The presence of various sites provides researchers with the opportunity to substitute or integrate different elements, leading to the development of new materials exhibiting significant photoelectrocatalytic activity. Moreover, the high electronic conductivity, elevated charge carrier mobility, and robust light absorption—spanning visible and ultraviolet spectra—position perovskite oxides as promising candidates for PEC devices [34].

2.2 Halide perovskites

In 2009, Kojima and his colleagues conducted pioneering work that sparked interest in halide perovskites as promising semiconducting materials. These materials gained attention for their potential application in photovoltaic devices [35]. Halide perovskites belong to a group of materials with the formula of ABX_3, A representing monovalent organic and/or inorganic cation, B is a divalent cation especially including IVA group elements (e.g., Pb^{2+}, Sn^{2+}), and X is a halide or mixed halide anions [36]. Mixed halogen-based perovskites exhibit numerous intriguing properties in contrast to single halogen-based perovskites, including enhanced thermal stability and a more manageable band structure. **Figure 1(c)** shows a crystal structure of a halide perovskite. In this arrangement, the B cation occupies the body center of an octahedron, which comprises six halide anions $[BX_6]^{4-}$ [37]. In cubic symmetry, the octahedral are interconnected by sharing their vertexes. As the temperature declines, the octahedral structure initiates to slant, leading to a sequence of lower symmetries, such as tetragonal, monoclinic, orthorhombic, and rhombohedral. When an organic cation, like methylammonium (MA^+) or formamidinium (FA^+), occupies the A-site in the halide perovskite structure, it is known as an organic-inorganic hybrid halide perovskite [10]. However, the presence of organic cations usually results in the instability of the structure. On the other hand, to enhance stability, the organic cation can be replaced with an inorganic cation, such as (Cs^+), resulting in an all-inorganic halide perovskite. The B site in the halide perovskites is commonly occupied by element Pb. Nevertheless, lead (Pb) is a recognized toxic metal capable of causing harm to both the environment and human health. As a result, there has been extensive research on using tin (Sn) as a partial or complete substitute for Pb. Nevertheless, tin-based halide perovskites are more prone to degradation compared to the lead-based perovskites. Additionally, other types of halide perovskites, such as halide double perovskites, layered halide perovskites, and Ruddlesden–Popper (RP) structures, have been studied as stable and environmentally friendly alternatives to lead halide perovskite [38].

Until now, halide perovskite has been extensively studied in various photocatalytic reactions, including hydrogen evolution, CO_2 reduction, and pollutant degradation [39–41]. The investigation into halide perovskites as a photoelectrocatalyst is motivated by their fascinating properties such as adjustable bandgap, high absorption coefficients, extensive carrier diffusion lengths, and long charge carrier lifetime. The stability of halide perovskites is the main challenge in various applications. Factors such as moisture, oxygen, light, and polar solvent can lead to rapid decomposition of halide perovskites. To address this issue, several strategies have been employed based on the intrinsic stability of halide perovskites. These strategies comprise: (i) substituting the organic MA or FA with inorganic ions like Cs^+; (ii) developing composites that expedite the swift extraction of photogenerated electrons and holes [11, 42, 43]; and (iii) The exploration of encapsulating halide perovskites has been undertaken as a method to avoid direct contact with degradable environments [44].

3. PEC processes

Sunlight is widely recognized as the most promising energy source due to its abundance, carbon-free nature, worldwide availability, non-polluting, and limitlessly renewable source of clean energy. However, solar energy is known to be intermittent and unstable, which limits direct use in practical applications. By converting solar irradiation into chemical fuels, it can be conveniently stored and transported, providing a reliable and versatile energy solution [45]. Photoelectrochemical processes serve as major means to harness solar energy and convert sunlight into storable chemical fuels. In a photochemical procedure, light energy is directly utilized to drive specific half-reactions, such as the oxygen evolution reaction (OER), and hydrogen evolution reaction (HER). When this process is combined with an additional potential bias, it is referred to as a photoelectrochemical process, also known as artificial photosynthesis [46].

Since the first PEC cells were described in 1972, extensive research has focused on PEC-based solar water splitting, which has been the most extensively studied application in this field [47]. As research on PEC cells advanced, scientists discovered that the potential of PEC extended beyond their initial applications, leading to the reporting of other applications of PECs, such as fuel production, CO_2 reduction, and pollutant degradation [3]. A schematic of a CO_2 reduction PEC cell in an acidic electrolyte has been presented in **Figure 2**. The PEC system comprises working and counter electrodes, with one or both of them being photoelectrodes. These photoelectrodes can be classified into various systems, including the photoanode/cathode system, anode/photocathode system, or the photoanode/photocathode system. In laboratory-scale setups, a reference electrode is typically included to monitor the precise reduction potential value of the working electrode's half-reaction within a cell. To design an efficient PEC system, it is crucial to select a light absorber that satisfies the following criteria: (1) it should possess the capability to absorb visible light within a band gap energy range of 1.5–2.5 eV, (2) the conduction band (CB) should have minimum levels, and the valence band (VB) should have maximum levels to provide the thermodynamic driving force for the desired reaction, and (3) it must demonstrate sufficient stability in the electrolyte solution to ensure the long-term performance and durability of the PEC system [48]. Therefore, the key component for PEC systems is photoelectrode, which directly regulates the energy conversion efficiency. In addition, it is worth noting that PEC systems have the potential to operate using available and stable materials, such as TiO_2, Fe_2O_3, and Cu-based metal oxides

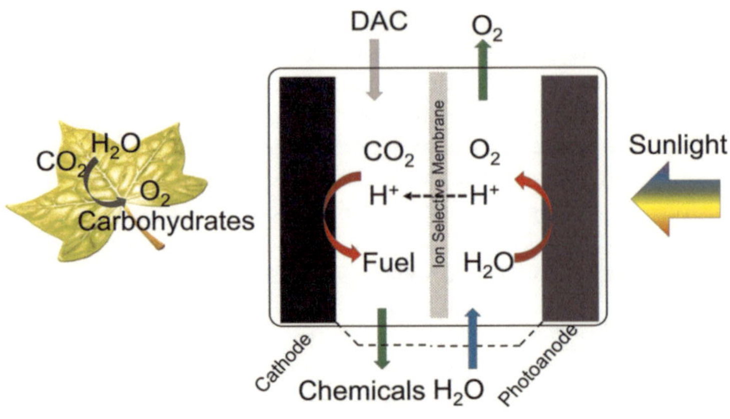

Figure 2.
Schematic of a CO_2 photoelectrochemical reduction cell (acidic electrolyte), resembling natural photosynthesize organisms for production of valuable chemicals and oxygen from direct air capture (DAC), water and sunlight.

[49]. However, as photoelectrodes in PEC systems, perovskite materials can be more attractive than simple oxides, by reason of tunable band structures, high electrocatalytic activity, enhanced light absorption, and efficient charge separation [50]. In addition, there has been significant interest in the energy conversion of perovskites through photovoltaic technology. The combination of high-performance photovoltaic cells that convert solar irradiation into electricity with electrochemical systems as a part that simultaneously stores or converts the generated electricity, has been widely studied. The integration of PV and EC in a single device, known as monolithic PV-EC systems, allows for continuous output and a high external voltage, which is essential for efficient solar reactions [51]. The mechanism of PV-EC devices consists of two parts: (1) solar energy is converted to electricity by the photovoltaics part, and (2) the electrochemical system converts the generated electricity into stored electricity or renewable fuels. Perovskite materials can be incorporated directly into the PV-EC device, permitting for a monolithic design (**Figure 3**). In fact, the perovskites as an absorber and the electrocatalytic cell are joined into a single device, eradicating the need for distinct components and simplifying the system design [52].

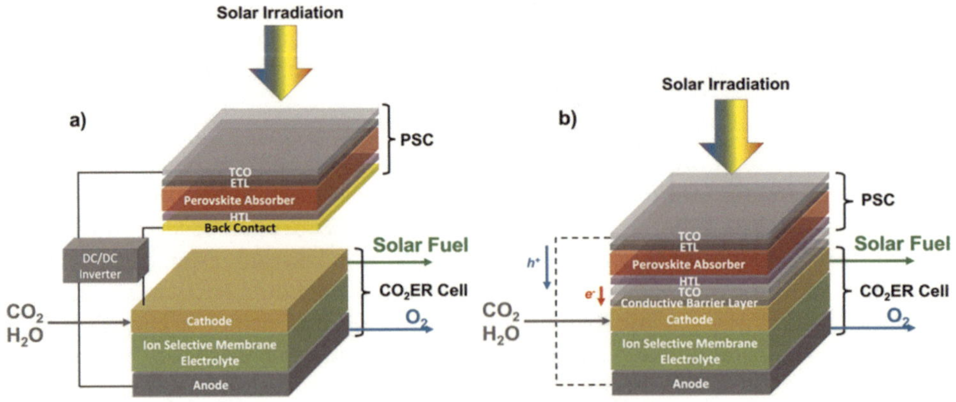

Figure 3.
Schematic of a PV-EC system containing perovskite solar cells and CO_2 electroreduction units in polylithic (a) and monolithic (b) device configuration.

3.1 Water splitting

Solar-driven water-splitting is considered as the most substantial and rapidly growing method for producing renewable H_2 to address the urgent energy crisis [53]. Hydrogen is widely recognized as a crucial clean fuel for the future. However, finding a cost-effective and sustainable approach to produce H_2 that can replace fossil fuels is a significant challenge. Inspired by the initial findings in 1967 by Honda and Fujishima, it was demonstrated that a photoelectrochemical cell comprising a single-crystalline TiO_2 (rutile) anode and a Pt cathode, combined with ultraviolet (UV) irradiation and an external bias, can effectively achieve overall water splitting (see **Figure 4**) [54].

Until now, many metal oxides as a photocatalyst for water splitting have been discovered [55]. However, most of them have large band gaps and can only be activated by ultraviolet (UV) light irradiation (l < 400 nm). It is worth noting that a significant portion of solar energy on Earth's surface falls within the visible light region (400 nm < l < 800 nm), and near-infrared (NIR) region (800–2500 nm) [56]. So, the efficient operation of visible light is crucial for the photocatalytic water splitting. In the PEC systems (acidic medium), the water splitting procedure consists of two half-reactions [57]. The oxidation reaction, known as, OER, occurs on the photoanode: $H_2O + 2\ h^+ \longrightarrow 2H^+ + ½O_2$ and the reduction reaction, known as HER, takes place on the photocathode: $2H^+ + 2e^- \longrightarrow H_2$.

In PEC cells, three processes are vital: (I) formation of photogenerated electron–hole pairs when the semiconductor materials are excited by light; (II) the separation and transfer of electron–hole pairs; (III) the participation of electrons or holes in the reduction or oxidation reactions. Due to the positive Gibbs free energy changes, water splitting is not a reaction that occurs spontaneously at atmospheric pressure and room temperature. According to theoretical calculation, a potential of 1.23 eV is needed to drive this reaction [58]. However, several factors lead to an increase in the theoretical energy quantities, including the recombination of photogenerated electron-hole pairs, the resistances of the electrodes and components, and voltage losses. To achieve efficient practical PEC- based water splitting, a semiconductor should have the following features: appropriate conduction and valence band positions, a band gap within the range of 1.0–2.2 eV in order to efficiently convert visible and NIR light, proper electron-hole pair separation, low cost, and high stability. Perovskite materials have recently emerged as promising candidates for photoelectrocatalytic performance in water splitting even under visible light excitation [59]. Besides, they can be

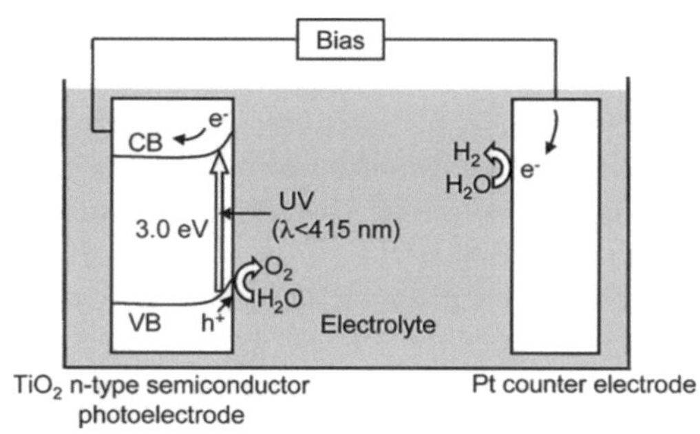

Figure 4.
PEC cell fabricated by Honda and Fujishima. Reproduced with permission from ref. [54].

integrated with other photovoltaic materials to create multi-junction devices.. The PV units offer a photovoltage for water splitting without needing additional bias.

A simple and efficient technique for the creation of porous $BiFeO_3$ as photoanodes, was investigated by Liu et al. [60]. Additionally, successive engineering procedures were applied to improve the photoelectrochemical performance in water splitting. A TiO_2 overlayer and a cobalt oxide/oxyhydroxide co-catalyst were prepared by atomic layer deposition and photo-assisted electrodeposition, respectively. Under 1 sun radiation (100 mW cm^{-2}, AM 1.5G spectrum), the engineered photoanode showed the photocurrent density of 0.16 mA cm^{-2} for oxygen evolution reaction (1 M NaOH). The Nyquist plots revealed that the charge-transfer resistance at the photoanode| electrolyte interface was reduced by the modification of $BiFeO_3$ by TiO_2 overlayer and CoO_x deposition. Furthermore, the deposition of CoOx on the photoanode surface amplified the capacitance approximately 15 times. These results indicated a higher charge density on the $BiFeO_3$ surface, which enhances charge transfer for the oxygen evolution reaction.

The mixed-cation perovskite $(HOOC(CH_2)_4NH_3)_x(CH_3NH_3)_{1-x}PbI_3$ ((5-AVA)$_x$(MA)$_{1-x}PbI_3$) was suggested for highly efficient water splitting [61]. The presence of 5-AVA cations can enhance the creation of perovskite crystals within the mesoporous oxide host, promoting their growth in the preferred direction. Compared to the $MAPbI_3$, the (5-AVA)$_x$(MA)$_{1-x}PbI_3$ exhibited lower defects and improved surface interaction with the TiO_2 surface. The obtained photoanode encapsulated with inexpensive and stable conductive carbon paste and waterproof silver conductive paint for PEC water splitting. The structure acts as a protective barrier, preventing the electrolyte from invading the perovskite. Besides, transportation of the generated holes from the perovskite layer to the electrolyte is facilitated by the tight interface between the conductive carbon paste layer, silver conductive paint layers, and the perovskite layer. This interface not only reduces resistance but also allows for efficient movement of the holes. Under AM 1.5G illumination in KOH electrolyte, the perovskite photoanode exhibited a noteworthy photocurrent density of 12.4 mA/cm^2 at 1.23 V versus the reversible hydrogen electrode (RHE) and maintained stability for 48 hours in a highly oxidized environment.

A conductive adhesive-barrier (CAB) that allows for an all-in-one, integrated assimilation and transformation of any photovoltaic system to a PEC cell was studied by Fehr et al. [62]. The discrete PEC cells constructed by p-i-n perovskite solar cell (PSC)|CAB|Pt and n-i-p PSC|CAB|IrO_x as photocathode and photoanode, respectively. The prepared system showed more than 99% conversion of the perovskite photovoltaic power into a chemical reaction. The unassisted water-splitting analyses on the photovoltaic cells were carried out with two different strategies. The first strategy involves connecting a photocathode and photoanode protected by CAB electrically in series, while illuminating them optically in parallel geometry, leading to a Solar-to-Hydrogen (STH) efficiency of 13.4%. Other system was fabricated by halide perovskite (HaP)/Si tandems presenting a hopeful low-priced photovoltaic system that showed a photovoltaic efficiency of about 30%. Water-splitting examines were conducted using a photoanode fabricated by monolithic HaP/Si tandem an IrO_x-coated CAB and Pt foil cathode, resulting in STH efficiency of 20.8% with 102 h of operation.

A metal-encapsulated photoanode based on black α-phase formamidinium lead triiodide ($FAPbI_3$) as an organic–inorganic metal halide perovskite (PSK) was suggested by Hansora et al. for efficient PEC water splitting (see **Figure 5**) [63]. To construct a photoanode with high efficiency and good durability for water splitting in

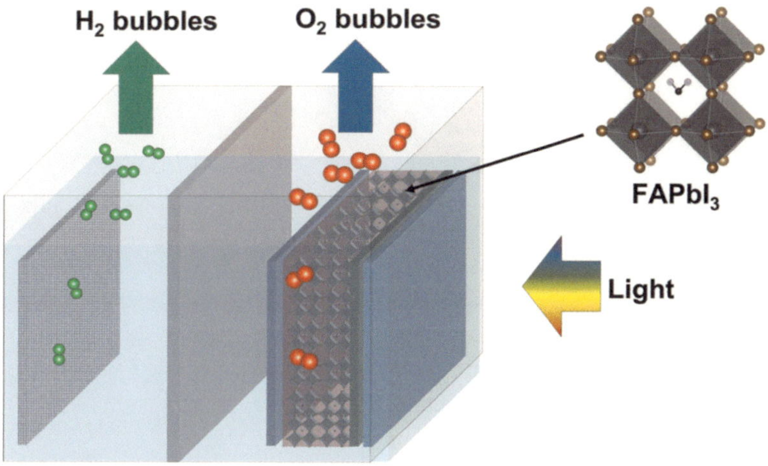

Figure 5.
Encapsulated $FAPbI_3$ photoanode for PEC water splitting.

an alkaline electrolyte, Ni foil as the passivation layer and nickel–iron oxyhydroxide (NiFeOOH) as an oxygen-evolving catalyst (OEC) were used to encapsulate $FAPbI_3$ and form NiFeOOH/Ni/$FAPbI_3$ photoanode. Actually, Ni metal foil was applied to enhance the activity and stability of the $FAPbI_3$ photoanode when immersed in water, and entirely block the infusion of the electrolyte. Without this protective layer, the photoanode becomes susceptible to degradation and loss of performance due to moisture, contaminants, and oxidation, which can negatively impact its functionality. To deposit NiFeOOH as an oxygen evolution reaction (OER) co-catalyst, the employed method was drop-casting of Ni and Fe precursor solutions on the Ni foil. Following the deposition of NiFeOOH on the photoanode, the most efficient charge transfer occurs during the mechanism of the OER. The synergistic effect of Ni and Fe within NiFeOOH has been thoroughly validated to enhance the OER kinetics in the NiFeOOH/Ni/$FAPbI_3$ photoanode. The Ag paste serves a critical role as a close ohmic connection metal between the Ni foil and the $FAPbI_3$ layer, crucial for maintaining optimal performance. In the absence of the Ag paste, both the Ni foil and the $FAPbI_3$ layer experience a significant drop in performance. The PEC water splitting performance of the NiFeOOH/Ni/$FAPbI_3$, photoanode was assessed in a setup employing a three-electrode configuration, with the photoanode as the working electrode, Hg/HgO (1 M NaOH) as the reference electrode, and a Pt wire as the counter electrode. The measurements were conducted under 1-sun illumination in 1 M KOH. At 1.23 V_{RHE} (the reversible hydrogen electrode), this photoanode demonstrated a photocurrent density of 22.8 mA cm^{-2} and exhibited superb stability for a duration of 3 days.

A PV-EC water splitting cell was made by incorporating the Pt/C/CFP||Ni-TFBDC/CFP (CFP: carbon fiber paper, TFBDC: tetrafluoroterephthalate) electrochemical cell with a tandem perovskite solar cell by Li et al. [64]. The OER part in the EC cell is joined to the cathode of the solar cell and the HER is coupled with the anode of the solar cell via a copper wire (see **Figure 6**). The PV cell was crafted by blending precursor solutions of FABr, FAI, CsBr, CsI, $PbBr_2$, and PbI_2 in DMF/DMSO with a ratio of 4:1 at a 1 M concentration. The Ni-TFBDC MOF was synthesized by substituting hydrogen atoms in terephthalate (BDC) with fluorine atoms, thereby altering the electronic structure of the MOF and enhancing its electrocatalytic activity. This system showed solar-to-hydrogen efficiency (η_{STH}) of 10.17% at 206.7 mA.

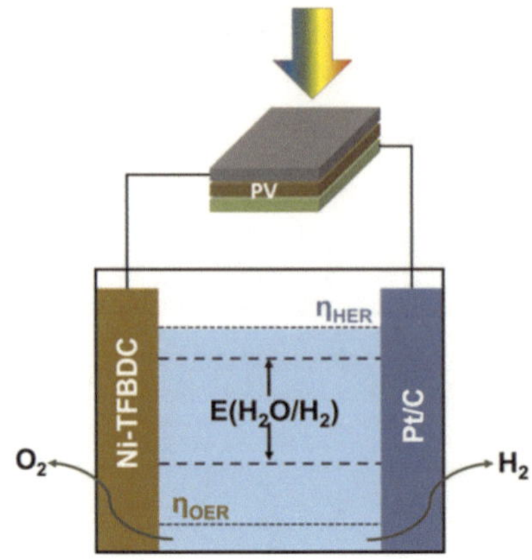

Figure 6.
Schematic of a PV-EC water splitting cell containing Pt/C cathode and Ni-TFBDC anode which is incorporated with a tandem perovskite solar cell.

3.2 Hydrogen production

The inimitable characteristics of hydrogen, including its great thermal properties, high energy density per unit mass (143 MJ kg^{-1}), and economic efficiency as a fuel, make H_2 a sustainable and crucial clean energy source [45]. Due to its lightweight nature (0.08988 g/L), hydrogen has the energy density much higher than that of other fuels like gasoline. One gram of hydrogen can release140 kJ of energy, almost four times the energy of methane (33 kJ/g) [65]. Currently, the promising usage of hydrogen still faces many challenges such as storage and handling, and an economically and environmentally clean production way. Photoelectrochemical hydrogen evolution offers a promising pathway toward a sustainable and low-carbon energy future by leveraging solar energy to produce clean and renewable hydrogen fuel. By developing suitable photocathode materials for PEC cells, researchers can enhance the efficiency, stability, and scalability of H_2 evolution processes driven by solar energy conversion [48]. Among different photocathodes, perovskites offer tailorable surface chemistry, versatility in operating conditions, and low cost for H_2 evolution reactions.

Water splitting for H_2 production is inherently slow and challenging, primarily due to the transfer of four electrons, necessitating a high overpotential. Additionally, the production of oxygen (O_2) through water oxidation raises safety concerns, requiring an additional separation process and resulting in increased costs [66]. In the past few years, H_2 evolution through PEC hydrohalic acids (HX, X = Cl, Br, I) splitting has attracted significant interest. PEC splitting of hydrohalic acids involves a process similar to water splitting, reduction of H^+ to H_2 and oxidation of X^- to X_2/X_3^-. Because of the solubility of X_2/X_3^- species in hydrohalic acid, the H_2 gas can be collected separately [38]. Lou et al. proposed an effective PEC system utilizing the $MAPbI_3$-TiO_2 nanorod array (TNA) for stable and efficient H_2 production in an aqueous HI solution [67]. **Figure 7** shows a schematic of the fabrication of the $MAPbI_3$-TNA photoelectrode for hydrogen evolution from HI splitting. During the PEC process, $MAPbI_3$ is capable of absorbing visible light with wavelengths up to 800 nm to create electron-hole pairs. The photogenerated electrons are efficiently injected into the TiO_2 nanorod array, where they are subsequently transported to the Pt counter electrode. At the counter electrode, the electrons participate in the

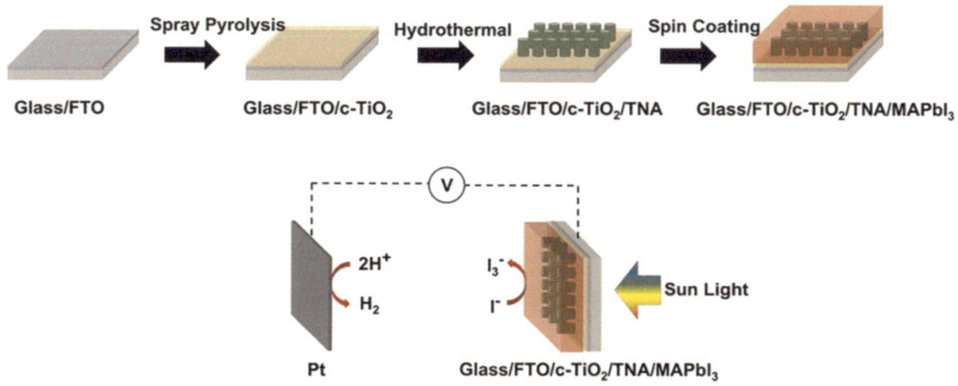

Figure 7.
$MAPbI_3$-TNA photoelectrode for PEC H_2 evolution from HI splitting.

reduction of protons, ultimately resulting in the evolution of hydrogen. In fact, TiO_2 nanorod array has a crucial role in blocking the transfer of holes from $MAPbI_3$ to Fluorine-doped Tin Oxide (FTO). In fact, the holes in $MAPbI_3$ take part in an oxidation process that converts I^- to I_3^-, which contributes to the stabilization of the $CH_3NH_3PbI_3$ crystals. Under AM 1.5 G illumination at 0.14 V (vs. Ag/AgCl), the optimized $CH_3NH_3PbI_3$-TNA PEC cell achieves a photocurrent density of 1.75 mA cm^{-2}, and remarkable stability in producing H_2 at a rate of 33.3 mmol cm^{-2} h^{-1} for over 8 h.

Choi et al. reported a PEC cell based on $Cs_{0.05}(FA_{0.83}MA_{0.17})_{0.95}(PbI_{0.83}Br_{0.17})_3$ (CsFAMA) perovskite as a photocathode and a CNT paper anode [68]. The photocathode was submerged in an acidic solution (0.5 M H_2SO_4 in water), while the anode was situated in a distinct 0.5 M H_2SO_4 solution containing reduced phosphomolybdic acid (PMA) as an electron mediator. PMA was chosen as a soluble catalyst for the extraction of electrons during biomass depolymerization because of its distinctive color transformation from yellow (PMA^{3-}) to dark green (PMA^{5-}) when reduced, especially with absorbance at 700 nm. Additionally, at a relatively low potential, PMA exhibits high solubility and reversible redox performance. Pre-reduction of PMA was carried out over biomass oxidation including lignin, hemicellulose, cellulose, or oak biomass. Furthermore, by selectively depolymerizing lignin in lignocellulosic biomass, they can produce value-added chemicals such as vanillin and acetovanillone (see **Figure 8**). This PEC system showed the current density of 19.8 $mAcm^{-2}$ under simulated AM1.5G 1-sun radiation with stability of more than 20 h without significant performance deprivation.

A method was proposed for high-efficiency hydrogen production via the oxidation of glucose using a photoelectrochemical approach. This method involves utilizing an ultrathin carbon-coated perovskite-modified TiO_2 nanotube photonic crystal (C@Cr-$SrTiO_3$/TiO_2 NTPC) as the photoanode [66]. Compared with water oxidation, the oxidation of biomass requires less external force. The oxidation of biomass derivatives through PEC process is a more favorable method for hydrogen generation, as it can be powered by solar energy. Glucose, a significant biomass component, is a key product of photosynthesis and is abundant in nature. Moreover, it serves as one of the primary sources of waste in agricultural and food industries. If not properly managed, this waste can cause significant environmental pollution. The PEC H_2 evolution from biomass derivatives oxidation process consists of two half-reactions:

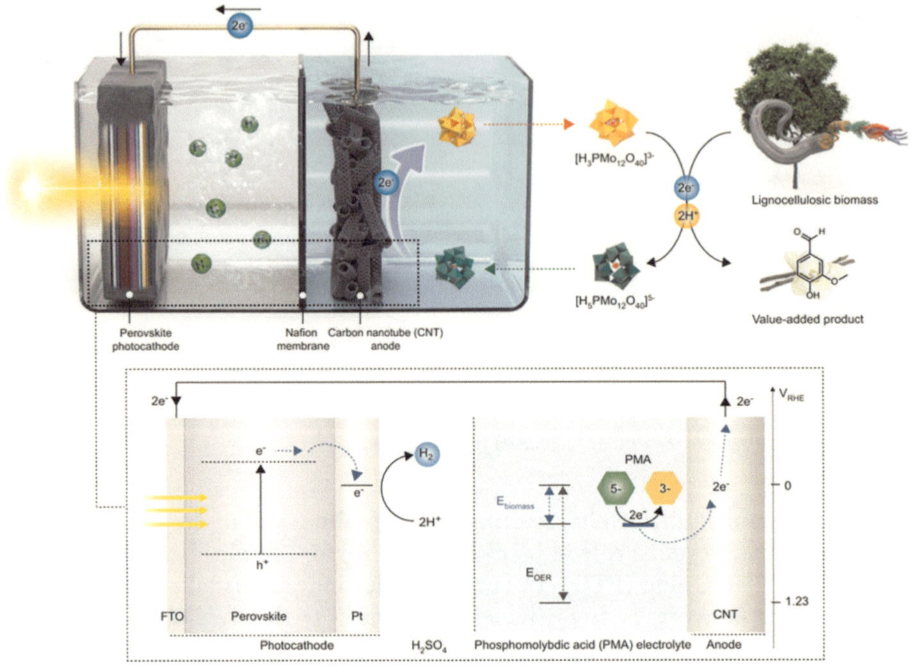

Figure 8.
A perovskite photocathode integrated with a Pt catalyst and an electron extraction system from biomass is utilized to achieve bias-free hydrogen evolution. Reproduced with permission from ref. [68].

Oxidation reaction:

$$C_xH_yO_z + (2x–z)\ H_2O \rightarrow xCO_2 + (4x–2z + y)(H^+ + e^-),\quad E_{anode} = E^{\theta}–0.059(pH)\ (vs.NHE). \tag{1}$$

Reduction reaction:

$$(4x–2z + y)H^+ + (4x–2z + y)\ e^- \rightarrow (2x–y + y/2)\ H_2,\quad E_{cathodic} = 0\ V–0.059(pH)\ V\ (vs.NHE). \tag{2}$$

Total:

$$C_xH_yO_z + (2x - z)\ H_2O \rightarrow xCO_2 + (2x–z + y/2)\ H_2. \tag{3}$$

In glucose electrolyte, the C@Cr-$SrTiO_3$/TiO_2 NTPC photoanode revealed the photocurrent density of 0.43 mA cm^{-2} at the potential of 0.6 V (vs. SCE).

3.3 CO_2 reduction

Due to the fast population growth and the rapid progress of the global economy, the global demand for fossil fuels is continuously growing. This leads to higher levels of greenhouse gases especially large amounts of CO_2 emissions and numerous environmental problems such as ozone depletion, climatic change, melting ice caps, acid rain, and rising sea levels. According to data from the National Oceanic and Atmospheric Administration (NOAA) research, in July 2017 the atmospheric CO_2

concentration reached approximately ~403.95 ppm, and it is anticipated to reach about 600 ppm by 2100 [69]. Converting CO_2 into valuable chemical feedstocks is a promising approach to address the problem of controlling CO_2 levels and growing the new energy storage. Since the late 1970s, there has been significant research on photoelectrochemical CO_2 reduction as "artificial photosynthesis". This process has garnered considerable attention for its potential for CO_2 reduction. CO_2 reduction reactions offer energy-saving capabilities and generate extensive products such as carbon monoxide (CO), formic acid (HCOOH), formaldehyde (HCHO), methane (CH_4), methanol (CH_3OH), olefins, and other hydrocarbons. **Table 1** shows the obtained products and the number of electrons and protons involved in the CO_2 reduction reaction (CO_2RR) potentials [70].

It should be considered that the nature of the electrocatalysts and the applied potential significantly impact the creation of final products. Inherently, the CO_2 molecule is highly stable because of its linear structure, which consists of two symmetric C=O bonds, and the C-O bond length in CO_2 is approximately 116.3 pm, requiring a significant energy of ~750 kJ/mol to break the initial bond in CO_2. Also, the reduction of protons can occur at the potential demand for CO_2 reduction reactions, leading to the low efficiency and low selectivity of a convenient reaction [71]. Moreover, most of the existing electrocatalysts still face several challenges such as poor stability, low conversion efficiency, low selectivity, and inability to conquer the hydrogen evolution as a competitive reaction.

In recent years, perovskite materials have garnered considerable interest in the field of photoelectrochemical CO_2 reduction. Crucially, the band positions of the majority of perovskites satisfy the essential thermodynamic criteria for CO_2 reduction [72]. As a result of these features, these materials have been effectively employed in PEC CO_2 reduction. Fundamentally, the PEC CO_2 cells contained the cathodic and anodic parts (**Figure 9**). They are separated by an ion-selective membrane. Considering the PEC CO_2 reduction in acidic solutions, in the cathodic compartment, the CO_2 reduction reaction occurs, while in the anodic compartment, the water oxidation reaction takes place. During water oxidation ($2H_2O + 4\,h^+ \rightarrow O_2 + 4H^+$) in the anodic section, protons are generated and subsequently transported to the cathodic part through the ion-selective membrane, propelled by a concentration gradient [73].

Zhang et al. investigated a PEC cell for the conversion of CO_2 to CO, employing a perovskite-based $CsPbBr_3$ (CPB) thin film as the photocathode, a Pt foil as the counter

Electrochemical thermodynamic half-reactions	Standard potentials (V vs. SHE)
$CO_2(g) + 4H^+ + 4e^- = C(s) + 2\,H_2O(l)$	0.210
$CO_2(g) + 2H^+ + 2e^- = HCOOH\ (l)$	−0.250
$CO_2(g) + 2H^+ + 2e^- = CO(g) + H_2O(l)$	−0.106
$CO_2(g) + 4H^+ + 4e^- = CH_2O(l) + H_2O(l)$	−0.070
$CO_2(g) + 6H^+ + 6e^- = CH_3OH(l) + H_2O(l)$	0.016
$CO_2(g) + 8H^+ + 8e^- = CH_4(g) + 2H_2O(l)$	0.169
$2CO_2(g) + 2H^+ + 2e^- = H_2C_2O_4(aq)$	−0.500
$2CO_2(g) + 12H^+ + 12e^- = CH_2CH_2(g) + 4H_2O(l)$	0.064

Table 1.
Different products under standard potentials of CO_2 reduction (versus standard hydrogen electrode, SHE).

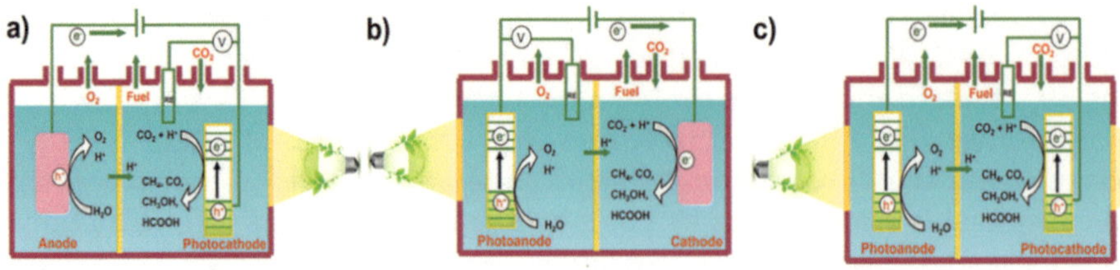

Figure 9.
Schematic representations of possible two-compartment PEC cells, designed for CO_2 reduction as: semiconductors employed as photocathodes (a), semiconductors utilized as photoanodes (b) and semiconductors serving as both photocathodes and photoanodes (c). Reproduced with permission from ref. [73].

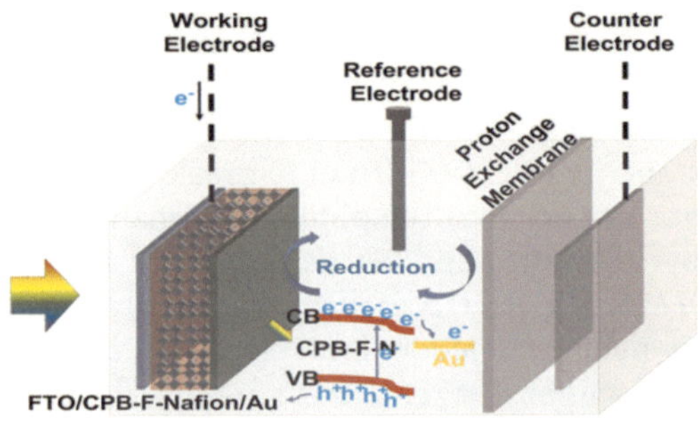

Figure 10.
A PEC cell for CO_2RR based on $CsPbBr_3$ (CPB) thin-film photocathode.

electrode, and an Ag/AgCl (3 M KCl) reference electrode [74]. **Figure 10** displays the schematic of this designed system. A two-step procedure was utilized to obtain perovskite photocathode. First, by a spin-coating process, a $PbBr_2$ layer film was made. Then, the solutions of the CsBr-based materials were spin coated onto the surface. Next, to prepare $CsPbBr_3$-F-N-Au thin films, fluorine, Nafion, and Au were added to the above solution. When F-ions are added in perovskite structure, they are able to promote the chemical interactions. Additionally, by Nafion addition, an adjustment in the band gap of $CsPbBr_3$ and a decreased electron-hole recombination rate were obtained. Besides, Au can accelerate separation of charge carriers in the photocathode. The photocathode exposed the best PEC performance at a potential of -0.5 V (vs Ag/AgCl), with the maximum photocurrent of -0.23 mA cm^{-2}. This investigation proved that a halide perovskite can be directly utilized to assist CO_2 reduction.

Moreover, PV-EC systems offer significant benefits in terms of scalability, durability and optimization compared to the PEC approach for CO_2 reduction. By construction of a monolithic PV-CE system, Schreier and colleagues demonstrated an efficient CO_2 to CO conversion with 90% faradaic efficiency [75]. **Figure 11** displays the designed PV-EC cell including an electrochemical unit with oxidized gold and IrO_2 as cathode and anode, respectively coupled to $CH_3NH_3PbI_3$-based photovoltaics. This study demonstrated the prolonged and consistent operation of a PV-EC cell utilizing perovskite photovoltaics. The current density remained stable at approximately 5.8 mA cm^{-2}, with only minor fluctuations observed for over 18 hours.

A $CsPbBr_3$-based PEC system was fabricated for CO_2 reduction into oxalic acid [76]. To improve the stability of $CsPbBr_3$ in water, it was coated with platinum and graphite (C). Graphite was selected for its stability under standard temperature and pressure conditions, as well as its affordability. Platinum was chosen due to its widely

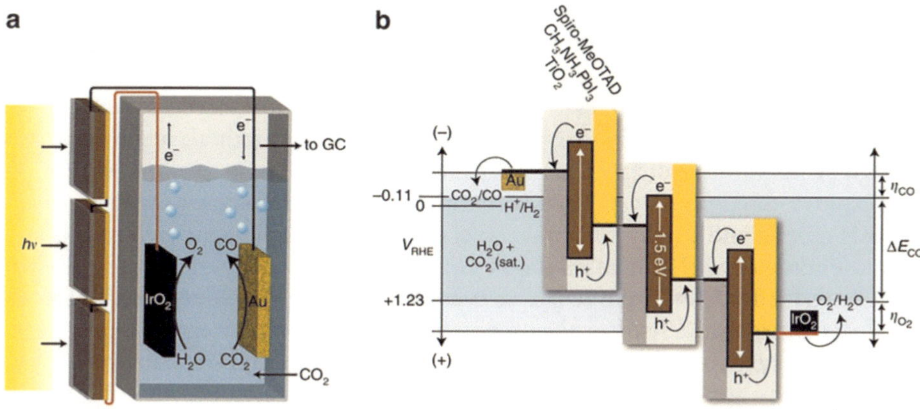

Figure 11.
(a) Schematic of the apparatus integrating photovoltaics with an electrochemical cell. (b) Comprehensive energy diagram depicting the conversion of CO_2 into CO utilizing three perovskite solar cells. The PSCs connected in series generate a voltage adequate to surpass the combined energy of the reaction free energy (ΔE) and the reaction overpotentials (η) at the electrodes. Reproduced with permission from ref. [75].

acknowledged high electrocatalytic activity. Chronoamperometric tests were conducted to study photoelectrochemical properties of $CsPbBr_3$/Pt and $CsPbBr_3$/C for CO_2 reduction at low external potentials applied (− 0.5 V and - 0.8 V vs. Ag/AgCl) under illumination. During CO_2 conversion, it was observed that the $CsPbBr_3$/Pt samples exhibited low charge transfer resistance to the electrolyte. As a result, these samples ultimately produced the highest photo-current (12.6 mA cm^{-2}). The $CsPbBr_3$/C showed the highest Faradic Efficiencies of 22% (at −0.5 V) and 44% (at −0.8 V) vs. Ag/AgCl, respectively. This behavior is attributed to the properties of the carbon surface, which catalyze the formation of oxalic acid more effectively than platinum. In contrast, platinum promotes the competing reaction of hydrogen formation.

Abarca and colleagues introduced a novel photoanode comprising a combination of perovskite-based calcium titanate ($CaTiO_3$) as an electron collector and $BiVO_4$ as a catalyst for visible-light-driven water oxidation, with the goal of enhancing PEC CO_2 conversion [77]. The designed photoanode is incorporated into an electrolyzer to continuously convert CO_2 into formate using visible light. The attained concentration was 63.8 g L^{-1}, with a Faradaic Efficiency of 79.1% and a solar-to-fuel (STF) conversion efficiency of 7.6%. The combination of $BiVO_4$ and $CaTiO_3$ creates a p-n junction, which enhances the separation of charges. This junction generates an interfacial electric field at the interface of both catalysts. Consequently, electrons can migrate from $CaTiO_3$ to the conduction bands (CB) of $BiVO_4$, while holes can move from $BiVO_4$ to the valence band (VB) of $CaTiO_3$. Moreover, the negative CB and VB levels in the electronic structure of $CaTiO_3$ lead to a spontaneous charge transfer from a thermodynamic perspective.

3.4 N_2 evolution

Nitrogen (N_2) making up about 78% of Earth's atmosphere, has the potential to be converted into valuable industrial products, profiting the environment and humanity. Conversion of N_2 to ammonia (NH_3) is crucial for industrial processes, and the production of pharmaceuticals, dyes, and fertilizers. Besides, NH_3 is highly valuable as a hydrogen carrier and is recognized for its ability to generate CO_2-free energy [78]. Considering the kinetic and energy aspects, N_2 Reduction Reaction (N_2RR) presents a challenging chemical process. This is primarily owing to the enthalpy of triple bonds between nitrogen atoms as 940.95 kJ mol^{-1}. At present, the industrial production of NH_3 continues to depend on the conventional Haber-Bosch method, which operates under high temperatures and pressures. This approach gives rise to considerable challenges, including excessive CO_2 emissions and high energy consumption [79]. Hence, it is crucial and urgent to devise an economically viable approach for N_2 reduction. Photoelectrochemical methods provide a more energy-efficient and environmentally friendly alternative to the conventional Haber-Bosch process. Photoelectrochemical cells potentially offer the advantages of both photochemical and electrochemical systems by employing solar energy, separating NH_3 from other byproducts and by applied electrical bias, these systems develop the reduction capability of photogenerated electrons. In a PEC N_2RR system, at the cathodic surface the N_2 reduction reaction occurs ($N_2 + 6H^+ + 6e - \longrightarrow 2NH_3$), while on the anodic part oxidation of H_2O takes place [80].

In 1977, Schrauzer and Guth converted N_2 to NH_3 using TiO_2 in UV light consequently, a number of photoelectrocatalytic systems including TiO_2, MoS_2, LDH, $V_2O_3{\cdot}nH_2O$, and g-C_3N_4 were studied for N_2 PEC reduction [81]. However, many of these materials still encounter numerous challenges, such as low efficiency, limited light absorption range, and instability, significantly restricting their applicability for industrial N_2 reduction. Consequently, there is a pressing need for the development of a suitable photoelectrode for N_2 conversion. In this regard, perovskite materials-driven photocathodes have been identified as a suitable and promising option.

For photocatalytic and photoelectrocatalytic N_2 reduction, a photocatalyst system composed of $CoTiO_3$/N-rGO p-n junction, was suggested by Paramanik et al. [82]. $CoTiO_3$ as a p-type semiconductor with band gap energy of about 2.25 eV, great charge carrier mobility, large absorption coefficient, and active in visible-light regions, reveals potential for well-organized PEC applications. In addition, the satisfactory N_2-adsorption sites can be generated by N-dopant in reduced graphene oxide

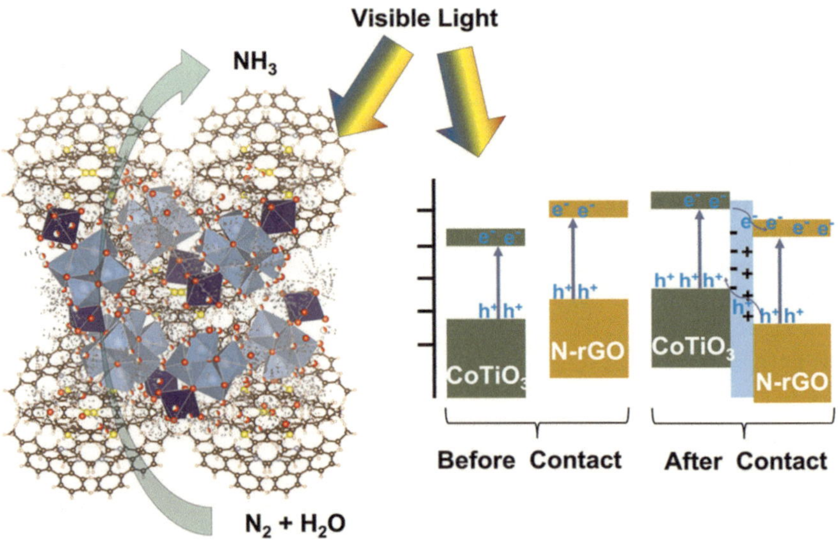

Figure 12.
Schematic of PEC N_2RR cell based on the $CoTiO_3$/N-rGO.

that creates a positive charge on the graphene surface, resulting in more N_2 activation. Under light irradiation, the produced electrons on the conduction band of $CoTiO_3$ transfer to CB of N-rGO, and the generated holes on the valence band of N-rGO flow to $CoTiO_3$ (**Figure 12**). In order to achieve the hydrogenation procedure of N_2 and conversion to NH_3, the activated N_2 molecules are attacked by photogenerated electrons and active H atoms from water to produce the final product. This system showed great performance with a faradaic efficiency of 18.05%.

3.5 Degradation of pollutants

The release of organic pollutants from industries such as textiles, dyeing, cosmetics, and food poses significant environmental challenges related to dangerous wastes, and polluted groundwater [83]. As a result, it is vital and desirable to adopt sustainable approaches to minimize the discharge of pollutants, protecting both the environment and human well-being. Overall, the PEC technique, with its integration of light response and electrochemical analysis, has opened up exciting possibilities for the degradation of pollutants [84]. The photoelectrocatalytic procedures are greater than traditional approaches owing to their capability to remove organic waste products through operation at room temperature and pressure, cost-effectiveness, and the avoidance of polycyclic compound generation. In a PEC system, the photogenerated electrons in conduction band (e_{CB}^-) and generated holes in the valence band (h_{VB}^+) can engage in several reactions, resulting in the production of highly Reactive Oxidation Species (ROSs) such as hydroxyl radical ($^{\bullet}OH$) and superoxide radical ($O^{2\bullet-}$). These species have the capability to react with organic contaminants in a non-selective manner. Ultimately, this process leads to the mineralization of organic contaminants, converting them into H_2O and CO_2 [85]. Semiconductors like TiO_2, CdS, ZnO, and MoS_2 have been widely used as photoelectrocatalyst in degradation via PEC systems. Nonetheless, their photoelectrocatalytic efficiencies are impeded by their wide band gaps, limiting their capacity to efficiently absorb visible light [86]. So, research has focused on the visible light-active electrocatalysts that possess narrow band gaps and exhibit minimal electron-hole recombination. Perovskite materials have demonstrated exceptional potential in the degradation of pollutants due to their remarkable catalytic properties. These materials exhibit high surface area and reactivity, enabling efficient conversion of harmful pollutants into less toxic byproducts.

The perovskite $BaSnO_3$ was synthesized using a chemical method with a cubic crystalline structure (average size of 63 ± 2 nm) by Sahmi et al. [87]. $BaSnO_3$ is both safe and cost-effective and exhibits chemical stability across a wide pH range. The active surface area is increased when it is synthesized through a chemical route, which leads to improved photo activity. The waste containing ibuprofen ($C_{13}H_{18}O_2$: propionic acid 2-(4-isobutylphenyl)) which classify as a hazardous content and the $BaSnO_3$ was successfully applied for ibuprofen degradation by its mineralization. Through electrocatalysis on $BaSnO_3$, ibuprofen has been observed to undergo elimination at a conversion rate of 55% when a current of 150 mA is applied. Furthermore, when subjected to UV illumination, the conversion rate has been increased to 68% through a PEC process. The electrons in conduction band of $BaSnO_3$ produce $O_2^{\bullet-}$ and H_2O_2. The H_2O_2 serves as an electron acceptor and reacts with the $O_2^{\bullet-}$ to form $^{\bullet}OH$ radicals. These $^{\bullet}OH$ radicals are used to oxidize ibuprofen. Additionally, the high potential of the $^{\bullet}OH/H_2O$ couple (2.27 V) allows for the production of $^{\bullet}OH$ radicals, which are essential for the mineralization of ibuprofen. These $^{\bullet}OH$ radicals are covalently bonded to $BaSnO_3$. Morcoso et al. investigated the photocatalytic and photoelectrochemical oxidation of 2-mercaptobenzothiazole (MBT) using $CsPbBr_3$ quantum dots [88].

The remarkable photoluminescence quantum yield displayed by perovskite quantum dots (QDs) provides compelling evidence of the notable reduction in nonradiative recombination procedures. As a result, when these QDs are photoexcited, a surplus of energized carriers is generated, presenting exciting possibilities for their utilization. They confirmed that complete degradation of MBT occurs as a result of hole injection from $CsPbBr_3$ to MBT. MBT is classified as a member of the heterocyclic aromatic compound group known as benzothiazole. It finds widespread application in various industrial sectors, serving as a rubber accelerator, corrosion inhibitor, fungicide, and antialgal agent. However, the poor biodegradability, aquatic toxicity, tumor-inducing properties, and role as a frequent allergen and human carcinogen make MBT a substance with significant concern. $CsPbBr_3$ QDs can be utilized for photoelectrocatalytic degradation of MBT, primarily due to their advantage of not overlapping with the characteristic band of MBT at 320 nm, which is monitored in PEC degradation. The presence of $CsPbBr_3$ QDs plays a significant role in the degradation of MBT under different lighting conditions. Without the existence of $CsPbBr_3$, no degradation occurs under visible illumination or in the dark and only UV light is effective in the MBT degradation. The addition of $CsPbBr_3$ QDs in the MBT solution has a remarkable impact on the degradation rate under different lighting conditions. Specifically, when $CsPbBr_3$ QDs are added, under UV-vis irradiation, the degradation rate doubles, while under visible light, it multiplies.

In another activity, Kuo et al. reported a photoelectrode utilizing $YFeO_3/CeO_2$ synthesized via a sol-gel method to tackle the challenge of recombination of photogenerated electron-hole pairs, thereby enhancing the degradation efficiency of the catalyst [89]. Due to its narrower bandgap relative to other perovskites, $YFeO_3$ has garnered considerable interest as a promising material for photoelectrocatalysis under visible light. Additionally, the ferromagnetic properties of $YFeO_3$ facilitate its magnetic retrieval post-utilization, rendering its reuse feasible. Hexagonal $YFeO_3$ exhibits superior photoelectrocatalytic activity compared to orthorhombic $YFeO_3$. However, metastable

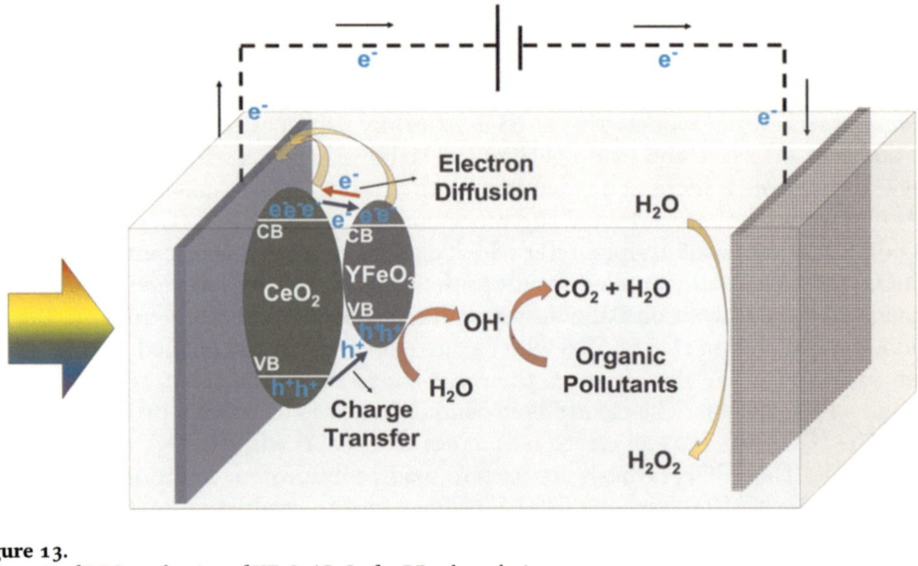

Figure 13.
Schematic of PEC mechanism of $YFeO_3/CeO_2$ for RB5 degradation.

nature and thermodynamically unstable crystal structure of hexagonal $YFeO_3$ present significant challenges for its formation. Therefore, employing a heterojunction electrocatalyst by combining $YFeO_3$ with another material presents an effective alternative strategy to enhance photoelectrochemical processes under visible light. The designed photoelectrode demonstrates a highly promising PEC efficiency of 75.2% for Reactive Black 5 dye (RB5) degradation under visible light irradiation, which is significantly higher compared to $YFeO_3$ and CeO_2 photoelectrodes with efficiency of 31.9 and 46.9%, respectively. As is observed in **Figure 13**, under visible illumination, electrons and holes were produced on the conduction band (CB) and valence band (VB) of $YFeO_3/CeO_2$ photoelectrode. Therefore, by reason of the reactions between the cumulative holes and OH ions, the creation of hydroxyl radical was elevated, which simplified degradation. The generated hydroxyl radical is capable of reacting with RB5 and degrading the pollutant.

3.6 Sensor devices

Photoelectrochemical systems have emerged as a highly valuable and efficient detection procedure in analytical chemistry owing to their superior recognition properties, stability, precision, portable apparatus and low cost. Because of the complete segregation between the excitation light source and the display current signal, they have emerged as a valuable tool with minimal background signal in trace analysis [90]. Photoactive materials play a critical role in influencing the sensitivity of a PEC device. These materials should possess specific characteristics including improved and steady photocurrent production, effective light harvesting, and high sensitivity. Recently, perovskite materials have been demonstrated outstanding analytical features in PEC sensors including distinctive carrier separation, fast response times, excellent robustness, and highly accurate detection capabilities.

A PEC aptasensor based on the "on-off-on" approach was investigated for ultrasensitive analysis of Prostate-Specific Antigen (PSA) as a target [91]. As it observed in **Figure 14**, principally, in order to development of the PEC system stability, Cd: Sb_2S_3 as a photosensitizer loaded on $La_2Ti_2O_7$ surface to produce the heterojunction related to the first "signal-on" state, and obtained a greater photocurrent because of the matched levels of band energy. Then, to realize the "signal-off" state, the V_2O_5 nanosphere as a H_2O_2 catalyst (electron donor in electrolyte solution) was utilized to mark aptamer DNA. The PSA was accurately detected by a DNA,

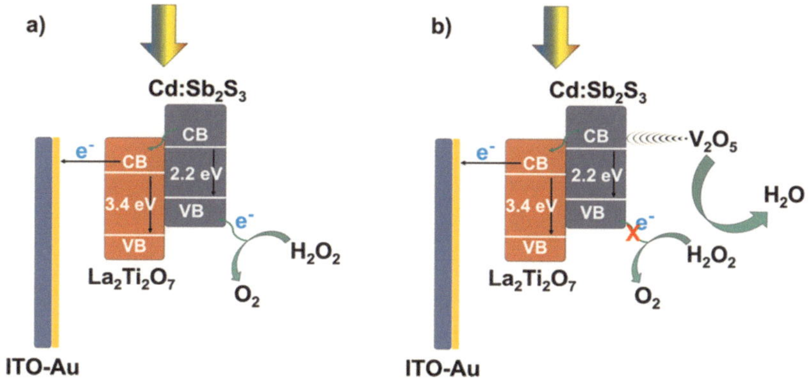

Figure 14.
Mechanism of Electron transfer (a) signal-on state and (b) signal-off state of a photoelectrochemical aptasensor based on $La_2Ti_2O_7/Sb_2S_3$ and V_2O_5 for effectively signal change strategy for cancer marker detection.

leading to the release of V_2O_5 from the electrode, thus reinstating the PEC "signal-on" state. The suggested PEC sensor for PSA detection exposed a wide linear range from 1.000×10^{5} to 500.0 ng/mL, as well as a low detection limit of 4.300 fg/mL. This sophisticated strategy not only progresses the sensor sensitivity but also expands the application of perovskites in the field of PEC sensing.

A novel highly stable, sensitive and selective PEC sensor for the recognition of dopamine (DA) was fabricated using $CsPbBr_{1.5}I_{1.5}$ QDs and immobilized them on the TiO_2 with porous structure [92]. The TiO_2 and $CsPbBr_{1.5}I_{1.5}$ QDs possess band gaps of approximately 3.2 and 2.1 eV, respectively. Their conduction band edges are at approximately −3.9 and − 3.5 eV, respectively, while their valence band edges are around −7.6 and − 5.6 eV, respectively. As depicted in **Figure 15**, under visible light excitation, the close proximity of the conduction band edge of $CsPbBr_{1.5}I_{1.5}$ QDs to that of TiO_2 suggests the potential for efficient transportation of photogenerated electrons from the conduction band of $CsPbBr_{1.5}I_{1.5}$ QDs to the FTO through the TiO_2 layer. This procedure efficiently conceals the direct electron-hole pair recombination that occurs in perovskite, resulting in an enhancement in the sensor sensitivity. In fact, it is easier for the photogenerated holes to transport from the valence band of perovskite to its surface. Additionally, the energy level of DA (−4.91 eV) is higher than the valence band (−5.6 eV) of perovskite, providing a likelihood for the detection of DA.

Leng et al. introduced a self-powered PEC-based biosensor composed of a BiOI as photocathode and a $WO_3/SnS_2/ZnS$ as photoanode [93]. Moreover, $CsPbBr_3$@COF-V was employed as a signal quenching agent for the quantitative detection of heart fatty acid binding protein (H-FABP) on the photocathode. In addition, the secondary antibody marker (Ab_2) was loaded to $CsPbBr_3$@COF–V to form $CsPbBr_3$@COF–V-Ab_2 for connection to FABP. This biosensor not only offers a continuous and powerful photocurrent response for FABP recognition, but also efficiently eradicates the interference of constituents. The photocurrent response of $CsPbBr_3$@COF–V is significantly influenced by the quenching source, which includes weak conductivity, steric interruption, and struggle against the substrate for dissolved oxygen as well as competition for the excitation light. Achieving multiple signal intensifications is possible with the intervention of this critical factor, unveiling an inventive approach to self-operating PEC biosensing devices. Actually, once FABP molecules were joined to the $CsPbBr_3$@COF–V-Ab_2 electrode, the response signal reduced because of steric

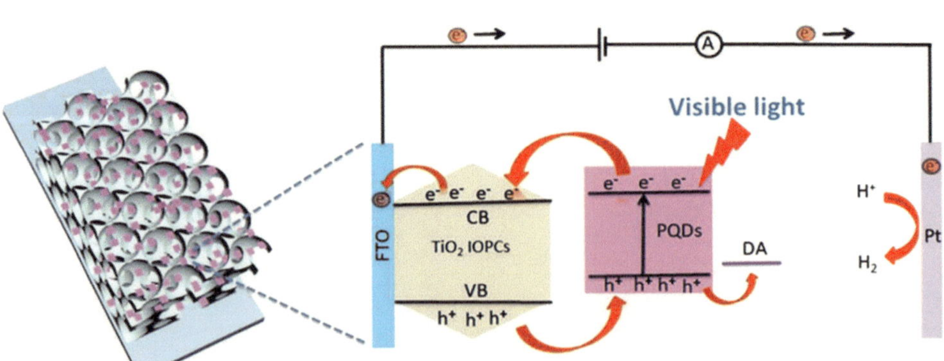

Figure 15.
Schematic of an all-inorganic perovskite quantum dot/TiO_2 inverse opal electrode platform for photoelectrochemical sensing of dopamine under visible light. Reproduced with permission from ref. [92].

hindrance and quenching process of Ab-2. This biosensor showed a wide-ranging from 0.0005 to 150 ng/mL with detection limit of 0.19 pg./mL for FABP detection.

In another research, Dong et al. reported a PEC sensor for trenbolone (TB) detection using $Cs_3Bi_2Br_9/BiOBr/Bi_2S_3$ as a working electrode with Ag-Au/CdS as a signal enhancer, an Hg/Hg_2Cl_2 reference electrode and a Pt wire as the counter electrode [94]. $Cs_3Bi_2Br_9/BiOBr/Bi_2S_3$ can efficiently avoid the electron-hole pairs recombination. Moreover, the working electrode was incubated with TB antibodies (Ab) of the same concentration. Subsequently, the designed sensor was applied for detection of various concentrations of TB. Trenbolone is a synthetic anabolic steroid that is used in veterinary medicine to promote muscle growth and appetite in livestock. It is known for its powerful effects on increasing muscle mass and strength, making it popular among bodybuilders and athletes looking to enhance their performance. Trenbolone is classified as a controlled substrate in many countries due to its potential for misuse and abuse, so, its detection is important. The concentration of TB can be assessed by monitoring the alteration in output current caused by the competitive binding of Ab with TB or Ag-Au/CdS. A decrease in TB concentration leads to an increase in connected Ag-Au/CdS on Ab. Therefore, the PEC sensor has the capability to reliably detect TB with a detection limit of only 29.7 fg mL^{-1}.

3.7 Synthesis of small organic compounds

PEC devices can be utilized not only for fuel production but also for the production of high-value simple organic molecules. PEC methods represent a cutting-edge approach to organic molecule synthesis that offers green and sustainable solutions, precise control over reactions, mild reaction conditions, versatility, rapid reaction kinetics, and compatibility with complex substrates [95]. 2,5-Dimethoxy-2,5-dihydrofuran (DMDF) is an important compound in organic chemistry due to potential applications in various synthetic processes as a vital intermediate for creating organic compounds such as benzenoid compounds, pyridines, pyridazines, and pyrroles. Using the conventional chemical synthetic method for DMDF, a significant quantity of bromine is utilized as an oxidizing agent, which poses concerns regarding environmental pollution [96]. PEC processes as an alternative method offer a sustainable and environmental approach to producing DMDF without additional catalyst consumption by harnessing renewable energy for driving chemical transformations.

Aimed at PEC synthesis of 2,5-Dimethoxy-2,5-dihydrofuran (DMDF) from furan, a thin film of methylammonium lead bromide ($MAPbBr_3$) single crystal (SCTF) as a photoanode was deposited on a FTO glass [97]. Then, an ultrathin Al_2O_3 layer was grown on the perovskite surface and a Ti layer was deposited on the Al_2O_3 layer. It was found that ultrathin Al_2O_3 layer efficiently inhibits surface defects of perovskite and Ti layer as a protective catalytic layer further improving PEC performance. Indeed, these layers significantly enhance the stability and photocatalytic activity of perovskite crystals. **Figure 16** displays the schematic of the PEC cell fabricated by $MAPbBr_3$ perovskite on FTO/TiO_2 substrate, and Pt mesh as the photoanode, and counter electrode, respectively, in an electrolyte including a Br^-/Br^+ in an acetonitrile/methanol mixed solution. Under light radiation, by photogenerated holes, the Br^- ions oxidize into Br^+. Generated Br^+ species act as oxidizing reagents to produce DMDF from furan. The PEC performance and stability of the designed photoanode were considerably enhanced resulting in 6 h constant DMDF evolution examination with a high Faraday efficiency of 93%.

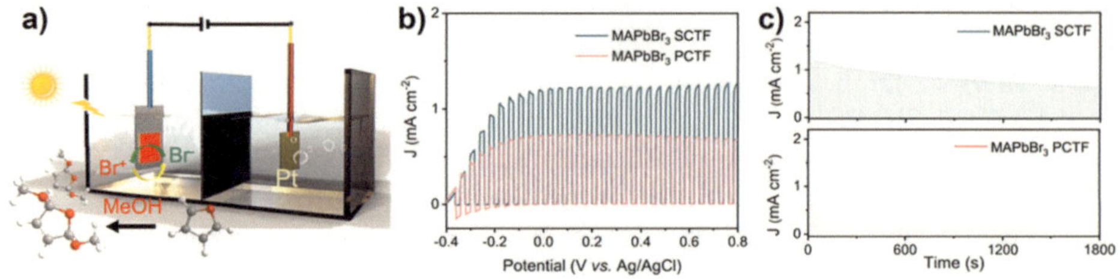

Figure 16.
(a) Schematic depicting the operational concept of $MAPbBr_3$ SCTF-based (or PCTF-based) PEC cells. Linear sweep voltammetry (LSV) (b) and Chronoamperometric trace (c) of $MAPbBr_3$ SCTF-based and PCTF-based photoelectrodes. Reproduced with permission from ref. [97].

4. Conclusions and prospective

The advantages of perovskites in photoelectrochemical applications profit from their exceptional properties, including appropriate band gap, high optical absorption coefficient, good electron mobility, high light-harvesting capability, simple manufacturing procedure, low cost, adaptable structure, and morphology. PEC application of perovskites holds great promise for advancing renewable energy technologies. While numerous chapters focus on the use of perovskite materials for solar cells, at present the performances of perovskites in PEC devices have not been completely investigated. This chapter thoroughly examines the significant impact of the perovskite structure on various photoelectrochemical (PEC) and monolithic Photovoltaic-Electrocatalytic (PV-EC) devices. Despite these advancements, the practical application of perovskites in PEC devices is still in its infancy stages. There are several challenges that need to be addressed in order to facilitate their practical use. One of the main challenges is the stability of perovskites under harsh operating conditions, which can limit their long-term performance. Crystal engineering or the encapsulation with appropriate semiconductor materials is an effective approach for developing stable perovskites for the practical applications. In addition, the toxicity of lead is a major concern when it comes to environmental issues which can be addressed by lead-free alternatives such as Ti, Bi, and encapsulating or coating of the lead-based perovskites to minimize the risk of lead exposure during the use of perovskite-based devices. Besides, the efficiency of perovskite-based PEC devices still needs to be improved to compete with other established technologies. Despite the obstacles, scientists are achieving rapid advancements, and there is no doubt that the current challenges will be overcome. The future of perovskite-based PEC devices is undeniably promising. To drive further progression in this realm, future research could be focused on designing perovskite architectures with various dimensions, integrating desired properties, and exploring the potential applications of perovskites in PEC processes.

Acknowledgements

M.Z. gratefully acknowledges the NEXTCCUS project (ID: 327327), which is funded through the ACT program (Accelerating CCS Technologies, Horizon2020; Project No. 691712), and financial contributions made by the Iritaly Trading Company

Srl and the Italian Ministry of Education, University and Research (MIUR). N.Y. gratefully acknowledges the P4SPACE project (ID: 101067838), funded by the European Union under HORIZON-MSCA-2021-PF-01.

Author details

Mina Ahmadi-Kashani[1], Mahmoud Zendehdel[2,3]*, Mohammad Mahdi Abolhasani[1] and Narges Yaghoobi Nia[4]*

1 Chemical Engineering Department, University of Kashan, Kashan, Iran

2 Kimia Solar Research Institute (K.S.R.I), Kimia Solar Company, Kashan, Iran

3 Iritaly Trading Company S.r.l., Rome, Italy

4 Laboratory of Photonics and Interfaces, Ecole Polytechnique Fédérale de Lausanne, Lausanne, Switzerland

*Address all correspondence to: m.zendehdel@iritalytrading.com and narges.yaghoobinia@epfl.ch

References

[1] Zhang D, Shi J, Zi W, Wang P, Liu SF. Recent advances in photoelectrochemical applications of silicon materials for solar-to-chemicals conversion. ChemSusChem. 2017;**10**(22):4324-4341. DOI: 10.1002/cssc.201701674

[2] Samuel E, Joshi B, Kim MW, Swihart MT, Yoon SS. Morphology engineering of photoelectrodes for efficient photoelectrochemical water splitting. Nano Energy. 2020;**72**(March): 104648. DOI: 10.1016/j.nanoen.2020.104648

[3] Prasad C, Tang H, Liu QQ, Zulfiqar S, Shah S, Bahadur I. An overview of semiconductors/layered double hydroxides composites: Properties, synthesis, photocatalytic and photoelectrochemical applications. Journal of Molecular Liquids. 2019;**289**: 111114. DOI: 10.1016/j.molliq.2019.111114

[4] Holmes-Gentle I, Tembhurne S, Suter C, Haussener S. Dynamic system modeling of thermally-integrated concentrated PV-electrolysis. International Journal of Hydrogen Energy. 2021;**46**(18):10666-10681. DOI: 10.1016/j.ijhydene.2020.12.151

[5] Lourenço AC, Reis-Machado AS, Fortunato E, Martins R, Mendes MJ. Sunlight-driven CO_2-to-fuel conversion: Exploring thermal and electrical coupling between photovoltaic and electrochemical systems for optimum solar-methane production. Materials Today Energy. 2020;**17**:100425. DOI: 10.1016/j.mtener.2020.100425

[6] Cen J, Wu Q, Liu M, Orlov A. Developing new understanding of photoelectrochemical water splitting via *in-situ* techniques: A review on recent progress. Green Energy & Environment. 2017;**2**(2):100-111. DOI: 10.1016/j.gee.2017.03.001

[7] Liu L, Zhang Y, Huang H. Junction engineering for photocatalytic and photoelectrocatalytic CO_2 reduction. Solar RRL. 2021;**5**(2):2000430. DOI: 10.1002/solr.202000430

[8] Chen D, Lu Y, Wu J, Li N, Zheng YZ, Tao X. Perovskite solar cells-TiO_2 tandem assembly for photoelectrocatalytic degradation of organic pollutants. Journal of Physics and Chemistry of Solids. 2019;**132** (September):204-212. DOI: 10.1016/j.jpcs.2019.03.025

[9] Sheng X, Xu T, Feng X. Rational design of photoelectrodes with rapid charge transport for photoelectrochemical applications. Advanced Materials. 2019;**31**(11):1-29. DOI: 10.1002/adma.201805132

[10] Sardashti MK, Zendehdel M, Nia NY, Karimian D, Sheikhi M. High efficiency $MAPbI_3$ perovskite solar cell using a pure thin film of polyoxometalate as scaffold layer. ChemSusChem. 2017;**10**(19):3773-3779. DOI: 10.1002/cssc.201701027

[11] Yaghoobi Nia N et al. Impact of P3HT regioregularity and molecular weight on the efficiency and stability of perovskite solar cells, ACS sustain. Chemical Engineering. 2021;**9**(14): 5061-5073. DOI: 10.1021/acssuschemeng.0c09015

[12] Yaghoobi Nia N, Méndez M, Paci B, Generosi A, Di Carlo A, Palomares E. Analysis of the efficiency losses in hybrid perovskite/PTAA solar cells with different molecular weights: Morphology

versus kinetics. ACS Applied Energy Materials. 2020;**3**(7):6853-6859. DOI: 10.1021/acsaem.0c00956

[13] Jeong J et al. Pseudo-halide anion engineering for α-FAPbI3 perovskite solar cells. Nature. 2021;**592**(7854): 381-385. DOI: 10.1038/s41586-021-03406-5

[14] Yaghoobi Nia N, Saranin D, Palma AL, Di Carlo A. Perovskite solar cells. In: Solar Cells and Light Management. Amsterdam, Netherlands: Elsevier; 2020. pp. 163-228. DOI: 10.1016/B978-0-08-102762-2.00005-7

[15] Zendehdel M, Yaghoobi Nia N, Yaghoubinia M. Emerging thin film solar panels. In: Reliability and Ecological Aspects of Photovoltaic Modules. London, UK: IntechOpen; 2020. DOI: 10.5772/intechopen.88733

[16] Zendehdel M, Yaghoobi Nia N, Paci B, Generosi A, Di Carlo A. Zero-waste scalable blade-spin coating as universal approach for layer-by-layer deposition of 3D/2D perovskite films in high efficiency perovskite solar modules. Solar RRL. 2021;**6**:2100637. DOI: 10.1002/solr.202100637

[17] Castriotta LA et al. Reducing losses in perovskite large area solar technology: Laser design optimization for highly efficient modules and mini panels. Advanced Energy Materials. 2022;**12**:1-12

[18] Yaghoobi Nia N et al. Solution-based heteroepitaxial growth of stable mixed cation/anion hybrid perovskite thin film under ambient condition via a scalable crystal engineering approach. Nano Energy. 2020;**69**:104441. DOI: 10.1016/J.NANOEN.2019.104441

[19] Ahmadi-Kashani M, Zendehdel M, Schirone L, Abolhasani MM, Yaghoobi Nia N. Recent Progress in the use of perovskites for electrochemical, photoelectrochemical, and photovoltaic–electrochemical CO_2 reduction. Energies. 2023;**16**(22):7632. DOI: 10.3390/en16227632

[20] Saparov B, Mitzi DB. Organic-inorganic perovskites: Structural versatility for functional materials design. Chemical Reviews. 2016;**116**(7): 4558-4596. DOI: 10.1021/acs.chemrev.5b00715

[21] Chouhan L, Ghimire S, Subrahmanyam C, Miyasaka T, Biju V. Synthesis, optoelectronic properties and applications of halide perovskites. Chemical Society Reviews. 2020;**49**(10): 2869-2885. DOI: 10.1039/c9cs00848a

[22] Ansari MIH, Qurashi A, Nazeeruddin MK. Frontiers, opportunities, and challenges in perovskite solar cells: A critical review. Journal of Photochemistry and Photobiology C: Photochemistry Reviews. 2018;**35**:1-24. DOI: 10.1016/J.JPHOTOCHEMREV.2017.11.002

[23] Navazani S, Yaghoobi Nia N, Zendehdel M, Shokuhfar A, Di Carlo A. Fabrication of high efficiency, low-temperature planar perovskite solar cells via scalable double-step crystal engineering deposition method fully out of glove box. Solar Energy. 2020;**206**: 181-187. DOI: 10.1016/J.SOLENER.2020.05.084

[24] Ermanova I et al. Crystal engineering approach for fabrication of inverted perovskite solar cell in ambient conditions. Energies. 2021;**14**(6):1751. DOI: 10.3390/en14061751

[25] Yaghoobi Nia N, Zendehdel M, Cinà L, Matteocci F, Di Carlo A. A crystal engineering approach for scalable perovskite solar cells and module

fabrication: A full out of glove box procedure. Journal of Materials Chemistry A. 2018;**6**(2):659-671. DOI: 10.1039/C7TA08038G

[26] Yaghoobi Nia N et al. Doping strategy for efficient and stable triple cation hybrid perovskite solar cells and module based on poly(3-hexylthiophene) hole transport layer. Small. 2019;**15**(49):1904399. DOI: 10.1002/smll.201904399

[27] Yaghoobi Nia N et al. Beyond 17% stable perovskite solar module via polaron arrangement of tuned polymeric hole transport layer. Nano Energy. 2021; **82**:105685. DOI: 10.1016/J.NANOEN.2020.105685

[28] Yaghoobi Nia N. Barrier layer for space. Nature Energy. 2023;**8**(2):117-118. DOI: 10.1038/s41560-023-01197-9

[29] Li W, Wang Z, Deschler F, Gao S, Friend RH, Cheetham AK. Chemically diverse and multifunctional hybrid organic-inorganic perovskites. Nature Reviews Materials. 2017;**2**(3):1-18. DOI: 10.1038/natrevmats.2016.99

[30] Kumar A, Kumar A, Krishnan V. Perovskite oxide based materials for energy and environment-oriented photocatalysis. ACS Catalysis. 2020; **10**(17):10253-10315. DOI: 10.1021/acscatal.0c02947

[31] Yin WJ, Weng B, Ge J, Sun Q, Li Z, Yan Y. Oxide perovskites, double perovskites and derivatives for electrocatalysis, photocatalysis, and photovoltaics. Energy & Environmental Science. 2019;**12**(2):442-462. DOI: 10.1039/C8EE01574K

[32] Wang K, Han C, Shao Z, Qiu J, Wang S, Liu S. Perovskite oxide catalysts for advanced oxidation reactions. Advanced Functional Materials. 2021; **31**(30):1-31. DOI: 10.1002/adfm.202102089

[33] Song HJ, Yoon H, Ju B, Kim DW. Highly efficient perovskite-based electrocatalysts for water oxidation in acidic environments: A mini review. Advanced Energy Materials. 2021;**11**(27): 1-13. DOI: 10.1002/aenm.202002428

[34] Xu X, Zhong Y, Shao Z. Double perovskites in catalysis, electrocatalysis, and photo(electro)catalysis. Trends in Chemistry. 2019;**1**(4):410-424. DOI: 10.1016/j.trechm.2019.05.006

[35] Di Carlo A, Lamanna E, Yaghoobi Nia N. Photovoltaics. EPJ Web of Conferences. 2020;**246**:00005. DOI: 10.1051/epjconf/202024600005

[36] Luo J et al. Halide perovskite composites for photocatalysis: A mini review. EcoMat. 2021;**3**(1):1-16. DOI: 10.1002/eom2.12079

[37] Quan LN, Rand BP, Friend RH, Mhaisalkar SG, Lee TW, Sargent EH. Perovskites for next-generation optical sources. Chemical Reviews. 2019; **119**(12):7444-7477. DOI: 10.1021/acs.chemrev.9b00107

[38] Xiao Z, Song Z, Yan Y. From lead halide perovskites to lead-free metal halide perovskites and perovskite derivatives. Advanced Materials. 2019; **31**(47):1-22. DOI: 10.1002/adma.201803792

[39] Jiang Y et al. *In situ* synthesis of lead-free halide perovskite Cs2AgBiBr6 supported on nitrogen-doped carbon for efficient hydrogen evolution in aqueous HBr solution. ACS Applied Materials & Interfaces. 2021;**13**(8):10037-10046. DOI: 10.1021/acsami.0c21588

[40] Singh M, Sinha I. Halide perovskite-based photocatalysis systems for solar-

driven fuel generation. Solar Energy. 2020;**208**(June):296-311. DOI: 10.1016/j.solener.2020.08.007

[41] Mollick S, Mandal TN, Jana A, Fajal S, Desai AV, Ghosh SK. Ultrastable luminescent hybrid bromide perovskite@MOF nanocomposites for the degradation of organic pollutants in water. ACS Applied Nano Materials. 2019;**2**(3):1333-1340. DOI: 10.1021/acsanm.8b02214

[42] Zendehdel M et al. Toward highly stable and scalable perovskite solar modules via Co-crystalline 2D perovskite passivation and blade-spin deposition. In: Proceedings of the International Conference on Hybrid and Organic Photovoltaics. València: Fundació Scito; 2022. DOI: 10.29363/nanoge.hopv.2022.185

[43] Yaghoobi Nia N, Zendehdel M, Reddy Sathy H, Schirone L, Di Carlo A. Super stable semitransparent perovskite solar modules for 4-terminal perovskite-silicon tandem modules via engineering of the polymeric hole transport layer and surface passivation. In: Proceedings of the International Conference on Hybrid and Organic Photovoltaics. València: Fundació Scito; 2022. DOI: 10.29363/nanoge.hopv.2022.187

[44] Du F et al. Improving the stability of halide perovskites for photo-, electro-, photoelectro-chemical applications. Advanced Functional Materials. 2023;**34**:2312175. DOI: 10.1002/adfm.202312175

[45] Wu H et al. Photocatalytic and photoelectrochemical systems: Similarities and differences. Advanced Materials. 2020;**32**(18):1-21. DOI: 10.1002/adma.201904717

[46] Jin L, Zhao H, Wang ZM, Rosei F. Quantum dots-based photoelectrochemical hydrogen evolution from water splitting. Advanced Energy Materials. 2021;**11**(12):1-28. DOI: 10.1002/aenm.202003233

[47] Li W et al. Origin of improved photoelectrochemical water splitting in mixed perovskite oxides. Advanced Energy Materials. 2018;**8**(31):1-7. DOI: 10.1002/aenm.201801972

[48] Wang S, Liu G, Wang L. Crystal facet engineering of photoelectrodes for photoelectrochemical water splitting. Chemical Reviews. 2019;**119**(8):5192-5247. DOI: 10.1021/acs.chemrev.8b00584

[49] Wang G, Ling Y, Wang H, Xihong L, Li Y. Chemically modified nanostructures for photoelectrochemical water splitting. Journal of Photochemistry and Photobiology C: Photochemistry Reviews. 2014;**18**(1):35-51. DOI: 10.1016/j.jphotochemrev.2013.10.006

[50] Guerrero A, Bisquert J. Perovskite semiconductors for photoelectrochemical water splitting applications. Current Opinion in Electrochemistry. 2017;**2**(1):144-147. DOI: 10.1016/j.coelec.2017.04.003

[51] Chatterjee P et al. Photovoltaic/photo-electrocatalysis integration for green hydrogen: A review. Energy Conversion and Management. 2022;**261**(April):115648. DOI: 10.1016/j.enconman.2022.115648

[52] Wang H, Wang X, Zhang H, Ma W, Wang L, Zong X. Organic–inorganic hybrid perovskites: Game-changing candidates for solar fuel production. Nano Energy. 2020;**71**:104647. DOI: 10.1016/j.nanoen.2020.104647

[53] Bard AJ, Fox MA. Artificial photosynthesis: Solar splitting of water to hydrogen and oxygen. Accounts of Chemical Research. 1995;**28**(3):141-145. DOI: 10.1021/ar00051a007

[54] Kudo A, Miseki Y. Heterogeneous photocatalyst materials for water splitting. Chemical Society Reviews. 2009;**38**(1):253-278. DOI: 10.1039/B800489G

[55] Concina I, Ibupoto ZH, Vomiero A. Semiconducting metal oxide nanostructures for water splitting and photovoltaics. Advanced Energy Materials. 2017;**7**(23):1-29. DOI: 10.1002/aenm.201700706

[56] Acar C, Dincer I. A review and evaluation of photoelectrode coating materials and methods for photoelectrochemical hydrogen production. International Journal of Hydrogen Energy. 2016;**41**(19): 7950-7959. DOI: 10.1016/j.ijhydene.2015.11.160

[57] Chen X, Zhang Z, Chi L, Nair AK, Shangguan W, Jiang Z. Recent advances in visible-light-driven photoelectrochemical water splitting: Catalyst nanostructures and reaction systems. Nano-Micro Letters. 2016; **8**(1):1-12. DOI: 10.1007/s40820-015-0063-3

[58] Yan K et al. Self-driven hematite-based photoelectrochemical water splitting cells with three-dimensional nanobowl heterojunction and high-photovoltage perovskite solar cells. Materials Today Energy. 2017;**6**: 128-135. DOI: 10.1016/j.mtener.2017.09.006

[59] Sabba D, Kumar MH, Wong LH, Barber J, Grätzel M, Mathews N. Perovskite–Hematite tandem cells for efficient overall solar driven water splitting. Nano Letters. 2015;**15**(6): 3833-3839

[60] Liu G et al. Enhancement of the photoelectrochemical water splitting by perovskite $BiFeO_3$ via interfacial engineering. Solar Energy. 2020;**202** (February):198-203. DOI: 10.1016/j.solener.2020.03.117

[61] Tao R, Sun Z, Li F, Fang W, Xu L. Achieving organic metal halide perovskite into a conventional photoelectrode: Outstanding stability in aqueous solution and high-efficient photoelectrochemical water splitting. ACS Applied Energy Materials. 2019; **2**(3):1969-1976. DOI: 10.1021/acsaem.8b02072

[62] Fehr AMK et al. Integrated halide perovskite photoelectrochemical cells with solar-driven water-splitting efficiency of 20.8%. Nature Communications. 2023;**14**(1):1-12. DOI: 10.1038/s41467-023-39290-y

[63] Hansora D et al. All-perovskite-based unassisted photoelectrochemical water splitting system for efficient, stable and scalable solar hydrogen production. Nature Energy. 2024;**9**(3): 272-284. DOI: 10.1038/s41560-023-01438-x

[64] Li X et al. Efficient solar-driven water splitting enabled by perovskite photovoltaics and a halogen-modulated metal-organic framework electrocatalyst. ACS Nano. 2023;**17**(23): 23478-23487. DOI: 10.1021/acsnano.3c05583

[65] Li Y, Zhang JZ. Hydrogen generation from photoelectrochemical water splitting based on nanomaterials. Laser & Photonics Reviews. 2010; **4**(4):517-528. DOI: 10.1002/lpor.200910025

[66] Zhang Y, Tang B, Wu Z, Shi H, Zhang Y, Zhao G. Glucose oxidation over ultrathin carbon-coated perovskite modified TiO_2 nanotube photonic crystals with high-efficiency electron generation and transfer for photoelectrocatalytic hydrogen production. Green Chemistry. 2016; **18**(8):2424-2434. DOI: 10.1039/c5gc02745d

[67] Luo J et al. Organic-inorganic hybrid perovskite – TiO_2 nanorod arrays for efficient and stable photoelectrochemical hydrogen evolution from HI splitting. Materials Today Chemistry. 2019;**12**:1-6. DOI: 10.1016/j.mtchem.2018.11.001

[68] Choi Y et al. Bias-free solar hydrogen production at 19.8 mA cm−2 using perovskite photocathode and lignocellulosic biomass. Nature Communications. 2022;**13**(1):1-9. DOI: 10.1038/s41467-022-33435-1

[69] Yang Y, Ajmal S, Zheng X, Zhang L. Efficient nanomaterials for harvesting clean fuels from electrochemical and photoelectrochemical CO_2 reduction. Sustainable Energy & Fuels. 2018;**2**(3): 510-537. DOI: 10.1039/c7se00371d

[70] Mao X, Hatton TA. Recent advances in electrocatalytic reduction of carbon dioxide using metal-free catalysts. Industrial and Engineering Chemistry Research. 2015;**54**(16):4033-4042. DOI: 10.1021/ie504336h

[71] Chang X, Wang T, Yang P, Zhang G, Gong J. The development of cocatalysts for photoelectrochemical CO_2 reduction. Advanced Materials. 2019;**31**(31):1-13. DOI: 10.1002/adma.201804710

[72] Shyamal S, Pradhan N. Halide perovskite nanocrystal photocatalysts for CO_2 reduction: Successes and challenges. Journal of Physical Chemistry Letters. 2020;**11**(16):6921-6934. DOI: 10.1021/acs.jpclett.0c00191

[73] Pawar AU, Kim CW, Nguyen-Le MT, Kang YS. General review on the components and parameters of photoelectrochemical system for CO_2 reduction with *in situ* analysis. ACS Sustainable Chemistry & Engineering. 2019;**7**(8):7431-7455. DOI: 10.1021/acssuschemeng.8b06303

[74] Zhang X et al. Synergistically interface-engineered inorganic halide perovskite photocathodes for photoelectrochemical CO_2 reduction. Energy and Fuels. 2023;**37**(23): 18163-18172. DOI: 10.1021/acs.energyfuels.3c02613

[75] Schreier M et al. Efficient photosynthesis of carbon monoxide from CO_2 using perovskite photovoltaics. Nature Communications. 2015;**6**:1-6. DOI: 10.1038/ncomms8326

[76] Bergamini L et al. CsPbBr3/platinum and CsPbBr3/graphite hybrid photoelectrodes for carbon dioxide conversion to oxalic acid. Solar Energy. 2023;**254**(March):213-222. DOI: 10.1016/j.solener.2023.02.043

[77] Abarca JA et al. Fabrication and optimization of perovskite-based photoanodes for solar-driven CO_2 photoelectroreduction to formate. Catalysis Today. 2023;**429**(November): 2024. DOI: 10.1016/j.cattod.2023.114505

[78] Yan Z, Ji M, Xia J, Zhu H. Recent advanced materials for electrochemical and photoelectrochemical synthesis of ammonia from dinitrogen: One step closer to a sustainable energy future. Advanced Energy Materials. 2020; **10**(11):1-35. DOI: 10.1002/aenm.201902020

[79] Li P et al. Amorphous Sn/crystalline SnS_2 nanosheets via *in situ* electrochemical reduction methodology for highly efficient ambient N_2 fixation.

Small. 2019;**15**(40):1-8. DOI: 10.1002/smll.201902535

[80] Zheng H, Zhang S, Liu X, O'Mullane AP. The application and improvement of TiO_2 (titanate) based nanomaterials for the photoelectrochemical conversion of CO_2 and N_2 into useful products. Catalysis Science & Technology. 2021;**11**(3):768-778. DOI: 10.1039/d0cy02048f

[81] Schrauzer GN, Guth TD. Photolysis of water and photoreduction of nitrogen on titanium dioxide1. Journal of the American Chemical Society. 1977;**99**(22):7189-7193. DOI: 10.1021/ja00464a015

[82] Paramanik L, Sultana S, Parida KM. Photocatalytic and photo-electrochemical ammonia synthesis over dimensional oriented cobalt titanate/nitrogen-doped reduced graphene oxide junction interface catalyst. Journal of Colloid and Interface Science. 2022;**625**:83-99. DOI: 10.1016/J.JCIS.2022.06.010

[83] Khaki MRD, Shafeeyan MS, Raman AAA, Daud WMAW. Application of doped photocatalysts for organic pollutant degradation - A review. Journal of Environmental Management. 2017;**198**:78-94. DOI: 10.1016/j.jenvman.2017.04.099

[84] Divyapriya G, Singh S, Martínez-Huitle CA, Scaria J, Karim AV, Nidheesh PV. Treatment of real wastewater by photoelectrochemical methods: An overview. Chemosphere. 2021;**276**:130188. DOI: 10.1016/j.chemosphere.2021.130188

[85] Kong J, Yang T, Rui Z, Ji H. Perovskite-based photocatalysts for organic contaminants removal: Current status and future perspectives. Catalysis Today. 2019;**327**:47-63. DOI: 10.1016/j.cattod.2018.06.045

[86] Liao W, Yang J, Zhou H, Murugananthan M, Zhang Y. Electrochemically self-doped TiO_2 nanotube arrays for efficient visible light photoelectrocatalytic degradation of contaminants. Electrochimica Acta. 2014;**136**:310-317. DOI: 10.1016/j.electacta.2014.05.091

[87] Sahmi A, Laib R, Omeiri S, Bensadok K, Trari M. Photoelectrochemical properties of the perovskite $BaSnO_3$ synthetized by chemical route. Application to electro-photocatalytic mineralization of ibuprofen. Journal of Photochemistry and Photobiology A: Chemistry. 2018;**364**:443-448. DOI: 10.1016/j.jphotochem.2018.06.034

[88] Cardenas-Morcoso D et al. Photocatalytic and photoelectrochemical degradation of organic compounds with all-inorganic metal halide perovskite quantum dots. Journal of Physical Chemistry Letters. 2019;**10**(3):630-636. DOI: 10.1021/acs.jpclett.8b03849

[89] Kuo YS, Lin MJ, Hsu CL, Liu YH. Heterostructured $YFeO_3/CeO_2$ as efficient photoelectrode for photoelectrochemical degradation of organic pollutants under visible light irradiation. Journal of Water Process Engineering. 2023;**56**(June):104394. DOI: 10.1016/j.jwpe.2023.104394

[90] Zhou Y, Yin H, Ai S. Applications of two-dimensional layered nanomaterials in photoelectrochemical sensors: A comprehensive review. Coordination Chemistry Reviews. 2021;**447**:214156. DOI: 10.1016/j.ccr.2021.214156

[91] Xu R et al. Photoelectrochemical aptasensor based on $La_2Ti_2O_7/Sb_2S_3$ and V_2O_5 for effectively signal change strategy

for cancer marker detection. Biosensors & Bioelectronics. 2021;**192**(March):3-8. DOI: 10.1016/j.bios.2021.113528

[92] Chen X et al. All-inorganic perovskite quantum dot/TiO_2 inverse opal electrode platform: Stable and efficient photoelectrochemical sensing of dopamine under visible irradiation. Nanoscale. 2018;**10**(22):10505-10513. DOI: 10.1039/c8nr02115e

[93] Leng D et al. A self-powered photoelectrochemical biosensing platform for H-FABP monitoring mediated by $CsPbBr_3$@COF–V. Biosensors & Bioelectronics. 2023;**241** (June):115710. DOI: 10.1016/j.bios.2023.115710

[94] Dong X et al. A competitive photoelectrochemical sensor based on the in situ formation of BiOBr/Bi_2S_3 by Cs3Bi_2Br_9 for the detection of trenbolone. Sensors and Actuators B: Chemical. 2023;**393**(June):134254. DOI: 10.1016/j.snb.2023.134254

[95] Wu YC, Song RJ, Li JH. Recent advances in photoelectrochemical cells (PECs) for organic synthesis. Organic Chemistry Frontiers. 2020;7(14): 1895-1902. DOI: 10.1039/d0qo00486c

[96] Tateno H, Miseki Y, Sayama K. Photoelectrochemical dimethoxylation of furan via a bromide redox mediator using a $BiVO_4$/WO_3 photoanode. Chemical Communications. 2017;**53**(31): 4378-4381. DOI: 10.1039/c7cc01190c

[97] Wang XD et al. Surface passivated halide perovskite single-crystal for efficient photoelectrochemical synthesis of dimethoxydihydrofuran. Nature Communications. 2021;**12**(1):1-9. DOI: 10.1038/s41467-021-21487-8

Chapter 4

Photoanode Applications of Polyene-Diphenylaniline Dyes Molecules Adsorbed on TiO_2 Brookite Cluster

Ife Elegbeleye, Eric Maluta, Regina Maphanga, Michael Walter and Oliver Brügner

Abstract

TiO_2 has excellent photoelectrochemical properties, which makes its suitable for photoanode applications. TiO_2 is widely utilized as semiconductor for dye-sensitized photoanode owing to its excellent stability and availability. The brookite polymorph of TiO_2 has been suggested to demonstrate better photocatalytic properties. In this work, we studied the adsorption of polyenediphenyl-aniline dyes on $(TiO_2)_{n = 8, 68}$ brookite clusters using density functional theory (DFT). We report our results on the UV–Vis absorption spectra of the dyes and dyes adsorbed on TiO_2 clusters, adsorption energies of the dyes adsorbed on TiO_2 clusters, electronic density of states and projected electronic density of states of the dyes adsorbed on TiO_2 complex, and electron density of the main molecular orbitals involved in photoexcitation for dye-sensitized solar cells application. The calculated adsorption energies of D5@$(TiO_2)_{68}$, D7@$(TiO_2)_{68}$, D9@$(TiO_2)_{68}$, and D11@$(TiO_2)_{68}$ are 4.84, 4.78, 4.66 and 4.92 eV, respectively. The results of the adsorption energies are in the order D11@$(TiO_2)_8$ > D9@$(TiO_2)_8$ > D7@$(TiO_2)_8$ > D5@$(TiO_2)_8$ and D11@ $(TiO_2)_{68}$ > D5@$(TiO_2)_{68}$ > D7@ $(TiO_2)_{68}$ > D9@ $(TiO_2)_{68}$. This implies that D11 dye molecule reacts more strongly with $(TiO_2)_8$ and $(TiO_2)_{68}$ brookite clusters than the other corresponding dye molecules.

Keywords: dye sensitized solar cells, TiO_2 brookite semiconductor, polyene-diphenyl-aniline dyes, density functional theory, optical properties

1. Introduction

The fascinating features of dye sensitized solar cells (DSSC) over conventional solar energy devices are their low manufacturing temperature, low fabrication cost, flexibility, and potential of high photon to current conversion efficiency [1–3]. DSSC photoanode is made up of dye molecules chelated to typical TiO_2 semiconductors surface. TiO_2 is widely used as semiconductor for DSSCs because it is readily available and has good stability when illuminated to sunlight [4, 5]. The ruthenium N3 complex has been extensively utilized in TiO_2-DSSCs architecture, reported to achieve 11–12% power conversion efficiency (PCE) [6–7]. However, these ruthenium dye complexes are relatively expensive, toxic, and require difficult synthetic and refinement processes. On the other hand, organic dyes have advantages of non-toxicity, low cost, and availability of raw materials and are environmentally friendly, which makes them more desirable than the ruthenium complexes [3, 8].

Polyenediphenyl-aniline dye (on TiO_2 semiconductor) based on ionic liquid electrolyte and iodide/triiodide has been reported as a promising candidate for DSSC applications with PCE of 6.5 and 5% [9, 10]. TiO_2 exists in three major polymorphs, namely, anatase, rutile, and brookite [4]. Anatase is mainly synthesized from TiO_2 powder by sol-gel method, while rutile can be obtained as a by-product of precipitation at low temperature in an acidic medium [1, 11]. Synthesizing pure brookite is challenging owing to the intergrowth morphology of anatase and rutile usually encountered during the synthetic process [12–14]. As a result of this difficult synthetic route, most of the studies on TiO_2 polymorphs are centered on anatase and rutile phases as prototypical models for DSSCs fabrication [15, 16]. There are very few studies on the photocatalytic properties of TiO_2 brookite. However, some recent studies have suggested that brookite may exhibit fascinating photocatalytic properties and may perform better as a semiconductor than the typically used anatase [17–21]. Zallen et al. reported on the optical absorption edge of brookite TiO_2 at room temperature using natural crystals [16]. The photon energy spanning from 2.1 to 3.54 eV, absorption coefficient of 2000 cm^{-1}, and optical gap of 1.9 eV were reported for brookite crystals. The absorption edge of brookite was observed to be broad and extends to the solar spectrum visible region, whereas steep edges were observed for rutile and anatase polymorphs of TiO_2 in the visible region [16]. The crystallographic description of brookite polymorphs is presented in **Table 1**.

Brookite has an orthorhombic crystalline structure with eight formula units in the orthorhombic cell; its unit cell is described by the space group *Pbca* [4]. Brookite formation may be envisioned as the joining of TiO_6 distorted octahedral sharing three edges, each with a titanium atom at its center and oxygen atoms at its corners.

Information about the interfacial reaction of TiO_2 brookite with the promising polyenediphenylaniline organic dye is crucial for the understanding and optimization of photo conversion function, light harvesting, and photocurrent densities in DSSCs.

In this study, we carried out density functional theory (DFT) and time-dependent DFT (TD-DFT) study on the adsorption of polyenediphenyl-aniline dyes on TiO_2 brookite in order to understand the interaction between TiO_2 brookite and

	Lattice parameter (Å)	Space group	System type	Energy bandgap (eV)
Rutile	a = 4.594 c = 2.958	$P4_2/mnm - D_{4h}^{14}$	Tetragonal	3.0
Anatase	a = 3.784 c = 9:515	$I4_1amd - D_{4h}^{19}$	Tetragonal	3.4
Brookite	a = 9.166 b = 5.436 c = 5.135	$Pbca - D_{2h}^{15}$	Orthorhombic	3.3

Table 1.
Properties of rutile, brookite, and anatase polymorphs of TiO_2 [4].

polyenediphenylaniline dyes. We report on the optical and electronic properties of the polyenediphenyl-aniline dyes and the dyes adsorbed on TiO_2 brookite clusters.

2. Materials and methods

The main part of density-functional theory calculations was performed within an atomic simulation environment (ASE) [22] and the GPAW package [23, 24]. GPAW implements the projector-augmented wave method with the smooth wave-functions and electron density represented on real space grids [21]. The exchange correlation energy was approximated within the generalized gradient approximation PBE [25]. The structures were relaxed without any symmetry constraints until all forces were found to be below 0.05 eV/Å. Their structure was visualized with Avogadro [26]. Optical spectra were obtained from linear response time-dependent DFT [27, 28].

The structures of the D5, D7, D9, and D11 dye molecules [10] were built in Gaussian 03 quantum chemical package [25, 29]. First geometrical optimization of the dye molecules was performed while employing the density functional theory and hybrid density functional theory B3LYP and 6-31G* basis set [30, 31] with the default integration grid as implemented in Gaussian 03. The structures were subsequently relaxed as described above and are depicted in **Figure 1(a-d)**.

The TiO_2 brookite models considered in this study are presented in **Figure 2**. The cluster models were imported using the crystallographic information mode (cif) from CASTEP material studio [32]. The brookite cluster models are brookite $(TiO_2)_8$ comprising eight titanium and 16 oxygen atoms, while brookite $(TiO_2)_{68}$ is a periodic supercell comprising 68 titanium atoms and 136 oxygen atoms [21]. The periodic structure is a bulk structure comprising unit cells that are repeated in x and y direction to form a cluster or supercell of an arbitrary size.

All the structures of the dye@TiO_2 were relaxed in vacuum using the PBE exchange correlation functional [32]. The structures were considered to have converged when the maximum forces that were acting on all the atoms were lower than 0.05 eV/Å. The periodic boundary conditions were implemented during the relaxation and the atoms of the cluster. The cluster were reoriented during the relaxation until the ground state structure was obtained where they become stable, and the forces converge to a value of 0.05 eV/Å.

The polyenediphenyl aniline dye molecules were adsorbed unto the TiO_2 surface *via* the carboxylic group in the bidentate bridging (BB) adsorption mode [33, 34], in which each of the oxygen atoms of the carboxylic moiety binds to titanium atom on the TiO_2 clusters. This mode was adopted because it was reported to be more energetically favorable [33, 34]. The adsorption energies and optical and electronic properties of the dye molecules absorbed on TiO_2 cluster were calculated upon adsorption.

2.1 Structures of polyenediphenyl-aniline dyes absorbed on (TiO2) n = 8, 68

Four polyenediphenyl-aniline dye molecules coded D5, D7, D9, and D11 were separately adsorbed by bidentate adsorption mode on TiO_2 clusters with stoichiometry $(TiO_2)_8$ and $(TiO_2)_{68}$ as showed in **Figures 3(a-d)** and **4(a-d)**. All dye@TiO_2 complexes were relaxed upon adsorption; the relaxed structures of the dye molecules adsorbed on $(TiO_2)_8$ and $(TiO_2)_{68}$ clusters are presented in **Figures 3** and **4**, respectively. Bidentate bridging adsorption mode was utilized to adsorb the polyenediphenyl-aniline dye molecules on the TiO_2 clusters *via* the carboxylic acid moiety. Each of

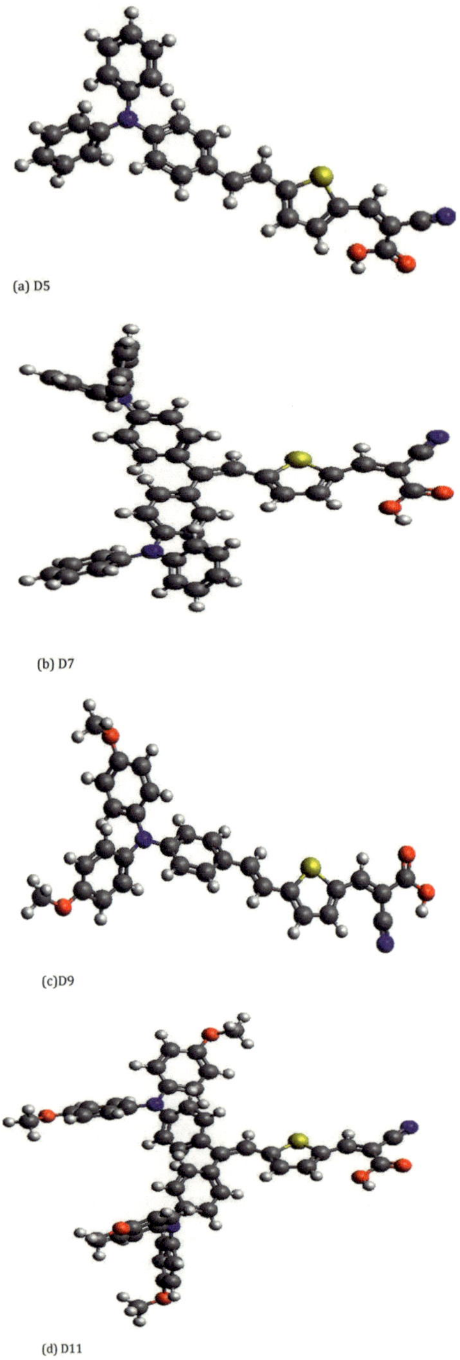

Figure 1.
Structures of the polyenediphenyl-aniline dye molecules in gas-phase.

the two oxygen atoms was adhered to the titanium atom on the TiO_2 cluster surface. In order to achieve a stable system for TD-DFT calculations, the atom of the hydrogen was transferred to a neighboring oxygen atom.

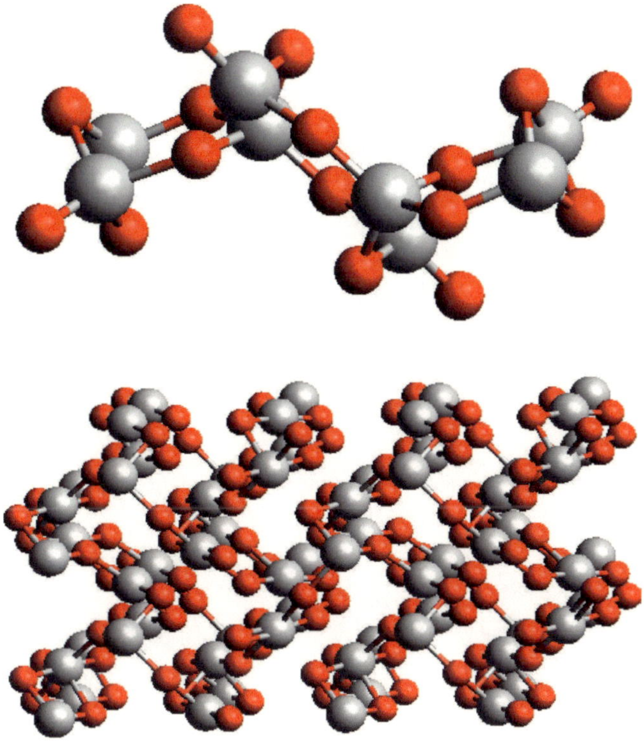

(a)

(b)

Figure 2.
(a-b) Broookite $(TiO_2)_{n\,=\,8,\,68}$ cluster periodic models considered for dye molecule adsorption in this study.

2.2 Adsorption energies of polyene-diphenylaniline dye molecules absorbed on (TiO2) $_{n\,=\,8,\,68}$ brookite clusters

The relaxation energies of the dye molecules, $(TiO_2)_8$ and $(TiO_2)_{68}$ brookite clusters, and polyenediphenyl-aniline dyes-TiO_2 complexes were computed the adsorption energies using (Eq. (1)).

$$E_{ads} = \left[E_{TiO_2} + E_{(\text{molecule})}\right] - \left[E_{(TiO_2+\text{molecule})}\right] \tag{1}$$

where E_{TiO_2} is the energy of the TiO_2 complex, $E_{molecule}$ is the energy of the adsorbate, and $E_{(TiO_2+\text{molecule})}$ is the total energy of the TiO_2 with the adsorbed molecule. The adsorption energy denotes the binding ability of the dye molecules [5, 21, 33, 35–37]. The adsorption energies of D5, D7, D9, and D11 dyes adsorbed on $(TiO_2)_8$ and $(TiO_2)_{68}$ clusters are listed in **Table 2**. All the calculated results show a positive adsorption energy, suggesting stable grafting of the dyes molecules unto the surface of TiO_2. The calculated adsorption energies of D5@$(TiO_2)_8$, D7@$(TiO_2)_8$, D9@$(TiO_2)_8$, and D11@ $(TiO_2)_8$ are 1.60, 1.62, 1.67, and 1.81 eV, respectively. The order of the adsorption energies is D11@$(TiO_2)_8$ > D9@$(TiO_2)_8$ > D7@$(TiO_2)_8$ > D5@$(TiO_2)_8$. The findings suggest that D11 dye molecule exhibits

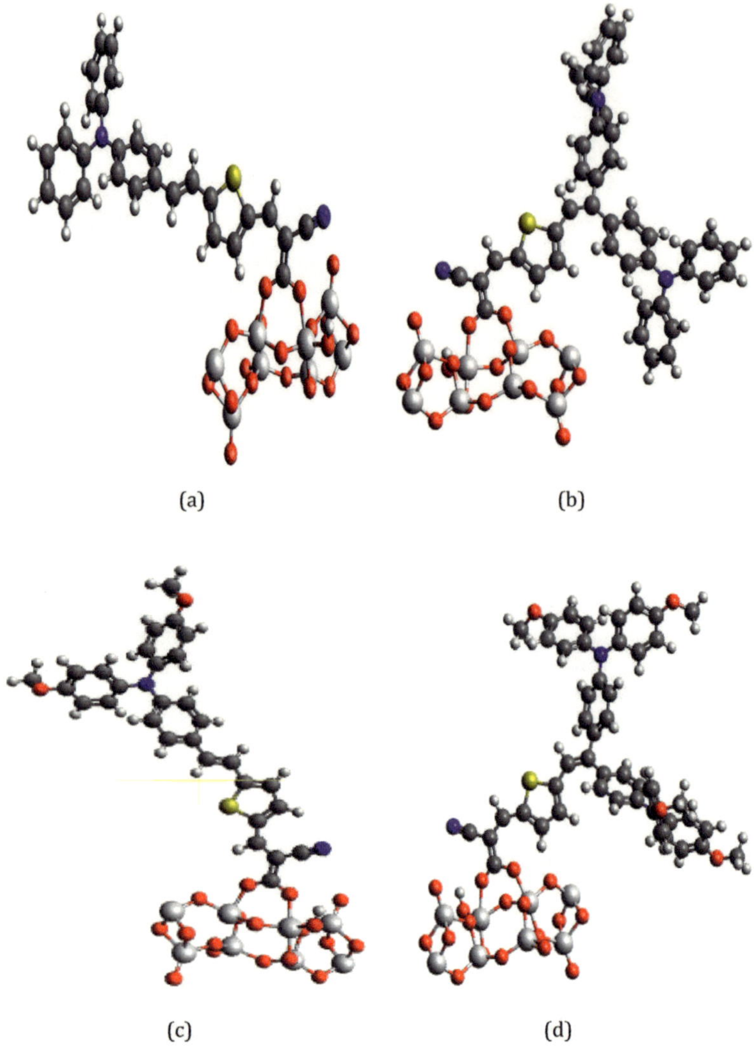

Figure 3.
Structures of polyenediphenyl-aniline dyes absorbed on $(TiO_2)_8$ (a) D5@$(TiO_2)_8$, (b) D7@$(TiO_2)_8$, (c) D9@$(TiO_2)_8$, and (d) D11@$(TiO_2)_8$. Gray spheres represent titanium atoms; red spheres represent oxygen atoms; white spheres represent hydrogen atoms; yellow spheres represent sulfur atoms; dark gray balls represent carbon atoms, and blue spheres represent nitrogen atoms.

stronger reaction with $(TiO_2)_8$ brookite cluster than the other dye molecules. Values of 4.84, 4.78, 4.66, and 4.92 eV were obtained for the adsorption energies of D5@$(TiO_2)_{68}$, D7@ $(TiO_2)_{68}$, D9@ $(TiO_2)_{68}$, and D11@ $(TiO_2)_{68}$, respectively. Unlike $(TiO_2)_8$, the order of the adsorption energies is not sequential but D11@$(TiO_2)_{68}$ > D5@$(TiO_2)_{68}$ > D7@$(TiO_2)_{68}$ > D9@$(TiO_2)_{68}$. The results show that D11 dye molecule demonstrates a stronger reaction with $(TiO_2)_{68}$ brookite surface than D5, D7, and D9 molecules. Variation in the obtained adsorption energies for the four dye molecules adsorbed on the different brookite surfaces indicates that the cluster size affects the binding ability of the molecules. The polyenediphenyl-aniline dye molecules are found to have stronger interaction with the $(TiO_2)_{68}$ cluster than $(TiO_2)_8$ cluster.

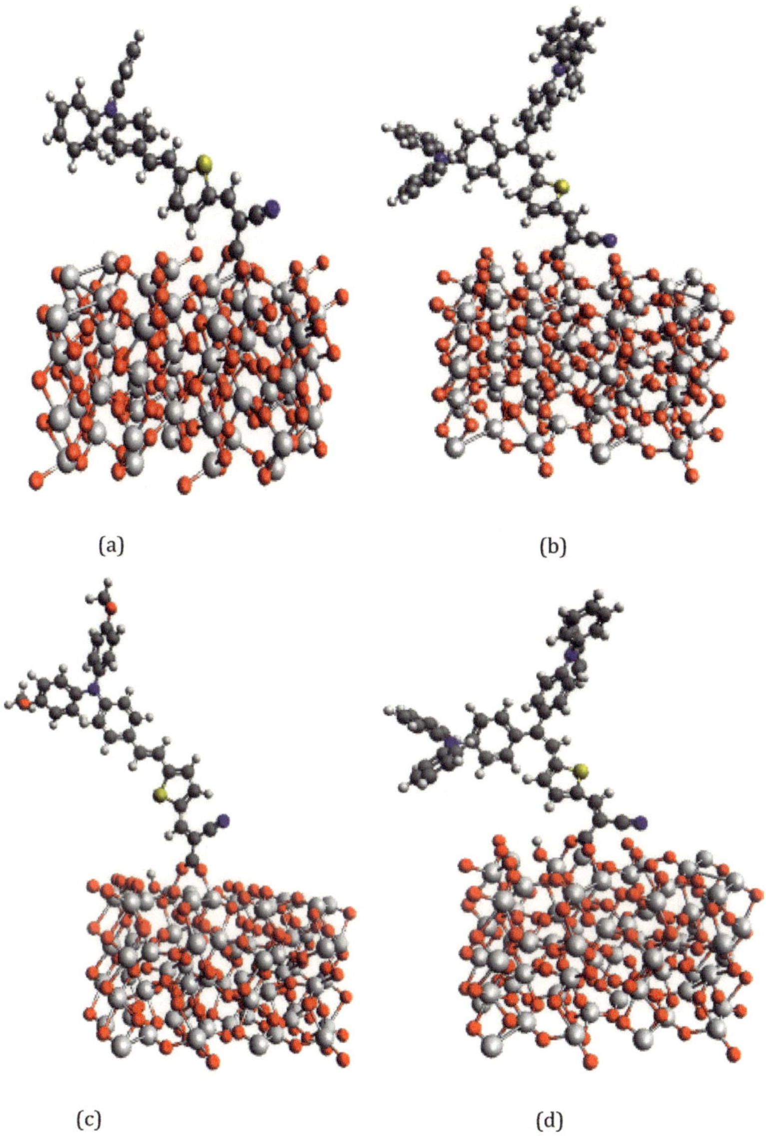

Figure 4.
Structures of polyenediphenyl-aniline dyes absorbed on $(TiO_2)_{68}$ (a) D5@$(TiO_2)_{68}$, (b) D7@$(TiO_2)_{68}$, (c) D9@$(TiO_2)_{68}$, and (d) D11@$(TiO_2)_{68}$. Same notation as in **Figure 4**.

Adsorption Energy of Dyes@$(TiO_2)_{n=8,68}$		**(eV)**
Dyes@$(TiO_2)_8$	D5@$(TiO_2)_8$ D7@$(TiO_2)_8$ D9@$(TiO_2)_8$ D11@$(TiO_2)_8$	1.60 1.62 1.67 1.81
Dyes@$(TiO_2)_{68}$	D5@$(TiO_2)_{68}$ D7@$(TiO_2)_{68}$ D9@$(TiO_2)_{68}$ D11@$(TiO_2)_{68}$	4.84 4.78 4.66 4.93

Table 2.
Adsorption energies of dyes @ $(TiO_2)_{n=8,68}$ clusters.

Balaganapathi et al. synthesized porous brookite and mixed brookite with rutile TiO_2 crystals using Polyethylene Glycol (PEG) assisted sol-gel synthesis and investigated their dye adsorption characteristics [38]. Reflectance measurements were used to determine the optical bandgap of both the brookite and brookite mixed with rutile samples as 2.26 and 2.48 eV, respectively. The reported binding energies ranged from 445 to 470 eV for pure brookite and from 520 to 540 eV for brookite mixed with rutile phase. In this work, the calculated bandgap of brookite was found to be 2.26 eV, which is consistent with the aforementioned study [38].

In another study, Corneliu et al. investigated the adsorption of coumarin dyes C343 on $Ti_nO_{2n+2}H_4$ (n = 14, 24) clusters using DFT/B3LYP with 3-21G (d) or LANL2DZ basis set in vacuum. The reported adsorption energies ranged between 0.429 and 3.254 eV. The findings concluded that the size of the cluster affects adsorption properties and revealed that adsorption onto n = 14 cluster leads to a significant distortion of the cluster and the molecule, a behavior that was not observed with a larger cluster of n = 24. This is in line with our findings and may be the reason why polyenediphenyl-aniline dye molecules are found to absorb more onto $(TiO_2)_{68}$ cluster than $(TiO_2)_8$ cluster [39].

2.3 Optical spectrum of polyenediphenyl-aniline dye molecules adsorbed on brookite $(TiO2)_{n=8,68}$

The excitation spectra of the dyes were calculated in gas phase as shown in **Figure 3**. All the dyes show a region of absorbance between 700 and 900 nm at higher energy region. The two dyes with pure phenyl groups D5 and D7 show two main strong features in their spectrum. The absorption peaks of the D5 dye are located at 450 and 750 nm, while that of D7 dye are located around 480 and 850 nm. D9 shows an absorption peak at 500 and 800 nm, while D11 shows double absorption peaks around 400–500 nm and absorption maximum at 900 nm. The largest absorption wavelength is notable for D11 containing methoxy and polyene groups, which is consistent with the experimental study by Kuang et al. [10], where D11 showed spectral red shifts and thus provided better current-voltage characteristics than the corresponding D5, D7, and D9.

Huy et al. carried out UV/Vis diffuse reflectance spectra measurements of brookite TiO_2/DC668 (styrene-acrylatethin) to infer the optical characteristics and to determine the optical bandgap energy of printed TiO_2 brookite thin film [19]. The study reported that TiO_2/DC668 thin film strongly absorbs light in the UV-A region at wavelength of 379 nm and below, and exhibited absorption between 400 and 700 nm [19]. Their findings is consistent with the near infrared absorption obtained for dye adsorption on brookite in this study.

Polyenediphenyl-aniline dyes D5, D7, D9, and D11 absorption spectra are presented in **Figure 5**. The excited state properties of the dyes bonded to TiO_2 clusters are illustrated by the absorption spectra. The computed UV/Vis spectra of $(TiO_2)_8$ brookite cluster using GPAW/TD-DFT exhibit absorption around 200–400 nm in the UV region. It was observed that polyenediphenly-aniline dye adsorption on $(TiO_2)_8$ brookite cluster results in shifting the absorption peaks to higher wavelength due to photoexcitation. Absorption maxima in the near infrared region were observed to be 750, 850–950, 850, and 950 nm for D5@$(TiO_2)_8$, D7@$(TiO_2)_8$, D9@$(TiO_2)_8$, and D11@$(TiO_2)_8$, respectively. The D11@ $(TiO_2)_8$ complex showed the highest absorption maximum and absorption peak in the near-infrared region. The results revealed that adsorption of polyenediphenly-aniline dye onto $(TiO_2)_8$ cluster improves the spectral responsivity, exhibiting whole spectra range absorption in the UV, visible and near-infrared region of the solar spectrum. This suggests higher photocurrents density in DSSCs.

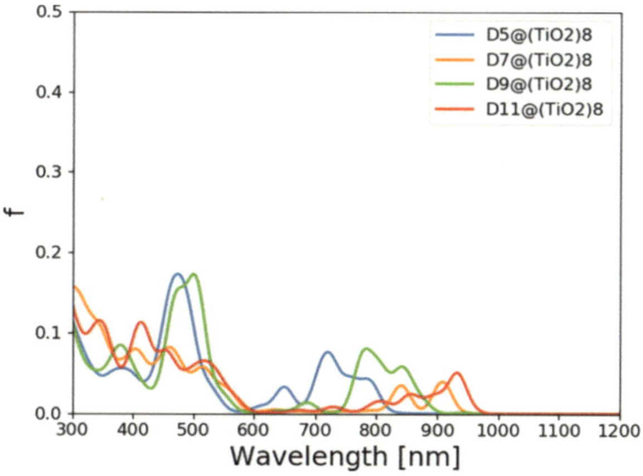

Figure 5.
Optical absorption spectrum of and polyene-diphenylaniline dye molecules adsorbed on $(TiO_2)_8$ cluster.

This makes the complex of this findings consistent with those reported by Huy et al. on a different dye molecule TiO_2/DC668 (styrene-acrylatethin) brookite thin film, exhibiting light absorption in the near-infrared region [19].

2.4 Electronic density of states polyenediphenyl-aniline dye molecules adsorbed on $(TiO2)_{n = 8, 68}$ brookite clusters

In order to gain further insights into the electronic structure of the dye-TiO_2 complex, GPAW and PBE exchange correction functional was utilized to compute the total density of states (DOS) for D5, D7, D9, and D11 dye molecules absorbed on $(TiO_2)_8$ and $(TiO_2)_{68}$. **Figure 6(a)-(d)** illustrate the total density of states of polyenediphenyl-aniline dye adsorbed onto $(TiO_2)_8$ and $(TiO_2)_{68}$. The DOS is composed of surface valence and conduction bands, separated by a wide bandgap. Upon the adsorption of the dye molecules on TiO_2 clusters, it was observed that sharp molecular energy levels were introduced by the dyes in the wide bandgap of TiO_2 as visible in **Figure 6(a)-(d)**. Thus, adsorption of dye onto the TiO_2 clusters resulted in the introduction of new occupied electronic orbitals in the conduction band and between the two states where there was an initial wide bandgap, thereby reducing the energy bandgap. Also, occupied orbital edge slightly shifted to higher energy levels; the shifting of the two states subsequently results in narrowing of the wide bandgap. This behavior explains the observation on red spectral shift upon dye adsorption discussed in the previous section.

The result of the PDOS for the clusters before the adsorption of the dyes shows that titanium and oxygen atomic orbitals contribute to the valence states; the major contributions of oxygen 2*p* atomic orbitals were mostly to the highest occupied valence band (VB), while the titanium 3*d* atomic orbitals contributions were dominant in the lowest unoccupied state of the conduction band.

Oxygen p atomic orbitals contributions dominated the valence band with minor contributions from *p* atomic orbitals of titanium. The conduction band major contributions emanate from the *d* and *p* orbitals of titanium.

The PDOS spectra of the dyes @$(TiO_2)_8$ and dyes@$(TiO_2)_{68}$ are presented in **Figure 6**. m2*p* orbitals of carbon, 2*p* orbitals of nitrogen, 3*p* orbitals of sulfur, 3*d* orbitals of titanium, and 2*p* orbitals of oxygen contributed majorly to the valence

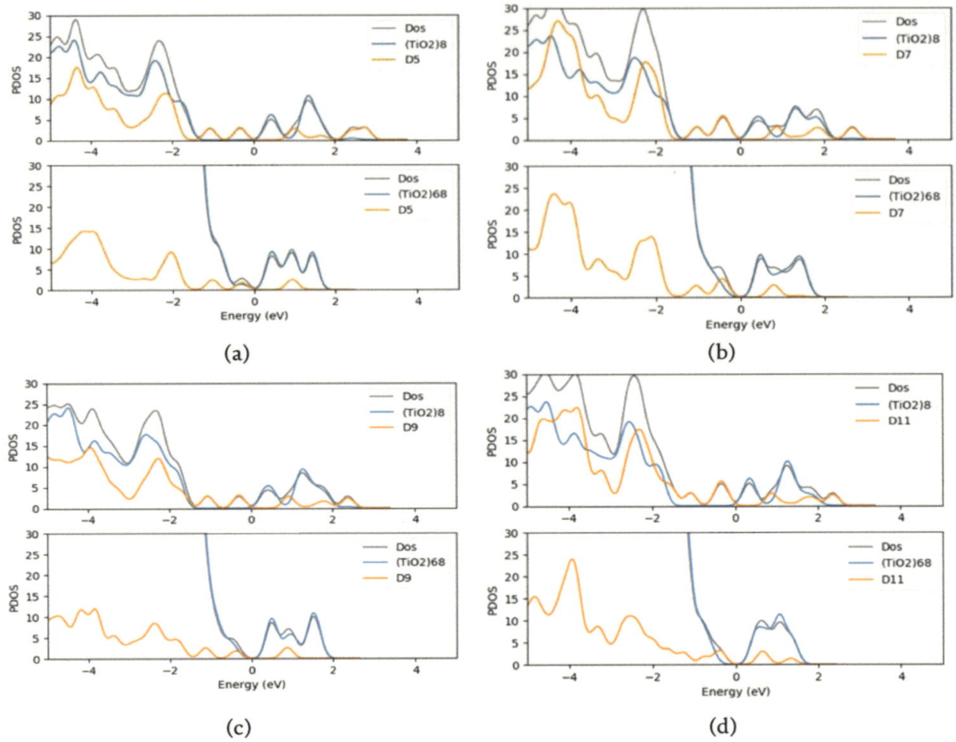

Figure 6.
(a) The total density of states for D5@ $(TiO_2)_8$ and D5@ $(TiO_2)_{68}$ (b) density of states for D7@ $(TiO_2)_8$ and D7@ $(TiO_2)_{68}$ DOS. (c) Total density of states for D9@ $(TiO_2)_8$ and D9@ $(TiO_2)_{68}$. (d) Total density of states for D11@ $(TiO_2)_8$ and D11@ $(TiO_2)_{68}$.

states, while there is a minor contribution of *p* orbitals of sulfur and oxygen to the conduction band.

The major contributions of *p* atomic orbitals of sulfur and oxygen are dominant in the valence band, while carbon and nitrogen *p* atomic orbitals has minor contributions to the valence states. Sulfur 3*p* orbitals contributions dominate the valence states. The dye molecules constitute the main atomic orbital contributing majorly to the valence states. This explains while the electronic densities of the highest occupied molecular orbital were concentrated on the dye molecules.

2.5 HOMO, LUMO isodensity surfaces of polyenediphenyl-aniline dye molecules adsorbed on $(TiO2)_{n = 8, 68}$ brookite nanocluster

The isodensity surfaces of the molecular orbital responsible for photoexcitation were created from cube files gotten from GPAW. The isodensity surfaces were visualized using the Avogadro software. The isodensity surfaces of the main molecular orbital that were responsible for the photoexcitation of the electrons for the adsorbed polyenediphenyl-aniline dyes@$(TiO_2)_8$ and $(TiO_2)_{68}$ brookite clusters are presented in **Figures** 7 and 8, respectively. The results of the highest occupied molecular orbitals (HOMO) clearly reveal the delocalization of the HOMO on the dye molecule. It is majorly located on the donor group constituting the occupied electronic states, while the lowest unoccupied molecular orbitals (LUMO) obtained

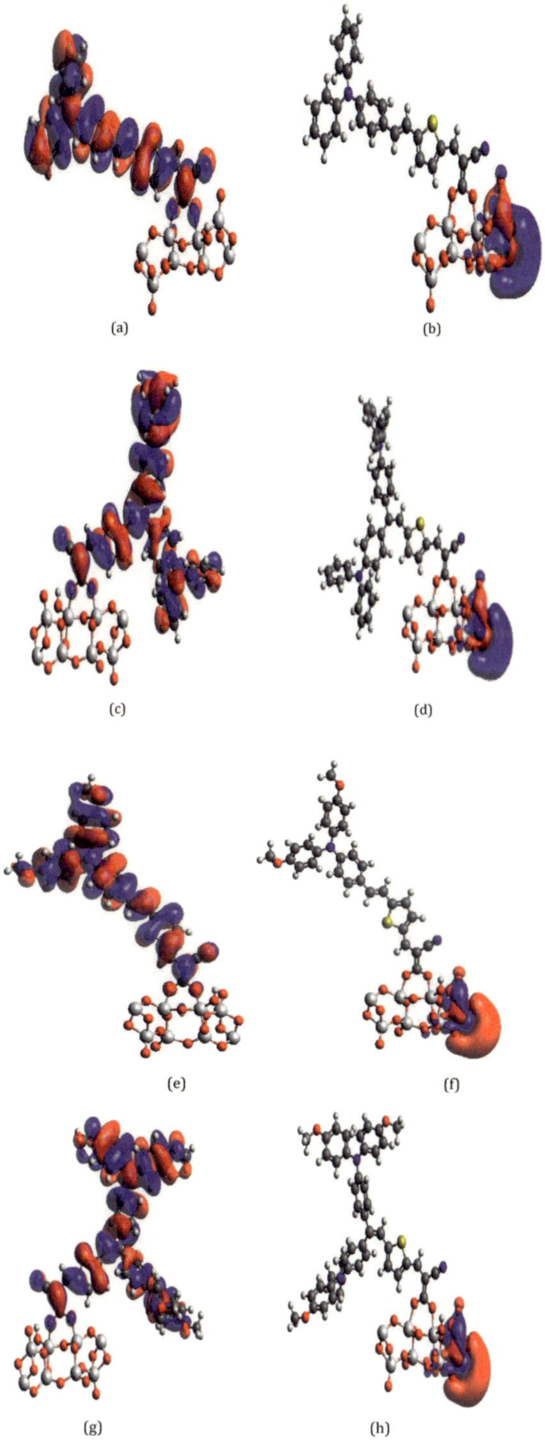

Figure 7.
Isodensity surfaces of molecular orbitals of dyes@ $(TiO_2)_8$ complex (a) D5@ $(TiO_2)_8$ HOMO, (b) D5@ $(TiO_2)_8$ LUMO, (c) D7@ $(TiO_2)_8$ HOMO, (d) D7@ $(TiO_2)_8$ LUMO, (e) D9@ $(TiO_2)_8$ HOMO, (f) D9@ $(TiO_2)_8$ LUMO, (g) D11@ $(TiO_2)_8$ HOMO, and (h) D11@ $(TiO_2)_8$ LUMO.

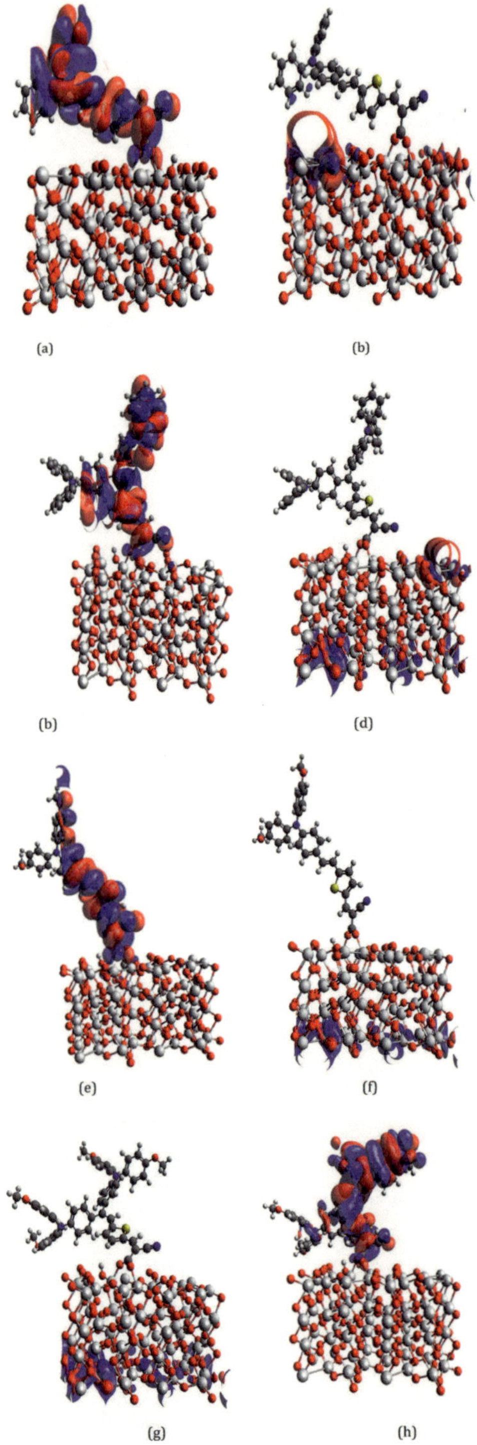

Figure 8.
Isodensity surfaces of molecular orbitals of dyes@$(TiO_2)_{68}$ complex (a) D5@$(TiO_2)_{68}$ HOMO, (b) D5@$(TiO_2$ 68 LUMO, (c) D7@$(TiO_2)_{68}$ HOMO, (d) D7@$(TiO_2)_{68}$ LUMO, (e) D9@$(TiO_2)_{68}$ HOMO, (f) D9@$(TiO_2)_{68}$ LUMO, (g) D11@$(TiO_2)_{68}$ HOMO, and (h) D11@$(TiO_2)_{68}$ LUMO.

reveal the concentration on the acceptor group, where the unoccupied electronic states are located. The distributions of the HOMO electron densities on the dye molecule and the distribution of the LUMO electron densities on the TiO_2 brookite indicate that there is a strong interaction between the dye molecules occupied excited state and the conduction band of the TiO_2 brookite semiconductor unoccupied acceptor levels. The concentration of the electronic levels of the HOMO on the dye molecules and the electronic levels of the LUMO on the TiO_2 clusters infer that electrons are injected from excited state of the dye molecules to the conduction band of the TiO_2 semiconductor.

2.6 Energy levels and bandgaps of polyenediphenyl-aniline dye adsorbed on brookite TiO2 complex

The energy levels of the HOMO, LUMO, and the energy gap of polyenediphenyl-aniline dyes, $(TiO_2)_8$, and the dyes adsorbed on $(TiO_2)_8$ clusters are presented in **Table 3**. This is to understand the effect of dye adsorption on the wide bandgap of the brookite TiO_2 clusters. The bandgaps of D5, D7, D9, and D11 polyenediphenyl-aniline dyes are situated at 1.32, 0.98, 1.08, and 0.85 eV, respectively. The lower bandgap values were found for D7 and D11 dyes molecules. To understand the effect of adsorption of the polyenediphenyl-aniline dyes and electron transfer mechanisms from the dye molecules to the TiO_2 semiconductor conduction band upon adsorption, we investigated the bandgap energy levels of the dye absorbed on $(TiO_2)_n$ clusters. The results show that the HOMO of $(TiO_2)_8$ brookite cluster without adsorption was situated at −6.92 eV while the LUMO is at −4.62 eV with a corresponding bandgap of 2.26 eV. It was observed that the adsorption of the dye molecules on $(TiO_2)_8$ brookite cluster introduces new molecular orbital emanating from the dyes into the wide bandgap of the brookite TiO_2 semiconductor cluster, consequently resulting in the narrowing of the bandgap. This new molecular orbital suggests interactions between the dye molecules occupied exited states and the brookite TiO_2 clusters unoccupied acceptor level. The results also suggest the transfer of electron from the dye molecules to the clusters upon adsorption.

3. Conclusions

In this work, we used DFT computational approach to explore the optical and electronic properties of polyenediphenyl-aniline dye adsorbed on TiO_2 brookite clusters for application in DSSCs. The aforementioned dye molecules were absorbed on brookite TiO_2 clusters in order to understand the interfacial properties between brookite TiO_2 semiconductor and the corresponding dye molecules. The adsorption energies, UV/Vis optical absorption properties, HOMO-LUMO energy gap, isodensity surfaces of key molecular orbital, electronic TDOS, and PDOS for various systems were presented and discussed. The presented results were based on polyenediphenyl-aniline, brookite $(TiO_2)_{n =, 8\ 68,}$ clusters, and the dyes absorbed on the clusters. The results suggest that the absorption of dye molecules on TiO_2 brookite cluster improves its spectra responsivity in the UV region and makes it possible to absorb over the whole spectra range, that is, the UV, visible, and near-infrared region of the solar spectrum. Our findings on the adsorption energy revealed stable grafting of the polyenediphenyl-aniline dyes on the surface of the TiO_2 brookite clusters with

	HOMO (eV)	LUMO (eV)	Fermi level	GAP (eV)
D5	−4.91	−3.59	−4.25	1.32
D7	−4.81	−3.83	−4.27	0.98
D9	−4.72	−3.64	−4.33	1.08
D11	−4.77	−3.92	−4.27	0.85
$(TiO_2)_8$	−6.92	−4.62	−5.75	2.26
D5@$(TiO_2)_8$	−5.09	−4.40	−4.76	0.69
D7@$(TiO_2)_8$	−5.06	−4.39	−4.72	0.67
D9@$(TiO_2)_8$	−4.94	−4.43	−4.64	0.51
D11@$(TiO_2)_8$	−4.73	−4.19	−4.46	0.54

Table 3.
HOMO, LUMO, and HOMO_LUMO GAP energy levels.

D11 dyes demonstrating more grafting on both clusters $(TiO_2)_{n =, 8\ 68}$. The analyses of the isodensity surfaces of the key molecular orbital, electronic TDOS, and PDOS and HOMO, LUMO energy gap also showed new molecular orbitals emanating from the dye molecules into the bandgap of TiO_2 upon absorption. Due to interactions of the electronic states between the orbitals of the TiO_2 and the adsorbed dye molecules, suggesting good electron injection kinetics from the dyes occupied donor level to TiO_2 brookite clusters unoccupied acceptor level. Generally, our findings suggest that TiO_2 brookite semiconductor and the corresponding polyenediphenyl-aniline dyes could perform well as photoanode material with promising potential for higher photocurrents density and open-circuit voltage and enhanced efficiency of DSSCs.

Acknowledgements

We would also like thank the University of Venda, Centre for High performance Computing (CHPC), South Africa, and Centre for Interactive Materials and Bioinspired Technology, Albert Ludwig's University Freiburg in Germany for their computing facilities and the support to carry out this research.

Conflicts of interest

The authors declare that they have no known competing financial interests or personal relationships that could have appeared to influence the work reported in this paper.

Author details

Ife Elegbeleye[1,2*], Eric Maluta[1,3], Regina Maphanga[3,4], Michael Walter[2] and Oliver Brügner[2]

1 Department of Physics, University of Venda, Thohoyandou, South Africa

2 Freiburg Centre for Interactive and Bioinspired Technology, Freiburg, Germany

3 National Institute for Theoretical Physics (NITheP), Gauteng, South Africa

4 Council for Science and Industrial Research, Pretoria, South Africa

*Address all correspondence to: ifelove778@gmail.com

References

[1] Mozhgan H, Zahra R. A review on flexible dye-sensitized solar cells as new sustainable energy resources. Pigment and Resin Technology. 2023;**52**:310-320. DOI: 10.1108/prt-01-2022-0010

[2] Abodunrin TJ, Obafemi O, Boyo AO, Adebayo TA, Jimoh R. The effect of electrolyte on dye sensitized solar cells using natural dye from mango (*M. Indica* L.) leaf as sensitizer. Advances in Material Physics and Chemistry. 2015;**5**:205-213. DOI: 10.4236/ampc.2015.56021

[3] Elegbeleye IF, Maluta NE, Maphanga RR. Density functional theory study of optical and electronic properties of $(TiO_2)_{n=5,8,68}$ clusters for application in solar cells. Molecules. 2019;**26**:1-19. DOI: 10.3390/molecules26040955

[4] Tong Z, Shang PG. The stability, electronic structure, and optical property of TiO_2 polymorphs. Journal of Physical Chemistry C. 2014;**118**:11385-11396. DOI: 10.1021/jp412462m

[5] Hao Y, Jia L, Gang Z, Sum WC, Hongda D, Lin G, et al. First principles study of ruthenium(ii) sensitizer adsorption on anatase TiO2 (001) surface. RSC Advances. 2015;**5**:60230-60236. DOI: 10.1039/C5RA06743J

[6] Mohammad KN, Filippo DA, Simona F, Annabella S, Guido V, Paul L, et al. Combined experimental and DFT-TDDFT computational study of photoelectrochemical cell ruthenium sensitizers. Journal of American Chemical Society. 2005;**127**:16835-16847. DOI: 10.1021/ja052467l

[7] Liyuan H, Ashraful I, Han C, Chandrasekharam M, Barreddi C, Shufang Z, et al. High efficient dye sensitized solar cell with novel co-adsorbent. Energy and Environmental Science. 2012;**5**:6057-6060. DOI: 10.1039/C2EE03418B

[8] Lu LL, Eric WGD. Porphyrin-sensitized solar cells. Chemical Society Reviews. 2013;**42**:291-304. DOI: 10.1039/C2CS35257E

[9] Boschloo G, Marinado T, Nonomura K, Edvinsson T, Agrios AG, Hagberg DP, et al. A comparative study of a polyene-diphenylaniline dye and $Ru(dcbpy)_2(NCS)_2$ in electrolyte-based and solid-state dye-sensitized solar cells. Thin Solid Films. 2008;**516**:7214. DOI: 10.1016/j.tsf.2007.12.035

[10] Daibin K, Pascal C, Shaik MZ, Daniel PH, Karl MK, Licheng S, et al. Stable dye-sensitized solar cells based on organic chromophores and ionic liquid electrolyte. Solar Energy. 2011;**85**:1189-1194. DOI: 10.1016/j.solener.2011.02.025

[11] Beltran A, Gracia L, Andres J. Density functional theory study of the brookite surfaces and phase transitions between natural titania polymorphs. Journal of Physical Chemistry B. 2006;**110**:23417-23423. DOI: 10.1021/jp0643000

[12] Maurizio A, Marianna B, Agatino DP, Leonardo P. Preparation and photoactivity of nanostructured anatase, rutile and brookite TiO_2 thin films. Chemical Communications. 2006;**47**:4943-4945. DOI: 10.1039/B612172A

[13] Damien D, Ilias B, Khalil A. Tailored preparation methods of TiO_2 anatase, rutile, brookite: Mechanism of formation and electrochemical properties. Chemical Materials. 2010;**22**:1173-1179. DOI: 10.1021/cm902613h

[14] Tarek AK, Armin F, Lars R, Ralf D, Detlef WB. Tailored titanium dioxide nanomaterials: Anatase nanoparticles and brookite nanorods as highly active photocatalysts. Chemical Materials. 2010;**22**:2050-2060. DOI: 10.1021/cm903472p

[15] Ken O, Bin L, Jin Z, Kenneth DJ, Jinlong Y, Hrvoje P, et al. Wet electrons at the $H_2O/TiO_2(110)$ surface. Science. 2005;**308**:1154-1158. DOI: 10.1126/science.1109366

[16] Zallen R, Moret MP. The optical absorption edge of brookite TiO_2. Solid State Communications. 2006;**137**:154-157. DOI: 10.1016/j.ssc.2005.10.024

[17] Pourjafari D, Reyes-Coronado D, Vega-Poot A, Escalante R, Kirkconnell-Reyes D, García-Rodríguez R, et al. Brookite-based dye-sensitized solar cells: Influence of morphology and surface chemistry on cell performance. Journal of Physical Chemistry C. 2018;**122**:14277-14288. DOI: 10.1021/acs.jpcc.8b02384

[18] Takayuki B, Ami H, Naoya A, Chika TY, Yutaka O. Hydrothermal synthesis of highly pure brookite-type titanium oxide powder from aqueous sols of titanate nanosheets. Advances Powder Technology. 2021;**32**:3601-3609. DOI: 10.1016/j.apt.2021.08.014

[19] Huy HD, Thi KCT, Thi DTU, Nguyen TD, Duc DN, Thu HT, et al. Controllable fabrication of photocatalytic TiO_2 brookite thin film by 3D-printing approach for dyes decomposition. Journal of Water Process Engineering. 2021;**43**:102319-1-10. DOI: 10.1016/j.jwpe.2021.102319

[20] Tingxin P, Jinlei X, Rong C. A novel multilayer brookite TiO_2 electrode for improved performance of pure brookite-based dye sensitized solar cells. Chemical Physics Letters. 2020;**738**:136902-1-6. DOI: 10.1016/j.cplett.2019.136902

[21] Gurenko VE, Popkov VI. Hydrothermal-assisted synthesis of sponge-like brookite for pseudo-fenton decomposition of crystal violet under UV–light. Nano-Structures and Nano-Objects. 2021;**28**(100801):1-6. DOI: 10.1016/j.nanoso.2021.100801

[22] Ask HL, Mortensen JJ, Jakob B, Ivano EC, Rune C, Marcin D, et al. The atomic simulation environment—A python library for working with atoms. Journal of Physics and Condensed Matter. 2017;**29**:1-59. DOI: 10.1088/1361-648X/aa680e

[23] Mortensen JJ, Lars BH, Karsten WJ. Real-space grid implementation of the projector augmented wave method. Physical. Review B. 2005;**71**:035109-1-11. DOI: 10.1103/PhysRevB.71.035109

[24] Enkovaara J, Rostgaard C, Mortensen JJ, Chen J, Dułak M, Ferrighi L, et al. Electronic structure calculations with GPAW: A real-space implementation of the projector augmented-wave method. Journal of Physics and Condensed Matter. 2010;**22**:253202-1-24. DOI: 10.1088/0953-8984/22/25/253202

[25] Perdew JP, Kieron B, Matthias E. Generalized gradient approximation made simple. Review Letters. 1996;**77**:3865-3868. DOI: 10.1103/PhysRevLett.77.3865

[26] Marcus DH, Donald EC, David CL, Tim V, Eva Z, Geoffrey RH. Avogadro: An advanced semantic chemical editor, visualization, and analysis platform. Journal of Chemical Information. 2012;**4**:1-17. DOI: 10.1186/1758-2946-4-17

[27] Mark EC. Time-dependent density functional response theory for molecules

in: Recent advances in density functional methods. Recent Advances in DFT Methods Workshop. 1995;**1**:155. DOI: 10.1142/9789812830586_0005

[28] Michael W, Hannu H, Lauri L, Martti P, Jussi E, Carsten R, et al. Time-dependent density-functional theory in the projector augmented-wave method. The Journal of Chemical Physics. 2008;**128**:244101-1-10. DOI: 10.1063/1.2943138

[29] Frisch MJ, Trucks GW, Schlegel HB, Scuseria GE, Robb MA, Cheeseman JR, et al. Gaussian 09, Revision B.01. Wallingford: Gaussian Inc.; 2010

[30] Becke AD. Density-functional exchange-energy approximation with correct asymptotic behavior. Physical Review A. 1988;**38**:3098-3100. DOI: 10.1103/PhysRevA.38.3098

[31] Chengteh L, Weitao Y, Robert GP. Development of the Colle-Salvetti correlation-energy formula into a functional of the electron density. Physical Review B. 1988;**37**:785-789. DOI 10.1103/PhysRevB.37.785

[32] Stewart JC, Matthew DS, Chris JP, Phil JH, Matt IP, Keith R, et al. First principles methods using CASTEP. Zeitschrift für Kristallographie—Crystalline Materials. 2005;**220**:567-570. DOI: 10.1524/zkri.220.5.567.65075

[33] Petre CIO, P, Fanica C, Marilena F, Mihai A. G. Density functional theory (DFT) study of coumarin-based dyes absorbed on TiO_2 nanoclusters—Applications to dye-sensitized solar cells. Materials. 2013;**6**:2372-2392. DOI: 10.3390/ma6062372

[34] Asif M, Salah UDK, Usman AR, Mudassir HT. Red shifting of absorption maxima of phenothiazine-based dyes by incorporating electron-deficient thiadiazole derivatives as π-spacer. Arabian Journal Chemistry. 2019;**12**:1447-1453. DOI: 10.1016/j.arabjc.2014.11.007

[35] Chiara A, Edoardo M, Mariachiara P, Enrico R, Filippo DA. Adsorption of organic dyes on TiO2 surfaces I,n dye-sensitized solar cells: Interplay of theory and experiment. Physical Chemistry Chemical Physics. 2012;**14**:15963-15974. DOI: 10.1039/C2CP43006A

[36] Avinash LP, Ramesh CK, Bhanuprakash K. Adsorption of croconate dyes on TiO_2 anatase (101) surface: A periodic DFT study to understand the binding of diketo groups. Journal of Chemical Science. 2012;**124**:301-310. DOI: 10.1007/s12039-012-0229-1

[37] Mariachiara P, Filippo DA. Computational modelling of TiO_2 surfaces sensitized by organic dyes with different anchoring groups: Adsorption modes, electronic structure and implication for electron injection/recombination. Physical Chemistry Chemical Physics. 2011;**14**:920-928. DOI: 10.1039/C1CP22663K

[38] Balaganapathi T, Kani Amuthan B, Vinoth S, Thilakan P. Synthesis, characterization and dye adsorption studies of porous brookite and mixed brookite with rutile TiO_2 using PEG assisted sol-gel synthesis. Material Research Bulletin. 2017;**19**:114-121. DOI: 10.1016/j.materresbull.2017.03.036

[39] Corneliu IO, Mihai AG. Structure and electronic properties of TiO_2 nanoclusters and dye–nanocluster systems appropriate to model hybrid photovoltaic or photocatalytic applications. Nanomaterials. 2019;**9**:1-31. DOI: 10.3390/nano9030357

Chapter 5

Photoelectrochemical Process on Semiconductor Electrodes

Yasuhisa Maeda

Abstract

For the purpose of conversion of solar energy to chemical energy and water purification, we have investigated the photoelectrochemical process on semiconductor electrodes. As the electrodes, we focused on TiO_2, ZnO, α-Fe_2O_3, and Cu_2O. The TiO_2 electrode was prepared from the anodic oxidation of Ti. The sintered ZnO and the electrochemically deposited film were used as ZnO electrodes. The α-Fe_2O_3 and Cu_2O electrodes were prepared from the deposited film by pulse electrolysis. The flat-band potential and photocurrent were checked for these semiconductor electrodes in the solution. The HPLC analysis was carried out for intermediates generated due to the photoelectrochemical reaction of organic materials. The performance of the apparatus for water purification was tested.

Keywords: semiconductor electrode, electrochemical preparation of semiconductor films, Mott-Schottky plot, photoelectrochemical reaction, HPLC analysis

1. Introduction

The earth can be regarded as a closed system from the point of view of thermodynamics. On the earth, solar energy comes in and heat energy goes out into the universe. Keeping energy balance is necessary to sustain life on Earth. The performance of conversion of solar energy to chemical energy is important to suppress carbon dioxide emissions. As the semiconductors have the property to generate holes and electron under irradiation of the light with a shorter wavelength than that corresponding to the band gap energy, they may be suitable for an effective utilization of solar energy. Thus, photoelectrochemical processes such as the photodecomposition of water on the titanium dioxide photoelectrode (Honda-Fujishima effect) have been attracting attention. There are a lot of reports of photoelectrodes and photocatalysis concerning water decomposition and environmental purification [1–9]. We have been investigating photoelectrochemical processes on semiconductor electrodes [10–17]. In the case of oxidation and reduction of chemical species in the solution, using a semiconductor electrode is preferable because reaction control may be possible by an applied voltage. For the semiconductor electrodes, it is necessary to prepare a large surface area. We carried out mainly the preparation of semiconductor films by electrochemical methods. They have the merits of simplicity, low cost, good reproducibility, and homogeneity for the formation of semiconductor films. We focused on TiO_2, ZnO, α-Fe_2O_3, and Cu_2O electrodes to investigate the preparation, characterization, and reactivity of organic materials in aqueous solution.

2. Processes of photocatalysis and photoelectrode

In order to cause photoreaction of chemical species by using functional semiconductors under light irradiation, there are two processes of photocatalysis and photoelectrode as shown in **Figure 1(A and B)**. In the photocatalysis system, the holes in the valence band and the electrons in the conduction band photogenerated due to light irradiation have the possibility to oxidize and reduce chemical species, respectively. In this case, the photooxidation and photoreduction may occur simultaneously on the surface of the photocatalysis particle. On a photoelectrode system consisting of a semiconductor electrode and a counter electrode, the oxidation and reduction of chemical species occur separately at each electrode. In the case of using an n-type semiconductor electrode, photoanodic oxidation proceeds on the surface of the photoelectrode. A p-type semiconductor photoelectrode causes a photocathodic reaction on its surface.

In the semiconductor electrode/electrolyte interface, the space charge layer may be formed from the surface to the inside of the semiconductor electrode. The band bending of the space charge layer can be controlled by the potential of the semiconductor electrode. An effective separation of hole-electron pairs photogenerated due to light absorption could occur in the presence of band bending. The flat-band potential and the photocurrent are important factors as the feature of the semiconductor electrode/electrolyte interface. The analysis of the chemical product generated on the photoelectrode may provide valuable information concerning the photooxidation and reduction process. The semiconductor photoelectrode system may be suitable for chemical synthesis and water purification.

3. Preparation and characteristics of semiconductor materials

The preparation and characteristics of TiO_2, ZnO, α-Fe_2O_3, and Cu_2O electrodes are mentioned below.

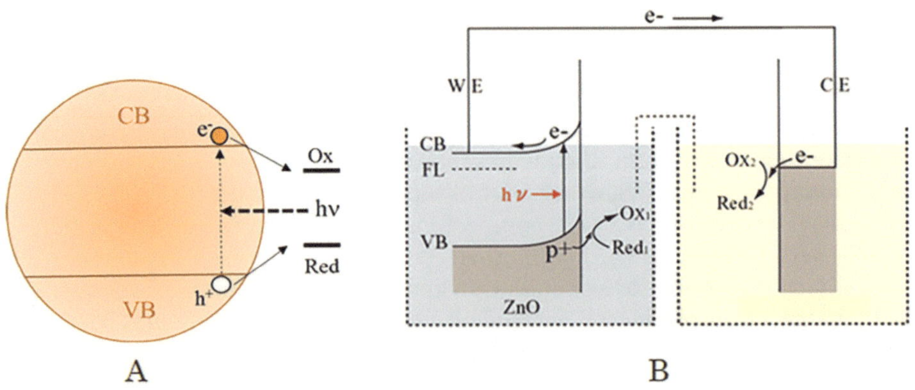

Figure 1.
Photocatalysis (A) and photoelectrode (B).

3.1 TiO_2 electrode

Titanium dioxide was prepared from the anodic oxidation of the titanium plate and its heat treatment in air. This method is useful in the aspects of simplicity, reproducibility, and low cost. The properties of titanium dioxide film may depend on the kinds of electrolytic solution for the oxidation of titanium plate. We used mainly nitric acid [12, 13] because the titanium dioxide from this solution showed a higher photocurrent response than those from sulfuric acid, chloric acid, and phosphoric acid.

Figure 2a shows the FE-SEM image of the surface of the sample prepared from anodic oxidation of titanium plate (size: $50 \times 50 \times 0.2$ mm) with a constant current of 300 mA for 60 min in aqueous 0.1 M HNO_3 solution. The porous surface with about 100 nanometer diameter was confirmed. **Figure 2b** shows the cross-section SEM of this sample. The boundary between titanium oxide and titanium substrate could not be distinguished. **Figure 2c** shows the relationship between the Ti:O ratio and the depth for the sample heat-treated at 500°C for 60 min. The thickness of the titanium oxide film was 1.5 μm. From the depth corresponding to the Ti: O ratio of 33: 67, the thickness of titanium dioxide was 0.2 μm. The XRD illustrated the dependence of the structure of the anodically oxidized sample on heat treatment temperature in air as shown in **Figure 3**. The peaks of titanium dioxide were not recognized on the sample without heat treatment. The anatase peaks of TiO_2 were observed on the samples

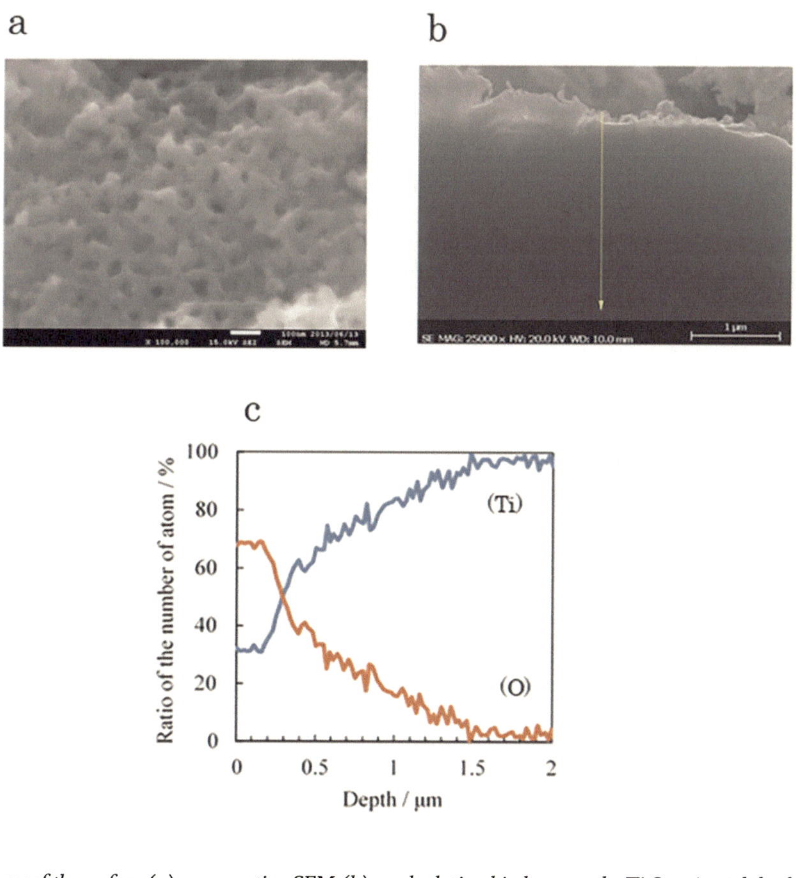

Figure 2.
FE-SEM image of the surface (a), cross-section SEM (b), and relationship between the Ti:O ratio and the depth (c) for the sample heat-treated at 500°C for 60 min after anodic oxidation of titanium in aqueous 0.1 M nitric acid.

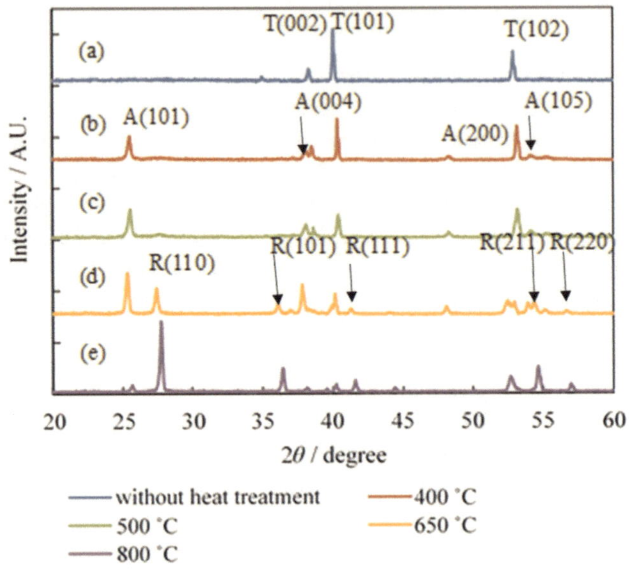

Figure 3.
XRD patterns of the samples (a) without heat treatment, and heat-treated at (b) 400°C for 60 min, (c) 500°C for 60 min, (d) 650°C for 60 min, and (e) 800°C for 10 min after anodic oxidation treatment in aqueous 0.1 M nitric acid A: Anatase, R: Rutile and T: Titanium.

heated at 400, 500, and 650°C The rutile peaks appeared clearly on that heated at 650°C, and the intensity of the rutile peaks increased instead of the much weaker anatase peaks on that heat treated at 800°C. Although there was no appearance of titanium dioxide peaks on the sample without heat- treatment, it might be amorphous because energy-dispersive X-ray spectroscopy (EDX) analysis showed the Ti: O ratio of 36.5: 63.5 for it. The anatase peaks were not recognized on the sample prepared from the titanium heat-treated only. Anodic oxidation of titanium may be necessary to obtain the anatase structure of titanium dioxide.

In order to characterize semiconductor electrodes, evaluation of the flat-band potential (E_{fb}) of semiconductor electrodes in electrolytic solution is important. The value of E_{fb} can be determined from the relationship between capacitance in the semiconductor electrode/electrolyte interface (C) and electrode potential (E) as shown in Eq. (1).

$$\frac{1}{C^2} = \frac{2}{eN\varepsilon\varepsilon^o}\left(E - E_{fb}\right) \tag{1}$$

e is the quantity of charge on an electron, N is the carrier density, ε dielectric constant, ε^o the permittivity of free space, and E_{fb} is the flat-band potential corresponding to the potential causing non-band bending inside the semiconductor electrode. In the linear plots of $1/C^2$ versus E (Mott-Schottky plots), the values of E_{fb} and N can be obtained by the intercept to the potential axis and the slope of the linear portion, respectively. The sign of slope is positive in the n-type semiconductor electrode, and that is negative in the p-type semiconductor electrode. The value of E_{fb} in the n-type semiconductor electrode may be close to the potential for the lower edge of the conduction band. Thus, the n-type semiconductor electrode with a more negative value of E_{fb} may have a higher reduction ability of photo-generated electrons. The potential for the upper edge of the valence band of the n-type semiconductor electrode could be regarded as the value of F_{fb} plus band gap energy. This potential implies the oxidation ability of a photo-generated hole. **Figure 4** shows the Mott-Schottky plot of TiO_2 electrode in an aqueous Na_2SO_4 solution. In this case, TiO_2 is prepared from anodic oxidation of Ti in aqueous HNO_3 solution and heat-treated at 500°C. The value of E_{fb} of -0.53 V vs. Ag/AgCl and that of N of 3.5 ×

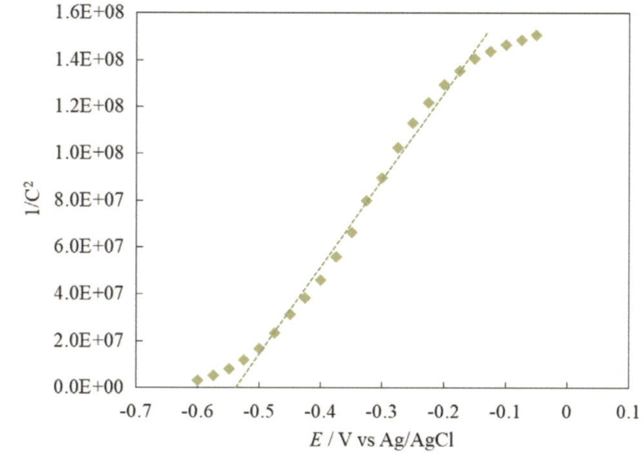

Figure 4.
Mott-Schottky plot for the TiO_2 (prepared from heat treatment at 500°C for 60 min after anodic oxidation treatment in aqueous 0.1 M nitric acid) electrode in aqueous 1 M Na_2SO_4 solution.

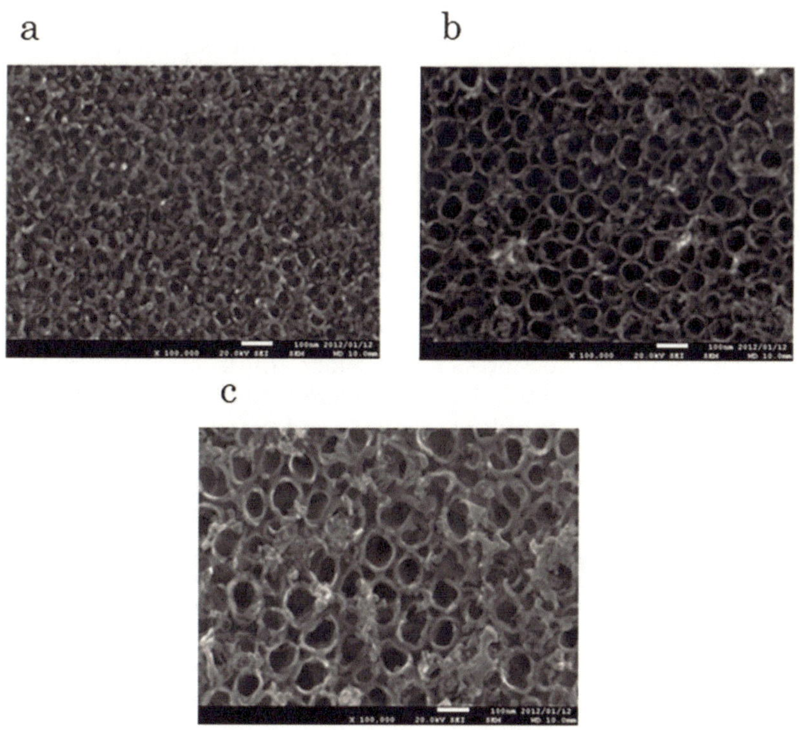

Figure 5.
FE-SEM images of the surface of the samples heat-treated at 500°C for 60 min after anodic oxidation of titanium with the voltage of 10 V (a), 15 V (b), and 20 V (c) in aqueous 0.1 M $(COOH)_2$ solution containing 0.2 Wt % NH_4F.

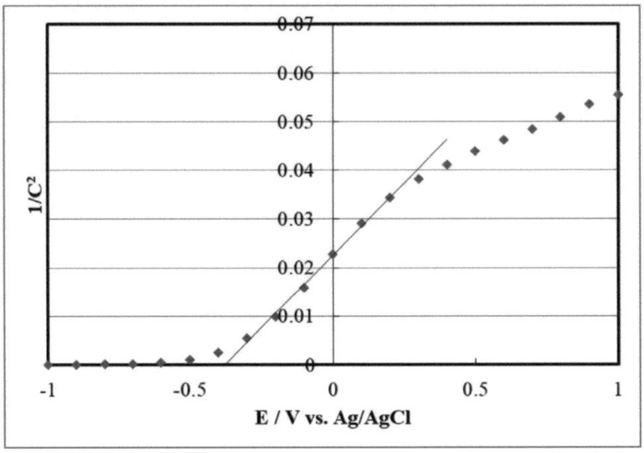

Figure 6.
Mott-Schottky plot for the TiO_2 (prepared from heat treatment at 500°C for 60 min after anodic oxidation treatment of titanium with the voltage of 10 V in aqueous 0.1 M $(COOH)_2$ solution containing 0.2 Wt % NH_4F) electrode in aqueous 1 M Na_2SO_4 solution.

10^{17} cm^{-3} were estimated from the extrapolation of the potential axis and slope of the linear potion, respectively. From the Mott-Schottky plots of the TiO_2 electrodes prepared from different heat-treatment temperatures of 400, 650, and 800°C the values of E_{fb} were − 0.54 V, −0.53 V, and − 0.54 V vs. Ag/AgCl, and the values of *N* were 1.9×10^{18}, 4.7×10^{16}, and 7.2×10^{15} cm^{-3}, respectively.

With regard to the anodic oxidation of titanium, it was reported that nano-porous titanium dioxide could be formed in an electrolytic solution in the presence of an F^- ion [18]. We carried out the anodic oxidation of Ti in aqueous 0.1 M $(COOH)_2$ solution containing 0.2 Wt % NH_4F. **Figure 5** shows the FE-SEM of the sample from the anodic oxidation of Ti plate in this solution with an applied voltage of 10 V (a), 15 V (b), and 20 V (c) for 60 min and then its heat treatment at 500°C for 60 min. The nano-porous structure was observed. The size of the nano-pore was dependent on an applied voltage. The porous surface of about 40, 70, and 100 nm pores was obtained by the anodic oxidation with an applied voltage of 10, 15, and 20 V, respectively, but the destruction of the porous structure was recognized in the case of higher voltage than 20 V. **Figure 6** shows the Mott-Schottky plot of the sample prepared from the anodic oxidation of Ti with 10 V in the same solution and its heat-treatment at 500°C in air. The value of E_{fb} was −0.38 V vs. Ag/AgCl and the value of *N* was 2.37×10^{19} cm^{-3}. From the Mott-Schottky plots for the nano-porous TiO_2 electrode prepared from the applied voltage of 15 V and 20 V, the values of E_{fb} were − 0.01 V and 0.06 V vs. Ag/AgCl, and the values of *N* were 1.89×10^{19} and 1.67×10^{19} cm^{-3}, respectively. In comparison with the TiO_2 electrode prepared from anodic oxidation of titanium in nitric acid, these nano-porous electrodes showed a more positive E_{fb} value.

3.2 ZnO electrode

It is previously known that photoanodic reaction proceeds on the surface of ZnO photoelectrode [19]. The simple preparation of the ZnO electrode we carried out is the sintering of ZnO powders [16]. In this case, the powders were pressed at a pressure of 2 MPa and heated at different temperatures for 60 min in air. The sample was connected to a lead wire through a Ga-In ally and covered with epoxy resin to protect leakage current. **Figure 7a** shows the SEM images of the samples heated at temperatures of 1000°C (A), 1100°C (B), and 1200°C (C). The aggregation and combination of particles proceeded with temperature and the grain with a larger size than 5 μm was observed on the sample heated at 1200°C. **Figure 7b** shows the relationship between electrode potential and photocurrent on the sintered ZnO electrode (geometric surface area: 1.54 cm^2) with heat-treatment temperatures of 1000°C (A), 1100°C (B), and 1200°C (C) under irradiation of UV light (wavelength: 360 nm and intensity: 30 mW/cm^2). The photocurrent on the 1200°C-treated electrode was much higher than that on the 1100°C- and 1000°C-treated electrodes, and it had a tendency to be saturated at more positive potential than 2.0 V vs. Ag/AgCl. This means that photogenerated holes and electron pairs could be separated smoothly on the 1200°C-treated electrode. The dependence of magnitude of photocurrent on electrode potential suggests that much more band bending may be necessary for the separation of photogenerated hole and electron pairs on the 1000°C- and 1100°C-treated electrodes.

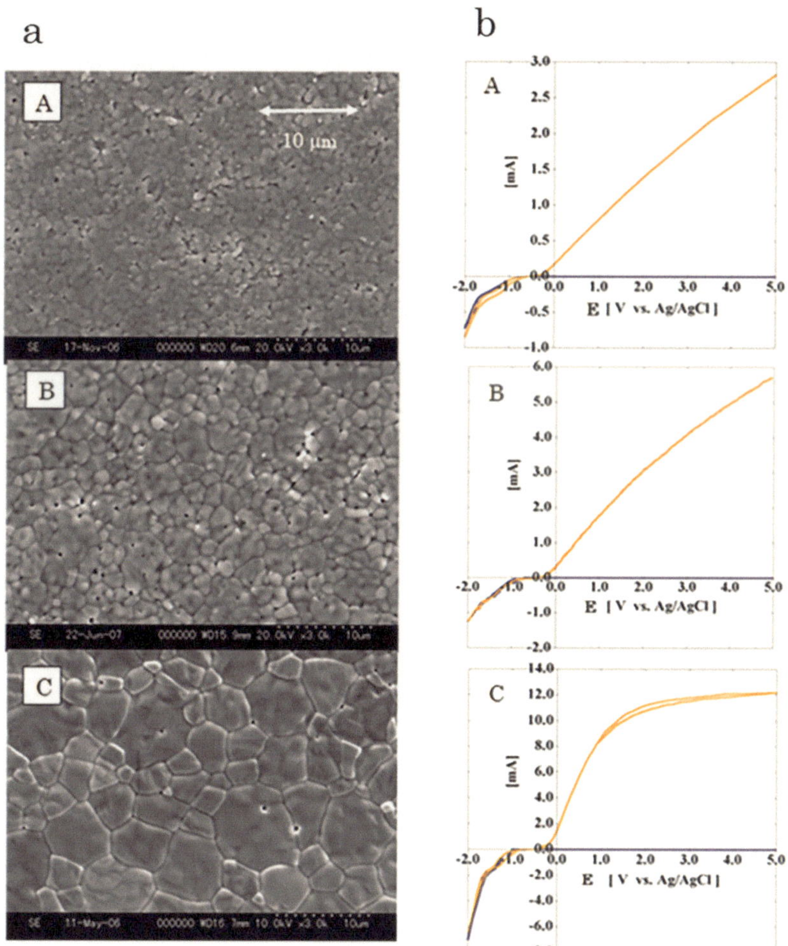

Figure 7.
SEM images of the surface (a) and relationship between photocurrent and electrode potential (b) for the ZnO electrode sintered at different temperatures of 1000°C (A), 1100°C (B) and 1200°C (C).

The photocurrent quantum efficiency ϕ is represented as the following equation.

$$\phi = \frac{N_e}{N_{h\nu}} \times 100 = \frac{\frac{I_{ph}}{e}}{\frac{WS}{h\frac{c}{\lambda}}} \times 100 \quad (2)$$

N_e is the number of electrons as photocurrent, $N_{h\nu}$ the number of photons of the incident light, I_{ph} the photocurrent, e the quantity of charge on an electron, W the intensity of light, S the surface area of the electrode, h the Planck constant, c the speed of light, and λ the wavelength of light. The values of photocurrent at 2.0 V vs. Ag/AgCl were 1.4, 3.1, and 10.8 mA for 1000°C-, 1100°C-, and 1200°C-treated electrodes, respectively. From Eq. (2), the values of ϕ were 10.4, 23.2, and 80.6% for each electrode. We had previously investigated the quantum efficiency by using the photothermal method (η_{pt}) for semiconductor photoelectrodes [20–22]. The value of 80.6% was close to the value of η_{pt} of 85% for the sintered ZnO electrode. **Figure 8** shows the Mott-Schottky plot of 1200°C-treated ZnO electrode in an aqueous 0.1 M Na_2SO_4 solution. The value of E_{fb} was –0.24 V vs. Ag/AgCl and the value of N was 2.2×10^{18} cm^{-3}. The 1100°C-treated electrode provided the values of −043 V vs. Ag/AgCl and 2.9×10^{17} cm^{-3} as E_{fb} and N, respectively. A remarked difference of N between 1200°C- and 1100°C-treated electrodes was recognized. The 1000°C-treated electrode did not show a linear relationship between $1/C^2$ and E clearly.

For the performance of water purification and photoelectrochemical synthesis on semiconductor electrodes, it is necessary to prepare the semiconductor films with a large surface area. Therefore, we carried out the electrochemical deposition of ZnO film on a titanium substrate by the following procedure. The electrode potential of a titanium working electrode (size: 50× 50 mm^2) was held at −0.9 V vs. Ag/AgCl for 3 hours in aqueous 0.1 M $Zn(NO_3)_2$ solution with O_2 gas bubbling. The temperature of the solution was kept at 62 °C. This electrolysis caused the deposition of the ZnO film on the surface of the titanium electrode in accordance with the following equations [23].

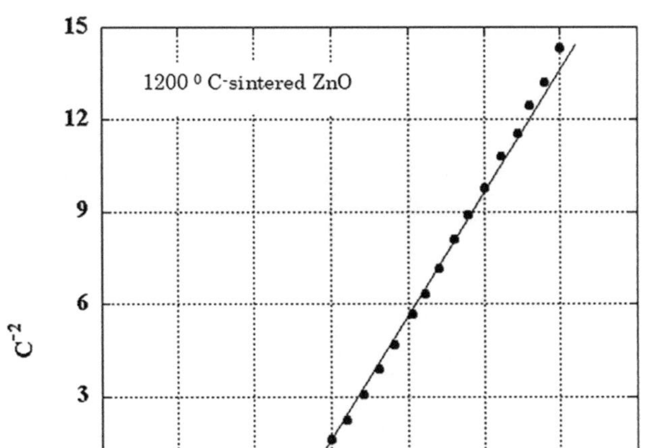

Figure 8.
Mott-Schottky plot for the 1200°C-sintered ZnO electrode in aqueous 0.1 M Na_2SO_4 solution.

$$NO_3^- + H_2O + 2e^- \rightarrow NO_2^- + 2OH^- \quad (3)$$

$$Zn^{2+} + 2OH^- \rightarrow Zn(OH)_2 \quad (4)$$

$$Zn(OH)_2 \rightarrow ZnO + H_2O \quad (5)$$

Figure 9a shows the FE-SEM image of the film deposited on the titanium substrate. From the cross-section analysis, the thickness of the film was about 20 μm (**Figure 9b**). **Figure 9c** shows the XRD of the film before (below illustration, A) and after (upper, B) heat treatment at 500°C for 60 min in air. The diffraction peaks due to ZnO were recognized on the as-deposited film, but the intensity of the peaks increased markedly after heat treatment. **Figure 10** shows the relationship between electrode potential and photocurrent on the ZnO electrode (geometric surface area: 25 cm^2) prepared from the electrochemical deposition and heat treatment under irradiation of UV light (wavelength: 365 nm and intensity: 2.35 mW/cm^2). The

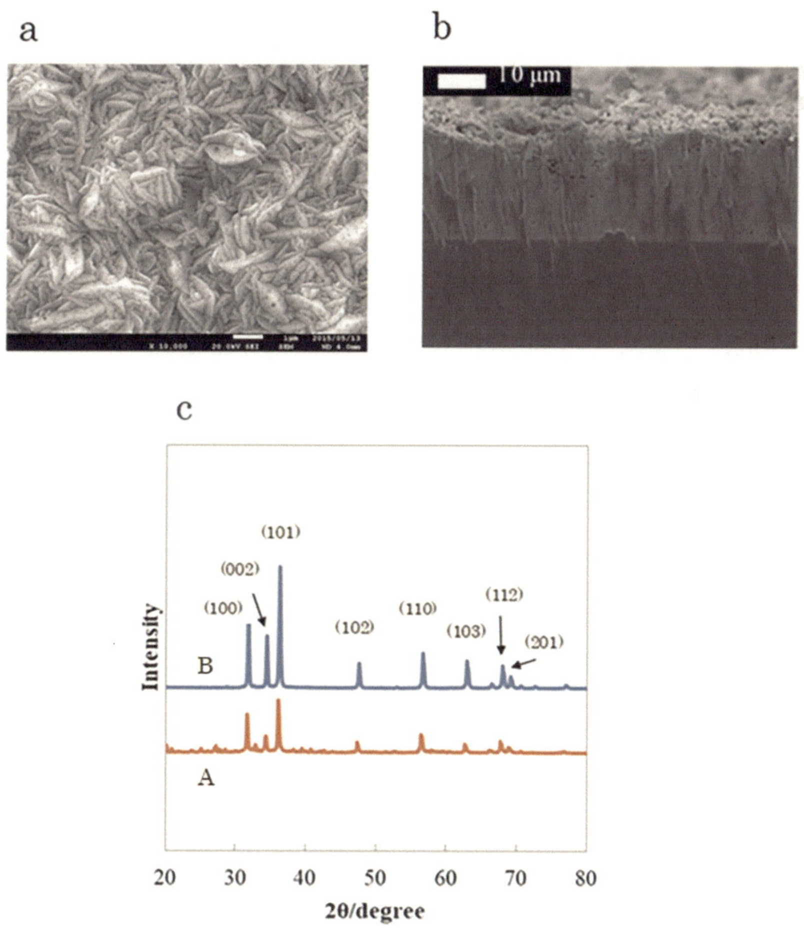

Figure 9.
FE-SEM image of the surface (a), cross-section SEM (b), and XRD patterns (c) for the film deposited electrochemically at the constant potential of −0.9 V vs. Ag/AgCl for 3 hours in aqueous 0.1 M $Zn(NO_3)_2$ solution with the temperature of 62°C. The below (A) XRD pattern is for the film before heat treatment, and the upper (B) for the film after heat treatment at 500°C for 60 min in air.

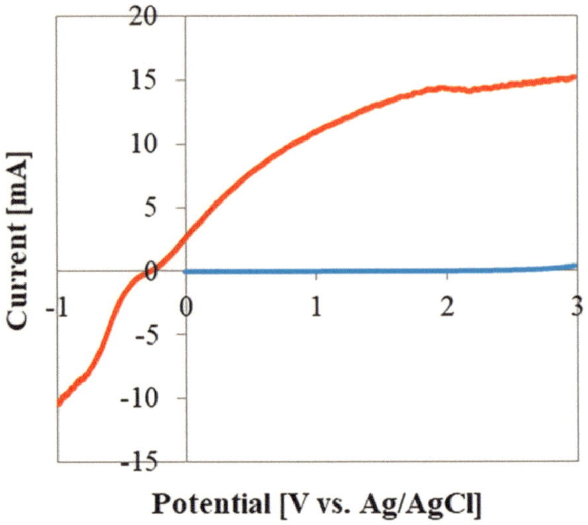

Figure 10.
Relationship between electrode potential and photocurrent on the ZnO electrode prepared from electrochemical deposition in aqueous $Zn(NO_3)_2$ solution and heat treatment at 500°C.

tendency to saturation of photocurrent was observed at more positive potential than 2.0 V vs. AG/AgCl. The photocurrent quantum efficiency at 2.0 V vs. Ag/AgCl for this electrode was estimated at 82.7% from Eq. (2) with I_{ph} of 14.3 mA, W of 2.35 mW/cm^2, S of 25 cm^2, and λ of 365 nm. The value of ϕ for the electro-deposited ZnO was high as well as that of ϕ for the 1200°C-sintered ZnO. This reflects a rapid separation of photo-generated holes and electron pairs on the ZnO electrode.

3.3 α-Fe_2O_3 and Cu_2O

As α-Fe_2O_3 (hematite) and Cu_2O are responding to visible light, they are expected as candidate semiconductors to play a part in artificial photosynthesis. Hematite electrode is prepared from thermal oxidation of iron easily [15]. We carried out the deposition of iron oxide film by current and potential pulse electrolysis on the titanium electrode in an aqueous KCl solution containing $FeCl_2$. The hematite electrode was obtained by heat treatment of the deposited iron oxide film at 500°C in air. The details of the preparation, characterization, and phtoelectrochemical response of hematite are summarized in the book [10].

For the preparation of Cu_2O, we used the electrochemical deposition of a film on the titanium electrode in an aqueous solution (pH = 10) of 10 mM $CuSO_4$, 0.2 M sodium lactate, and 20 mM NaOH. The electrolysis in this case was done for 60 min by the potential pulse method represented by the positive potential (E_p), the negative potential (E_n), and pulse width (W) as shown in **Figure 11**. **Figure 12a** shows the FE-SEM of the film deposited in the case of E_p = 0.0 V, E_n = −0.4 V vs. Ag/AgCl and W = 0.5 sec. The granular deposits with a size of about 80 nm were observed. From the cross-section analysis (**Figure 12b**), the thickness of the film was about 0.8 μm. **Figure 12c** shows the XRD of this film. The peaks of Cu_2O were confirmed in addition to Ti peaks. With regard to the electrochemical deposition of Cu_2O film in the basic solution containing lactate ion (Lac^-), the following process could be

DOI: http://dx.doi.org/10.5772/intechopen.1004844

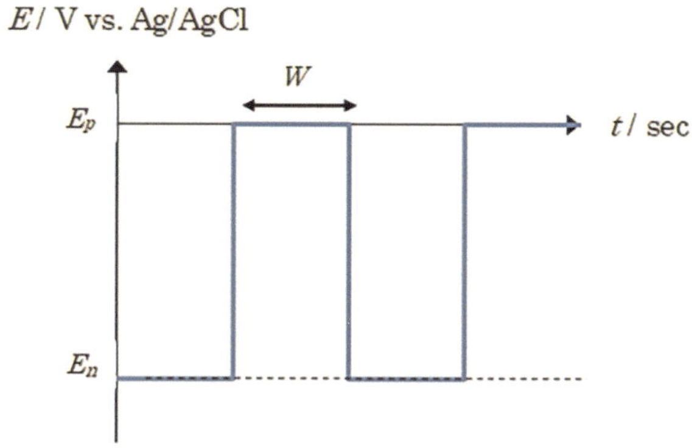

Figure 11.
Potential pulse electrolysis with negative potential (E_n), positive potential (E_p), and pulse width (W).

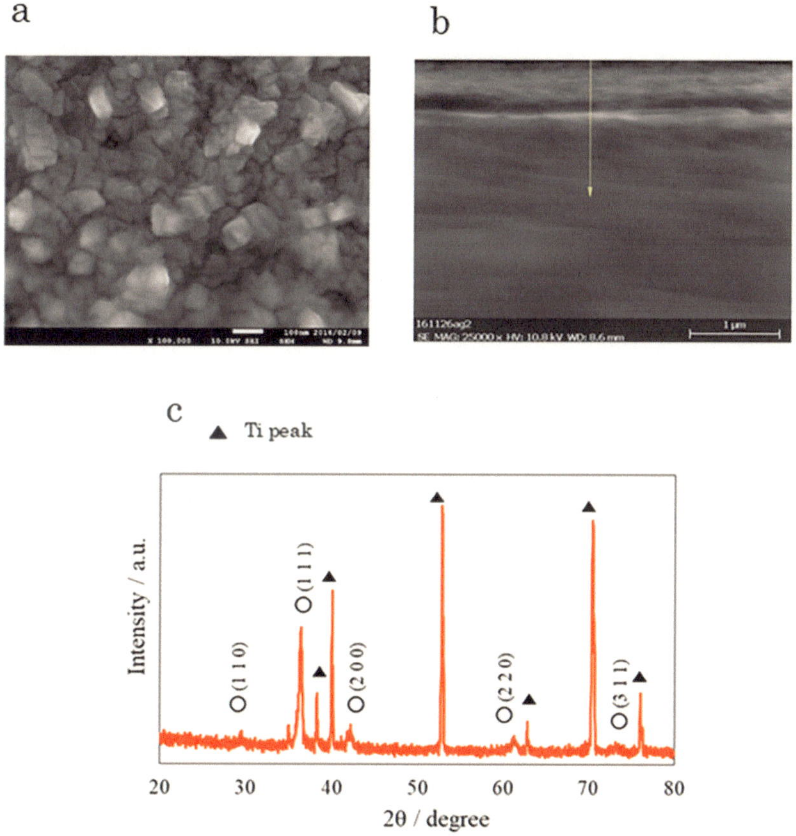

Figure 12.
FE-SEM image of the surface (a), cross-section SEM (b) and XRD pattern (c) for the film deposited from potential pulse electrolysis (E_n: −0.4 V vs. Ag/AgCl, E_p: 0.0 V vs. Ag/AgCl and W: 0.5 sec) for 60 min in aqueous basic solution of 10 mM $CuSO_4$, 0.2 M sodium lactate and 20 mM NaOH.

considered. The cu^{2+} ion forms the complex of Cu^{2+} $(Lac^{-})_2$. This complex is reduced electrochemically to form Cu_2O through an intermediate of CuOH as shown below.

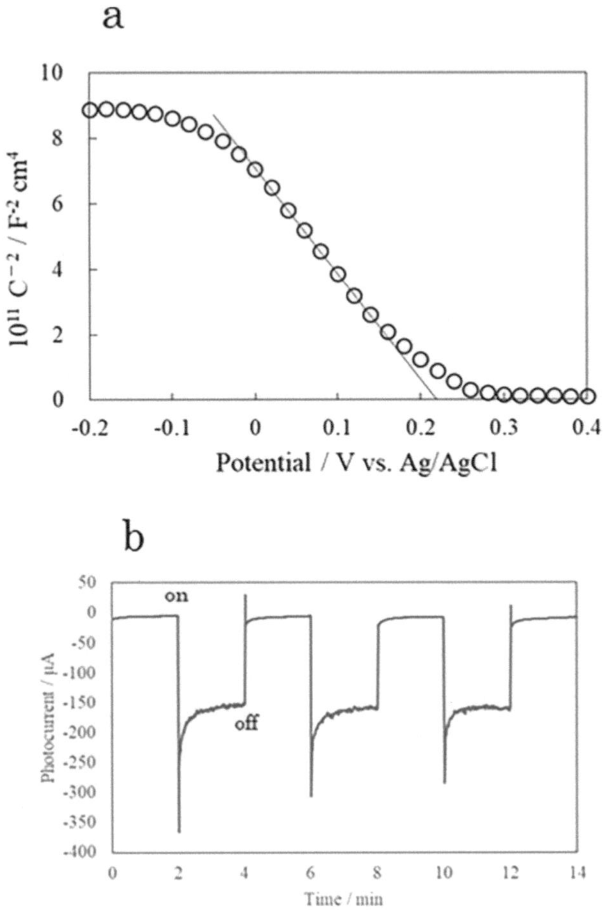

Figure 13.
Mott-Schottky plot for the Cu_2O electrode in aqueous 1 M Na_2SO_4 solution (a) and photocurrent response of the Cu_2O electrode to the irradiation of visible light (wavelength: 490 nm, intensity: 4.8 mW/cm^2) in aqueous 0.1 M Na_2SO_4 solution (b).

$$Cu^{2+}(Lac^-)_2 + e^- + OH^- \rightarrow CuOH + 2Lac^- \tag{6}$$

$$2CuOH \rightarrow Cu_2O + H_2O \tag{7}$$

Figure 13a shows the Mott-Schottky plot of the Cu_2O electrode in aqueous 1 M Na_2SO_4 solution. The negative slope of the linear portion indicated the p-type semiconductor electrode. The values of E_{fb} and N were 0.22 V vs. Ag/AgCl and 5.73×10^{18} cm^{-3}, respectively. **Figure 13b** shows the photocurrent response of the Cu_2O electrode to the irradiation of visible light (wavelength: 490 nm and intensity: 4.8 mW/cm^2). The cathodic photocurrent of the feature of the p-type semiconductor electrode was clearly observed.

4. Photoelectrochemical reaction on semiconductor electrodes

For the progress of organic synthesis and water purification based on photoelectrochemical reactions, it is necessary to make clear the reactivity of photo-generated

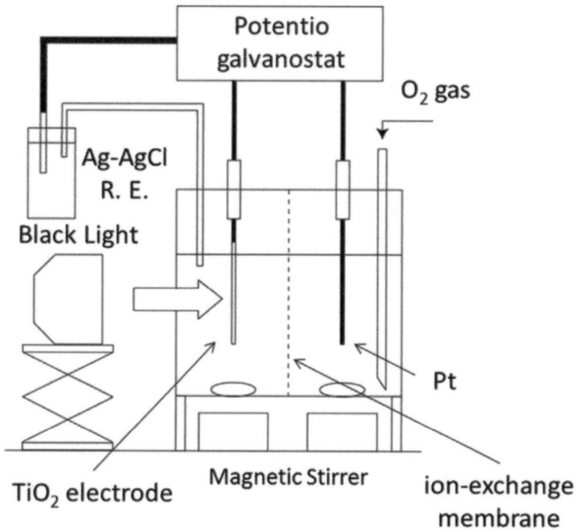

Figure 14.
Cell assembly for photoelectrochemical reaction of organic material in aqueous solution consisting of photoelectrode and counter electrode sides separated by a cation-exchange membrane.

holes and electrons on semiconductor electrodes to chemical species. We carried out an HPLC analysis of the products generated from a starting organic material on the photoelectrode by using a cell shown in **Figure 14**. In this cell, the solution of photoelectrode and that of the counter electrode were separated by a cation-exchange membrane (Selemion CMVm AGC Engineering). We investigated the photoanodic oxidation process of Veratryl Alcohol (VA, **Figure 15**) on the semiconductor electrode in an aqueous solution from the aspects of intermediates detected by applying the HPLC method to the photoelectrode solution.

In the case of using the TiO_2 electrode (geometric surface area: 25 cm^2) prepared from anodic oxidation of Ti in HNO_3 solution and then heat-treatment at 500°C, the intermediates due to oxidation of VA were clearly detected [12]. In this case, the electrolytic solution (volume: 420 ml) of two electrodes was aqueous 10 mM NaCl solution containing 50 μM VA. The solution of the Pt counter electrode (geometric surface area: 25 cm^2) was bubbled with oxygen gas. The light source was a black light (wavelength: 365 nm, intensity: 3.0 mW/cm^2). **Figure 15** shows the chromatograms for the solution of the TiO_2 electrode before and after irradiation of UV light. The potential of the TiO_2 electrode was held at 2.0 V vs. Ag/AgCl. Before irradiation, only the peak A due to VA was observed at a retention time of 6.9 min. After irradiation for 6 hours, the new peaks of Veratraldehyde (B), Veratric Acid (C), Vanillil Alcohol (D), Vanillin (E), Vanillic Acid (F), Protocatechualdehyde (G), Protocatechuic Acid (H), and Benzentriol (I) appeared. The structure of these intermediates is shown in **Figure 16**. Protocatechuic Acid and Benzentriol are highly oxidized intermediates. **Figure 17** shows the relationship between the concentration of intermediates and irradiation time, and between the relative concentration of VA and irradiation time on the TiO_2 photoelectrode. In addition to the main intermediates of Vanillil Alcohol, Veratraldehyde, and Vanillin, the formation of Benzentriol and Protocatechuic Acid is shown clearly. The HPLC analysis for organic acid revealed the formation of ketomalonic acid. This suggests the occurrence of aromatic ring cleavage on the TiO_2 photoelectrode. High response to VA with the

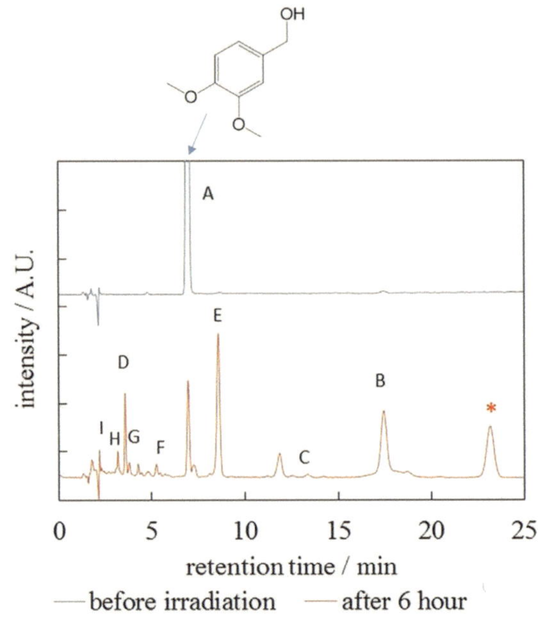

Figure 15.
Chromatograms before and after irradiation (wavelength: 365 nm, intensity: 3.0 mW/cm²) for 6 hours to the TiO_2 electrode in aqueous 10 mM NaCl solution containing initial concentration of 50 μm Veratryl alcohol. A) Veratryl alcohol B) Veratraldehyde C) Veratric acid D) Vanillil alcohol E) Vanillin F) Vanillic acid G) Protocatechualdehyde H) Protocatechuic acid I) Benzentriol.

formation of ketomalonic acid was also recognized on the TiO_2 prepared from the anodic oxidation of Ti in $(COOH)_2$ $-NH_4F$ solution.

In the photoanodic oxidation of 50 μM VA on the ZnO electrode in aqueous NaCl solution under UV irradiation (wavelength: 365 nm, intensity: 2.6 mW/cm^2), the main intermediates were Veratraldehyde and Vanillil Alcohol. The concentration of Vanillin was much lower than that of Veratraldehyde and Vanillil Alcohol. The formation of Veratric Acid, Isovanillin, Protocatechuic acid, and Benzentriol was recognized on the ZnO photoelectrode. Except for alkaline solution, the photoanodic dissolution may occur in an aqueous solution under UV irradiation as below.

$$ZnO + 2h^+ \rightarrow Zn^{2+} + \frac{1}{2}O_2 \tag{8}$$

Because of this side reaction, a difference of intermediate between ZnO and TiO_2 might appear.

On the α-Fe_2O_3 electrode in an aqueous NaCl solution, the visible light irradiation (wavelength: 490 nm, intensity: 4.8 mW/cm^2) caused the photoanodic oxidation of VA to generate Veratraldehyde, Vanillil Alcohol, and Vanillin. The HPLC peaks of Protocatechuic Acid and Benzentriol were not observed. After the photoanodic oxidation of VA with the initial concentration of 100 μM for 9 hours, 24.7 μM of VA decreased, and 17.5 μM of Veratraldehyde, 5.0 μM of Vanillil Alcohol, and 1.1 μM of Vanillin were formed. The ratio of the total concentration of these intermediates to the concentration of oxidized VA was 95.5%. This implies a possibility of selective oxidation of organic materials by α-Fe_2O_3 photoelectrode.

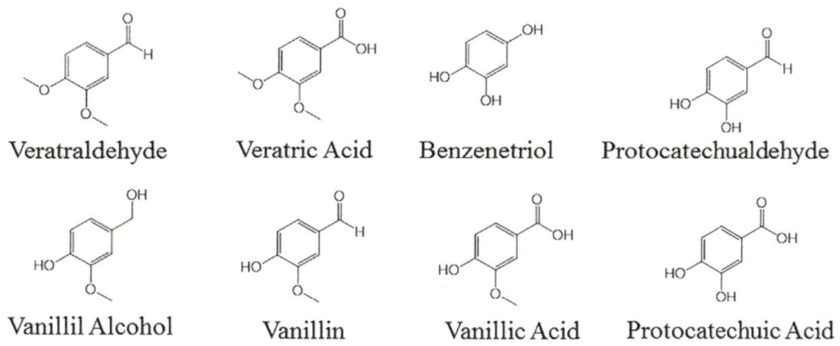

Figure 16.
Intermediates from Veratryl alcohol.

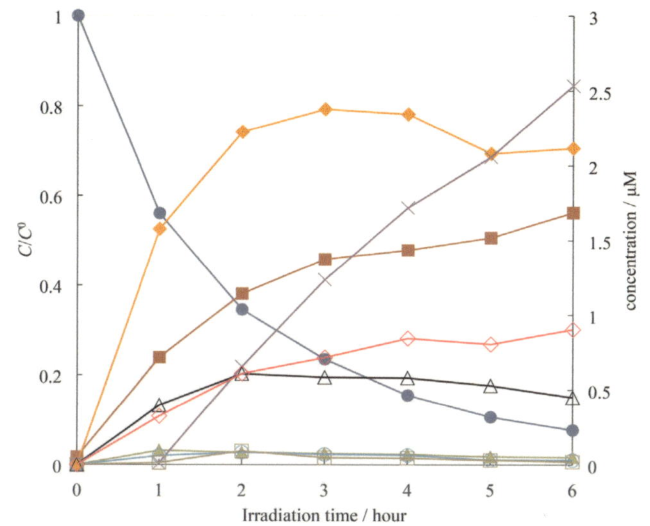

Figure 17.
Relationship between concentration of intermediate and irradiation time on the TiO_2 electrode Veratryl alcohol (●) Veratraldehyde (■) Veratric acid (▲) Vanillil alcohol (◆) vanillin (×) Vanillic acid (○) Protocatechualdehyde (□) Protocatechuic acid (△) Benzentriol (◇).

The E_{fb} of the p-type semiconductor electrode is close to the upper edge of the valence band. As the E_{fb} of the Cu_2O electrode we prepared is 0.22 V vs. Ag/AgCl, namely 0.42 V vs. NHE in an aqueous Na_2SO_4 solution, the lower edge of the conduction band can be regarded as −1.68 V vs. NHE assuming that the band gap of 2.1 eV. Thus, the photo-generated electrons could reduce the chemical species with more positive redox potential than −1.68 V vs. NHE. **Figure 18** shows the chromatograms of aqueous 10 mM Na_2SO_4 solution containing an initial concentration of 100 μM p-benzoquinone on the Cu_2O photoelectrode for different irradiation times. In this case, the potential of the Cu_2O electrode (geometric surface area: 1 cm^2) was held at −0.2 V vs.Ag/AgCl. The wavelength of irradiation light was 550 nm and the intensity was 4.8 $mW/cm^{2.}$. The solution was bubbled with N_2 gas. As shown in this figure clearly, the concentration of p-benzoquinone decreased and that of hydroquinone increased with irradiation time. The photocathdic reduction of p-benzoquinone to hydroquinone was confirmed. The current efficiency of this process was 95.8%.

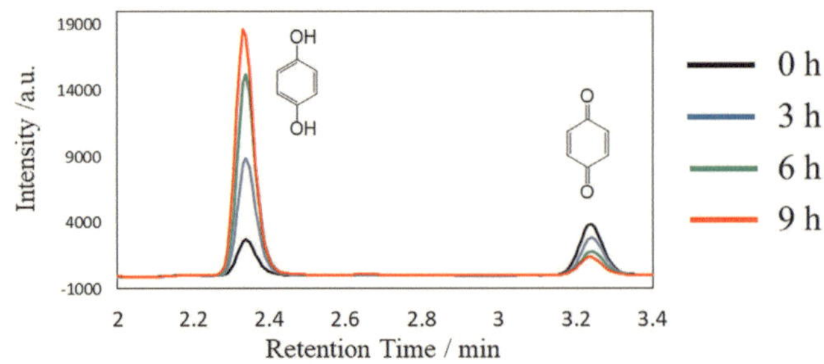

Figure 18.
Change of chromatogram with irradiation time on the Cu_2O electrode.

The occurrence of photocathodic reduction of 1,4-Naphthoquinone to 1,4-Dihydroxynaphthalene on the Cu_2O photoelectrode was also confirmed by the HPLC analysis. In this case, the current efficiency was 83.5%.

Lastly, we would like to mention the apparatus for water purification [13] as shown in **Figure 19a**. The photoelectrode is 30 sheets of the TiO_2 electrode (5×5 cm^2) prepared from anodic oxidation of Ti in nitric acid solution and heat treatment at 500°C. The counter electrode is a carbon cloth. The light source is four black lights

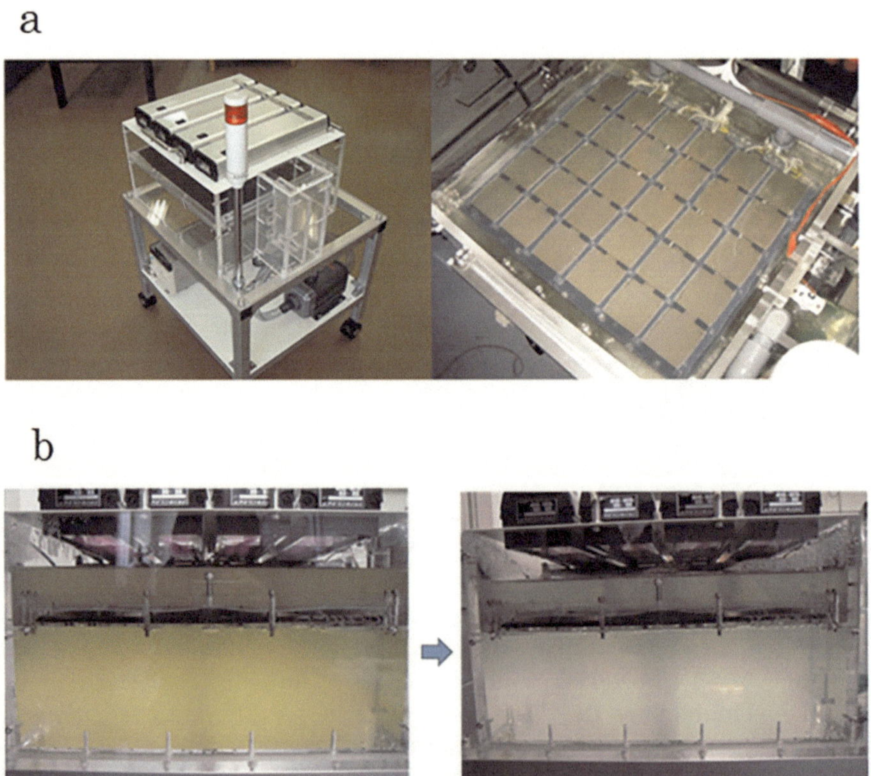

Figure 19.
The apparatus for water purification consisting of 30 sheets of titanium dioxide electrodes, carbon cloth counter electrode, four black lights, a power supply, and a stirring implement (a) and its performance for 20 L wastewater after 4 days of treatment (b).

(wavelength: 365 nm, intensity: 3.0 mW/cm^2). **Figure 19b** shows the test of performance of this apparatus for the waste liquid (volume: 20 L). After 4 days with an applied voltage of 3.0 V, the color of the liquid disappeared and The COD value of 120 decreased to a value of 20. Further application such as water purification in the field of food and medicine is expected.

5. Conclusion

We have been investigating the preparation, characteristics, and photoelectrochemical response to chemical species for semiconductor electrodes. The semiconductor films of TiO_2, ZnO, α-Fe_2O_3, and Cu_2O were prepared by electrochemical methods. From the Mott-Schottky plot for semiconductor electrodes in aqueous Na_2SO_4 solution, the values of flat-band potential and carrier density of each electrode were estimated. In the case of the Cu_2O electrode, a negative slope of the linear plot indicating the p-type semiconductor electrode was recognized. The Cu_2O electrode showed a cathodic photocurrent response under visible light irradiation. With regard to the photocurrent quantum efficiency of the ZnO electrode, both the 1200 °C-sintered electrode and the electrochemically deposited film electrode showed a high efficiency of over 80%. On the TiO_2, ZnO, and α-Fe_2O_3 electrodes, the photoanodic reaction of Veratryl Alcohol was analyzed by the HPLC method. On the TiO_2 and ZnO electrodes, the highly oxidized intermediates of Protocatechuic Acid and Benzentriol were formed, but these intermediates were not detected on the α-Fe_2O_3 electrode. There was also a difference in the appearance of intermediates between TiO_2 and ZnO electrodes. These reflect a difference in the reactivity of photo-generated holes to chemical species on the TiO_2, ZnO, and α-Fe_2O_3 electrodes. The phocathodic reduction of p-benzoquinone to hydroquinone and of 1,4-Naphthoquinone to 1,4-Dihydroxynaphthalene was confirmed on the Cu_2O electrode. The photoelectrochemical system by using α-Fe_2O_3 and Cu_2O is expected to apply to artificial photosynthesis.

Acknowledgements

I am grateful to Dr. D. Kodama, Mr. M. Itono, Mr. H. Mitsuyoshi, Mr. T. Kitagawa, Mr. A. Konemura, Mr. T. Nakamura, Mr. Y. Kanada, Mr. K. Miura, and Dr. Y. Kohno for experimental contribution in Photoelectrochemical Process on Semiconductor Electrodes.

Author details

Yasuhisa Maeda
Shizuoka University, Hamamatsu, Japan

*Address all correspondence to: tykmaeda@tees.jp

References

[1] Fujishima A, Honda K. Electrochemical evidence for the mechanism of the primary stage of photosynthesis. Bulletin Chemical Society of Japan. 1971;**44**:1148-1150

[2] Ohnishi T, Nakato Y, Tsubomura H. The quantum yield of photolysis of water on TiO_2 electrodes. Berichte der Bunsen-Gesellschaft fuer Physikalische Chemie. 1975;**79**:523-525

[3] Kennedy JH, Anderman M. Photoelectrolysis of water at α-Fe_2O_3 electrodes in acidic solution. Journal of the Electrochemical Society. 1983;**139**: 848-852

[4] Gerischer H, Heller A. Photocatalytic oxidation of organic molecules at TiO_2 particles by sunlight in aerated water. Journal of the Electrochemical Society. 1992;**139**:113-118

[5] Hoffmann MR, Martin ST, Choi W, Bahnemann DW. Environmental applications of semiconductor photocatalysis. Chemical Reviews. 1995; **95**:69-96

[6] Fujishima A, Hashimoto K, Watanabe T. TiO_2 Photocatalysis Fundamentals and Applications. Tokyo, Japan: Bkc. Inc.; 1998

[7] Bandara J, Mielczarski JA, Lopez A, Kiwi J. 2. Sensitized degradation of Chlorophenols on iron oxides induced by visible light comparison with titanium oxide. Applied Catalysis B Environmental. 2001;**34**:321-333

[8] Zhang Z, Hossain MF, Takahashi T. Self-assembled hematite (α-Fe_2O_3) nanotube arrays for photoelectrocatalytic degradation of azo dye under simulated solar light irradiation. Applied Catalysis B: Environmental. 2010;**95**:423-429

[9] Kment S, Hubicka Z, Krysa J, Zlamal M, Olejnicek J, Cada M, et al. On the improvement of PEC activity of hematite thin films deposited by high-power pulsed magnetron sputtering method. Applied Catalysis B: Environmental. 2015;**165**:344-350

[10] Maeda Y. Photoelectrochemistry of hematite. InrechOpen, Iron Ores and Iron Oxide Materials Chapter. 2018;**8**: 149-176

[11] Maeda Y, Itoh Y, Kodama D, Kohno Y. Photoanodic response of hematite electrode to citric acid in aqueous solution. Journal of Electroanalytical Chemistry. 2017;**785**: 166-171

[12] Kodama D, Uchiyama E, Kohno Y, Tomita Y, Kobayashi K, Maeda Y. Photodegradation of veratryl alcohol on a titanium dioxide photoelectrode in aqueous solution containing chloride ions. Journal of the Surface Finishing Society of Japan. 2015;**66**:420-424

[13] Maeda Y, Tsukakoshi H. Japan Patent No. 5093801. Manufacturing Method for Titanium Oxide Molded Material, Photocatalysis, Photoelectrode and Water Purification Device. 2012

[14] Morinaga Y, Kohno Y, Tomita Y, Kobayashi K, Maeda Y. Photoelectrochemical characteristics of iron oxide/polyaniline in aqueous acidic soluiton. Electrochemistry. 2011;**79**: 168-171

[15] Maeda Y, Morinaga Y, Tomita Y, Kobayashi K. Photoanodic response of iron oxide electrode in aqueous solution and its application to Pb^{2+} removal under visible light irradiation. Electrochimica Acta. 2009;**54**:1757-1761

[16] Maeda Y, Kitagawa T, Gotoh K, Tomita Y, Kobayashi K. Photoelectrochemical behavior of zinc oxide sintered electrode in aqueous solution. Transactions of the Materials Research Society of Japan. 2008;**33**: 997-1000

[17] Maeda Y, Morinaga Y, Kobayashi K. Photoelectrochemical behavior of iron oxide electrode prepared from thermal oxidation of iron. Journal of the Surface Finishing Society of Japan. 2007;**58**: 376-378

[18] Zhou M, Glushenkov AM, Kartachova O, Li Y, Chen Y. Titanium dioxide nanotube films for electrochemical Supercapacitors: Biocompatibility and operation in an electrolyte based on a physiological fluid. Journal of the Electrochemical Society. 2015;**162**:A5065-A5069

[19] Gomes WP, Freund T, Morrison SR. Chemical reactions involving holes at the zinc oxide single crystal anode. Journal of the Electrochemical Society. 1968;**115**: 818-823

[20] Fujishima A, Maeda Y, Honda K. Intrinsic quantum efficiencies of semiconductor photoelectrodes obtained by temperature measurement. Bulletin Chemical Society of Japan. 1980;**53**: 2735-2741

[21] Maeda Y, Fujishima A, Honda K. The investigation of energy efficiency of the ZnO semiconductor photoelectrode by photothermal measurement. Bulletin. Chemical Society of Japan. 1982;**55**: 3373-3376

[22] Maeda Y, Kazuyuki K, Kato T. Evaluation of the quantum efficiency of the photoanodic reaction on titanium dioxide in aqueous solution containing hydroxyl acid salts by a photothermal method. Journal of Electroanalytical Chemistry. 1997;**424**:213-216

[23] Izaki M, Omi T. Electrolyte optimization for cathodic growth of zinc oxide films. Journal of the Electrochemical Society. 1996;**143**: L53-L55

Chapter 6

Polymer Dispersed Liquid Crystal Smart Film Technologies: Overview

Canhan Sen, Berk Alkan, Omid Mohammadmoradi and Alpay Taralp

Abstract

Liquid crystal (LC)-based research and its technological output vary from daily-use personal electronics and flat panels to switchable optical devices such as sensors. Optical and dielectric anisotropy is a key attribute of LCs, imparting functionality and broadening the scope of smart film systems to such products. Among LC smart films, the polymer dispersed liquid crystal (PDLC) smart film depicts an electro-optical (EO) composite sandwiched by transparent conductive oxide electrode-coated polyethylene terephthalate (PET) films. LC orientation and optical transparency in the composite are readily tuned by altering the electric field. The competitiveness of such PDLC devices reflects its favorable response time, energy conservation potential, and manufacturing convenience, all attributes that are readily endorsed by smart home appliances and areas of architecture and the automotive industry. In response to unrelenting market demands, sustainable, energy-efficient, and "greener" PDLC variants have appeared. Particularly worthy of mention are systems featuring transparency at zero field (reverse-mode). Others boast very high energy efficiencies (%80). In this chapter, the science and technology of PDLC, reverse-mode PDLC, and related LC smart films will be reviewed with a highlight on fabrication methods and operating principles. Market potential and research prospects compared to non-LC smart film technologies will also be touched upon.

Keywords: liquid crystal (LC), smart glass, electro-optical (EO) properties, polymer dispersed liquid crystal (PDLC), polymer stabilized liquid crystal (PSLC)

1. Introduction

Since the discovery of liquid crystallinity in 1888 by Friedrich Reinitzer [1], potential applications and developments in science and technology have led to a thorough understanding of this remarkable phenomenon and its specific properties [2]. In contrast to the classic traits of soft condensed matter, LC behavior arises from mesophases existing between the solid crystalline form and isotropic liquid phase, which exhibit optical and dielectric anisotropy as a function of position and orientation. A plethora of registered examples [3] show such phase properties, including tobacco, many proteins, and even various soap solutions. Among these, only a selected few LCs specifically used in PDLC manufacturing and upcoming PDLC technologies and potential technologies will form the subject of this chapter.

Modern PDLC designs may be regarded as having evolved in response to economic growth, consumer demands, and improved LC manufacturing technologies. The first LC light valve device was patented in 1936 [4]. LC displays (LCD), flat screens, sensors, and many other LC-based technological products followed in the late 1960s [4, 5]. Among those LC-based smart films entering the market, the technology underlying PDLC-based films in particular was patented four decades ago [6–8]. With essentially no fundamental revisions thenceforth in terms of operating principle, modern PDLC designs can be physically described as a nematic LC droplet encapsulated in a polymeric matrix. The resultant film is sandwiched by plastic substrates coated along their inner face with transparent conductive oxides (TCO). Indium tin oxide (ITO) is the most common TCO coating used to contact the film, enabling potential use of the film as an electro-optically active device. A typical electro-optically active composite formed accordingly generally works under 60 V_{rms} AC bias at 50–60 Hz, hence bypassing infrastructural energy management issues and facilitating convenient deployment and integration into standard voltage platforms. Improvements to PDLC design constraints are still being shaped by novel interdisciplinary LC manufacturing techniques [5, 9], thin film deposition methods, transparent conductive oxide coating techniques [10–12], modern platform constraints, and customer needs.

After passing a period of relative inactivity due to patent disputes [13–15], PDLC manufacturing has experienced a substantial jump in worldwide interest. In particular, an increased smart glass market demand for esthetic, visually "comforting" products was encouraged by the strong performance of various manufacturers. Global climate/energy expenditure mitigation policies as well as the parallel development of complementary technologies such as electrochromic [16–19] or suspended particle devices [20–22] also fueled activities and growth. Indeed, the inevitable continual advancement of LC smart film technologies, both academic and industrially, has proven attractive and lucrative.

In keeping with global urbanization, %67 of world population will move to cities by 2050 (approximately 6.2 billion people). To meet housing demands, the current building count will likely double [23–25]. Already >30% of the global energy consumption and > 35% of carbon dioxide emissions are attributable to the routine operation of residential and nonresidential buildings (i.e., heating, ventilation, air-conditioning (HVAC), and artificial lighting). These values may exceed 50% by 2050 [26–30].

Given the obvious need to mitigate environmental as well as urban growth sustainability issues, the stringent implementation of environmental and energy policies, actions, and sanctions [26, 31–33] is clearly foreseen on a global scale. In addressing both environmental (e.g., climate change) and urban energy use concerns (i.e., HVAC energy consumption [34, 35]), it follows that old habits must die, making room for new construction concepts, innovative building designs, and mitigated material solutions, particularly along building exteriors [36, 37]. Indeed, the routine, if not legislatively mandatory practice of constructing energy-efficient buildings using recyclable construction materials and utilizing renewable energy, seems just on the horizon. A substantial part of this overall global strategy will be to incorporate active and passive smart glass solutions [38, 39] to selectively control the entry of daylight. For instance, when configured for cold weather, sunlight striking the smart glass surface will serve not only as a lighting source but also as a source of infrared heating. Conversely, when solar heating is unwanted, smart glass systems may be adjusted to block out the infrared part of the light spectrum from entering windows. In this way, dynamic light control technologies will permit not only the fine-tuning of transmitted light for visual and working comfort but also the selective passage or blockage of solar heating, thereby reducing energy consumed for HVAC [40, 41] purposes.

Windows clearly represent the weakest construction element of an exterior wall in terms of thermal insulation traits. Indeed, the heat flux ratio of an insulated wall compared to a single-pane window is on the order of 1:100 [42]. In spite of such compelling data encouraging the limitation of windows along external walls (and rooftops), the reality is that exterior window utilization nonetheless continues to proliferate both in terms of number and relative surface area. From an architectural standpoint, windows are esthetically pleasing and permit the exploitation of day illumination and reduction of artificial lighting, the latter consuming 14% of the total energy of EU countries and 19% of countries worldwide [43]. Moreover, sunlight, and an ability to view the outdoors, yields a strong psychological benefit on human behavior, health, and job performance [43–45]. Hence, window elements will continue to partake in building exteriors as a major design feature. Already certification and regulations such as the Green Building Rating system has led to the utilization of smart window techniques. It follows to suppose that future smart window systems will target the optimization of heat transfer, among other issues.

Among all commercially available, energy-efficient, and environmentally green smart glass technologies, the PDLC-based window can be manufactured and installed most cost-effectively into building exteriors. Indeed, modern variants of PDLC technologies fall into the scope of the European Union Net Zero Emission Building (NZEB) Directive as well as various International Energy Agency policies [46, 47]. Thanks to significant scientific and technical advances in smart and/or specialty nanomaterials, current PDLC systems feature improved dynamic light and heat flux control traits. Complementary smart glass systems such as electrochromic (EC) and suspended particle device (SPD) technologies also provide visual comfort and energy regulation [48, 49] features in buildings (see **Table 1**). Both are intrinsically limited, however, in that neither can be manufactured as cheaply or simply as PDLC systems.

The commercial smart glass systems mentioned in **Table 1** fundamentally differ in their mode of light control. While PDLC enables tint/haze control via light scattering, electrochromism (EC) and suspended particle devices (SPD) permit light control by light absorption and/or reflection. U-value and solar heat gain coefficients (SHGC) are two parameters describing energy transfer at window panes. Both coefficients are strongly related to the light control mechanism underlying SPD, EC, and PDLC. U-value, meaning thermal transmittance, describes the rate of nonsolar heat transfer through a material. It carries the units of W/m^2K. In contrast, SHGC describes the fraction of total solar energy transmitted through a window into a building. A low SHGC value arising from limited solar heat transmittance reflects strong shading traits. Such window attributes would be desired in hot climates in order to mitigate air conditioning costs during the day. In contrast, high SHGC values, reflecting weak shading traits, would be configured in devices used to facilitate the heating of building interiors in cold climates [50]. **Table 2** highlights the experimental and simulated thermal data of various PDLC window configurations used throughout the world. As shown, PDLC systems with differing characteristics permit the adjustment of near infrared (NIR) transmission, yielding different SHGC and u-values. The results also support the substantial capacity of PDLC windows to reduce energy consumption and CO_2 emissions compared to conventional window systems. In addition to mitigating energy consumption, another major benefit of PDLC is the economic advantage of

Attribute	Polymer dispersed liquid crystal (PDLC)	Electrochromic devices (EC)	Suspended particle devices (SPD)
Compatibility to Green Building Standards (LEED, BREEAM, etc.)	Yes – in progress	Yes	Yes
Color	Clear	Blue/Green	Dark Blue
Switching speed	5–15 ms	2–15 min	1–5 s
Privacy in "Off" state	Yes	No	Yes
Energy need for "On" state	Yes	No	Yes
Operating temperature	−20 – 70°C	−20 – 70°C	- 20 - 80 °C
Operating voltage	48–65 V AC	12 V DC	~100 V AC
Power consumption for switching	5 W/m^2	2.5 W/m^2	5–10 W/m^2
Power consumption for clear state	5 W/m^2	~0.2–0.4 W/m^2	5–10 W/m^2
Price range (USD)	250–600	1200–2400	1800–2500

Table 1.
A comparison of active smart glass systems sandwiched in glass panes.

downsizing the air-conditioning infrastructure and reducing the frequency of periodic maintenance.

Interestingly, the economic gains, merit, and sustainability of smart window systems have rarely been analyzed. In fact, this subject generally lacks sufficient recognition and understanding in the public eye. One study claimed that the high price and luxury perception of PDLC technologies compared to those of conventional glass systems are among the reasons why smart glass systems are not more widespread in the market [59]. Ironically, the costs of nonrenewable energy and environmental remediation will likely continue to increase for as long as fossil fuel combustion continues. Hence, the energy savings potential and short amortization period of PDLC is such that smart glass technologies should be viewed by consumers and builders as a necessity as opposed to a luxury.

Studies pitting the monetary and environmental costs of nonrenewable energy against the return on investment (ROI) period and life-cycle costs of PDLC [54] and EC [60–62] are limited. This lack of dissemination is unfortunate, as such studies would be required to inform and emphasize the importance and necessity of PDLC. It is hoped that this review may contribute to a better visibility and appreciation of PDLC, hence expediting its deployment and scope and prompting the routine and worldwide use of PDLC devices as an element of exterior building walls.

Compared to PDLC, EC fabrication plants with a typical capacity of 4 million m^2 cost over 100 million USD to construct and carry along high operational expenses (i.e., energy, maintenance, and qualified workforce). The use of energy-intensive high technology manufacturing equipment such as large physical vapor deposition, high-temperature annealing, and top-tier characterization/quality-control (QC) systems is another discouraging trait of EC, which markedly contrasts against the rapid setup capability and cheap and technically simple nature of PDLC manufacturing. In fact, PDLC manufacturing plants of the same output capacity cost less than 1 million USD. The low operational costs of solution-processed roll-to-roll manufacturing, which lies at the heart of PDLC, is equally compelling. The sales price of SPD technologies is

SHGC		u-value (W/m^2-K)		Configuration	Summary	Ref.
ON	OFF	ON	OFF			
0.252	0.22	1.666	1.666	8 mm PDLC - 12 mm air −6 mm low-e	PDLC double glazing reduced building energy consumption by 25.1% and raised daylight performance by 149.1% compared to a low-e double glazed building.	[51]
0.541	0.536	1.758	1.758	6 mm clear glass - 14 mm air - 6 mm clear glass - 23.1 mm air −13 mm PDLC laminated glass	PDLC overlaid along a conventional glass window showed a 22.4% energy reduction.	[52]
0.68	0.63	2.79	2.44	4 mm glass PDLC film −4 mm glass	The studies showed that the PDLC glazing yields low NIR transmission (44%) in the translucent state.	[53]
0.25	0.23	N/A	N/A	6 mm window glass-0.4 mm PDLC film	PDLC film glass accounts for 101.76 USD/m^2 annual air-conditioning cost mitigation.	[54]
0.63	0.42	N/A	N/A	Double glazed PDLC window	PDLC double glazing reduced indoor temperatures 1.2–3°*C* compared to conventional double glazing	[55]
0.68	0.63	2.79	2.44	4 mm glass-PDLC film-4 mm glass	PDLC showed potential to reduce HVAC energy loads (4.8–12.8%) and enhanced daylight performance	[56]
0.562	0.363	2.41	2.57	Double glazed PDLC window	NIR transmission varied from 42 to 24% depending on the choice of transparent and opaque phases. CO2 emissions were also reduced (32.5–89 kg CO_2/m^2).	[57]
0.53/0.64	0.39/0.49	5.1	5.0	Double glazed PDLC window	PDLC window systems reduced HVAC energy usage %22.35. CO_2 emissions fell.	[58]

ON: Transparent state of PDLC (electrically active); OFF: Opaque state of PDLC (electrically inactive); Ref: References; ***N/A:*** *Not available.*

Table 2.
Experimental and simulated dataset of PDLC glazing systems optimized for operation in different climate zones.

even more than EC, but the root cause of this high expense is likely related to the very low number of available suppliers utilizing propriety knowledge as opposed to the costs of establishing and running a manufacturing plant. Shown in **Table 1**, the current retail sales price of PDLC is 1/3 to 1/5 of the price of competing technologies. Given that PDLC production methods and raw materials are cheaply and relatively accessible, given that PDLC manufacturers are firmly and broadly established compared to EC and SPD manufacturers, and given that PDLC products can be easily installed without necessitating special training or equipment, it follows to reason that future PDLC technologies will contribute substantially to the energy efficiency of buildings. In fact, price trend comparisons strongly suggest that the market competitiveness of PDLC compared to the aforementioned other two commercial dynamic light and heat influx modulating systems will continue to increase even more over the course of the short and medium term. Factors underlying the favorable future prospects of PDLC include but are not limited to the perception of PDLC as a sustainable element of green growth and its potential for even better performance through continued R&D. Even now, PDLC performance has been further enhanced by the contribution of auxiliary constituents such as dyes, inorganic and organic particle substitutes, adhesive layers, and specific mesogens in the overall formulation. These more recent efforts have opened a door to specialized PDLC devices, which typically fall under one of the categories of polymer stabilized liquid crystal (PSLC) devices [63, 64], reverse-mode LC devices [65–68], dye and/or particle-doped LC devices [69–72], or guest-host systems [73, 74].

In the following passages, the fabrication of conventional and modern variants of PDLC systems as well as their chemical and EO operating principles will be reviewed. Additionally, specialized PDLC systems and their constituents will be overviewed and implications relating to performance will be assessed. It follows that a better visibility and appreciation of the science underlying the function of PDLC may expedite the deployment and scope of PDLC devices in energy conscious, "green building" projects.

2. Structure and characteristics of LCs as well as other constituents of PDLC technologies

2.1 General remarks

The LC journey sparked by Friedrich Reinitzer and Otto Lehman in 1888 was revitalized by fresh ideas rife with potential applications at a conference in 1965. The chemical structure, synthesis, application, and research of materials aimed at yielding LC products have undergone many revisions since then [75]. Among these, the chemical structure of LC has been the primary focus of research ever since the early years of investigation. LC materials have been subjected to a variety of methods to better understand aspects of their synthesis, characterization, modification, and performance. Many chemical compounds can exist in one or more LC phases. While each compound is chemically distinct in its own right, all LC materials share some common physico-chemical-optical traits. For instance, in spite of structural differences, the original and nearly most cited LC compound cholesteryl benzoate [76] responds and performs similarly to the currently most studied compound N-(4-methoxybenzylidene)-4-butylaniline (MBBA) (**Figure 1**) [77].

This section outlines and groups the various chemical compounds being used in PDLC smart glass production and comments on their function and interrelationship.

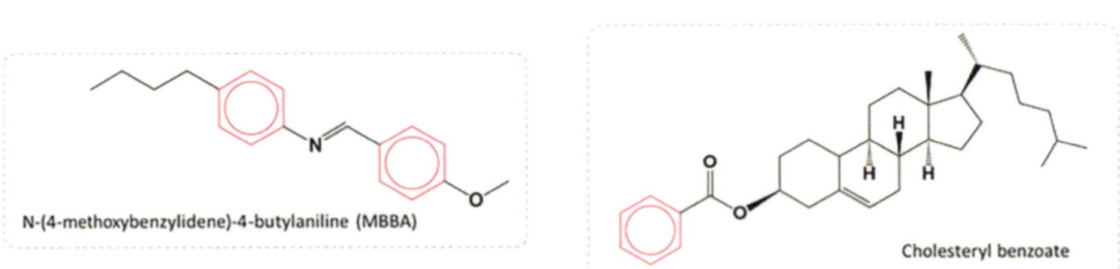

Figure 1.
Structures of N-(4-methoxybenzylidene)-4-butylaniline (MBBA) and cholesteryl benzoate.

2.2 Definition and distinguishing general traits of an LC

The LC is a thermodynamically stable 4th state of matter residing within a moderately broad temperature range. As might be surmised by the name, the LC state of matter relates to those special compounds, which can intrinsically display traits intermediate to that of ordered solids and isotropic liquids. On the lower end of this temperature continuum exists a phase transformation from LC to solid crystal while at the higher end is a second phase transformation from LC to isotropic liquid. Understandably, the LC state of matter has appropriately been dubbed "mesophase" where meso denotes "middle". Not every compound can behave like an LC, as most directly transform from solid to liquid. Compounds proceeding indirectly from the solid to the liquid phases via a mesophase are said to be mesophasic or to display mesophasic traits. These are dubbed "mesogens". The LC state is distinguishable in that the molecules (i.e., mesogens) tend to become aligned along a common axis. This axis, termed the "director" (*vide infra* for further discussion), shows the preferred direction of molecular orientation in mesophases.

As a point of reference, the isotropic liquid is a fluid displaying short-range order of its constituent molecules and a statistically random spatial distribution at longer ranges, whereas the solid crystal displays 3D long-range order with respect to the spatial position and orientation of its constituent molecules. Indeed, in true solids, molecules are highly ordered and display minimal translational freedom. As might be anticipated for matter lying intermediate to these states, the mesophase displays imperfect positional and/or orientational long-range ordering of its constituent molecules. The tendency of LC molecules to orient themselves contrasts with molecules in the liquid phase, which have no intrinsic order beyond the short range. As the orientational ordering of the LC state lies midway (i.e., 1D or 2D ordering) between the traditional solid and liquid phases, this state is appropriately described as "mesogenic" along the same logic underlying the origin of the term mesophase. Mesogenic and LC states are synonymous.

Compounds with an inherent ability to exist as a mesophase can be made to behave fluid-like, LC-like, and solid crystal-like by varying appropriate parameters, particularly temperature and pressure; depending on the conditions, the observable traits of an LC compound can vary substantially. That being said, most commercial applications deliberately select LC-active compounds, which retain their mesophasic traits throughout the operating temperature range of the device. It follows that in selecting LC compounds to meet a particular purpose, the thermal transition temperature is just one feature to consider, among many other constraints. The unusual traits of mesophasic compounds and their exploitation in LC devices stem not only from their ability to simultaneously display anisotropic crystal and liquid/fluid properties but also from their ability to be modulated by electric fields. It follows to speculate that the continued advancement of the PDLC industry will depend, in part, on developing and selecting intrinsically advantageous LC-active compounds, which microscopically prompt advantageous macroscopically observable traits in a particular glaze formulation.

2.3 Physical properties of the LC

The physical, or put more technically, the stereo-electronic properties of the LC, and to a substantial extent the non-LC environment surrounding the LC molecule (e.g., additives, residual solvent, etc.), all contribute in determining the energetics, behavior, and hence performance of a particular LC system. Since LC systems typically utilize high concentrations of LC-active species, the principle influence on performance generally reflects the intrinsic physical properties of the LC molecule itself. Given the ability of LC compounds to readily undergo various liquid crystalline phase transitions, a key criterion underlying the design of LC applications is the order parameter, S, which defines the degree of order across the boundaries of a phase transition system. Put more simply, S quantifies the arrangement or orientational order of the system, meaning the extent to which molecules deviate from the overall weight-averaged orientation of the system. S values typically range from 0.3 to 0.9, where S = 1 is perfect order. As might be anticipated, S values are highly temperature-sensitive. Expressed mathematically (Eq. (1)), the order parameter S is a second-rank symmetric traceless tensor usually ranging from 0 in one phase to nonzero in another. Eq. (1) requires the measureable parameters $\hat{n}$ and θ as input.

$$S = \frac{1}{2} < 3\cos^2\theta - 1 > \quad (1)$$

Exemplified in **Figure 2**, the angle θ between the principle axis of an individual LC molecule and the local director vector, $\hat{n}$, provides a simple means to quantify the orientational order of the LC phase. Again, in a hypothetically perfectly ordered phase, there would be no deviations from the local director and so S = 1. Conversely, S = 0 for a perfectly disordered situation, that is, isotropic phase. The phase order of nematic LCs will typically lie within the range of S = 0.5–0.7.

2.4 Anisotropic behavior of LC systems

A typical rod-like LC possesses uniaxial symmetry around the director, which produces shape anisotropy. When coupled to surrounding effects (e.g., an externally applied electric field), shape anisotropy indirectly permits the adjustment of many

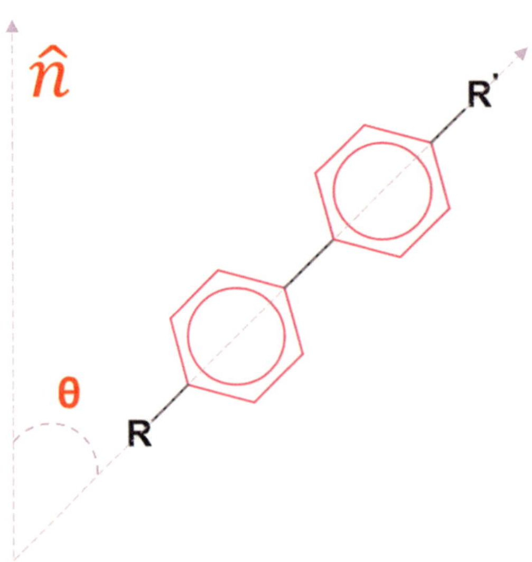

Figure 2.
Typical representation of an LC director vector and its angular deviation, θ, compared to the local director, $\hat{n}$, of the collection.

intrinsically anisotropic LC physical properties such as refractive index (n), dielectric permittivity ($\bar{\varepsilon}$), magnetic susceptibility (χ), conductivity, and viscosity.

2.4.1 Optical anisotropy

LCs are inherently optically anisotropy. Consequently, LCs possess optical birefringence; this is to say that LCs yield one refractive index for light polarized along the director and a second refractive index for light polarized perpendicularly to the director. One is termed the ordinary refractive index (n_0), whereas the second is the extraordinary refractive index (n_e). In the case of the ordinary refractive index, light travels with a constant velocity regardless of its polarization direction. In the case of the extraordinary refractive index, light travels at different velocities depending on its polarization orientation within the material. The overall effect is due to the anisotropic nature of the crystal structure, which results in two different refractive indexes for different polarizations. The difference between the ordinary and extraordinary refractive indexes allows researchers to understand the unique optical properties of anisotropic materials. By knowing these parameters, one can predict how light will interact with the material; for example, in the case of LC, one may surmise how light will be bent or polarized upon entering or passing through the material. This information is crucial in designing optical components like LC-based waveguides and polarizers, which exploit the birefringence trait of LCs to manipulate light in specific ways. Both inputs are needed to assess the optical birefringence, Δn, of an LC system (*vide infra* for further discussion). These inputs may also be used to define the average refractive index (n_{av}) of an LC (Eq. (2)):

$$n_{av} = \sqrt{\frac{1}{3}\left(n_e^2 + 2n_0^2\right)} \tag{2}$$

As defined in Eq. (3), the birefringence, Δn, is the difference between n_0 and n_e. This difference may be positive or negative, so the sign of Δn physically has meaning. In particular, a negative or positive value reports on the LC orientation with respect to the optical axis. As might be surmised, the birefringence criterion is important in many aspects, and it is used to classify LCs into different categories. For instance nematic (e.g., rod-like) LCs yield positive Δn values (i.e., $n_e < n_0$). These typically lie between Δn = 0.02 and 0.4. In contrast, chiral nematic/columnar and discotic LCs yield negative Δn values (i.e., $n_0 < n_e$).

$$\Delta n = n_0 - n_e \tag{3}$$

Shown in **Figure 3**, two geometric constructions, termed optical indicatrix or index ellipsoid, schematically represent opposite scenarios in which Δn takes on positive or negative values. Similar to a uniaxial crystal, the indicatrix herein is an ellipsoid in which the rotational and optical axes are also superimposed.

2.4.2 Dielectric anisotropy

The dielectric properties of LCs relate to the response of the LC molecule toward externally applied electric fields. As with many compounds, the permittivity of an LC is a material property, which determines the strength of interaction between a dielectric medium, here the LC, and an externally applied electric field. This extent of interaction

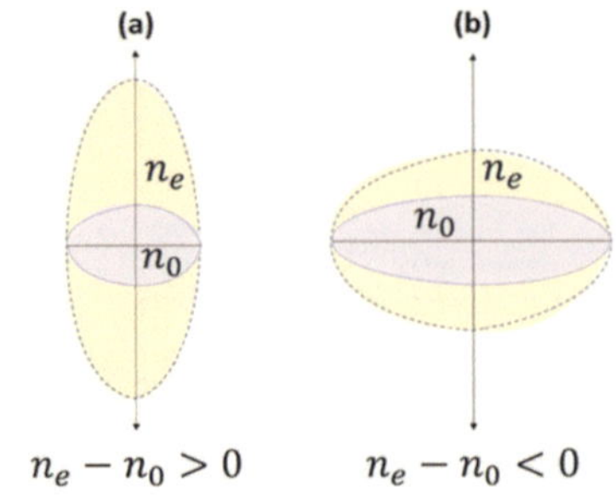

Figure 3.
Indicatrix depiction of optically (a) positive and (b) negative birefringence.

is determined by the inherent capability of the material to polarize upon exposure to an electric field. The observed polarization of the LC will thus reflect the vector sum of the external electric field and an opposing induced electric field of lesser magnitude inside the material. For LC materials comprised of nonpolar groups, electronic (notable at optical frequencies) and ionic polarization generally exists. For LC materials comprising polar groups, orientational polarization is also observed. Dielectric anisotropy is arguably the most important physical property of liquid crystalline compounds because it essentially determines the lower threshold voltages of LC displays.

Two principle permittivity coefficients can be defied by arbitrarily orienting the LC director, N, along the z axis of a macroscopic three-dimensional Cartesian system. The first principal permittivity obtained is $\epsilon_{/} = \epsilon_{zz}$, which is parallel to the director. The second director is $\epsilon_{\perp} = 1/2(\epsilon_{xx} + \epsilon_{yy})$, and it is perpendicular to N. The difference between the first and second principal permittivity coefficients gives the anisotropic permittivity value, $\Delta\epsilon$ (Eq. (4)):

$$\Delta\epsilon = \epsilon_{|} - \epsilon_{\perp} \tag{4}$$

When the dielectric anisotropy value is positive, the LC is oriented along the electric field, whereas when the value is negative, the orientation is perpendicular to the electric field (**Figure 4**). Quite simply, LC dielectric anisotropy is what led to the creation of PDLC smart windows, as LC molecules in the glaze could be reoriented in a controlled manner using electric fields. Indeed, given an appropriately chosen LC with positive/negative birefringence and suitable dielectric anisotropy, the alignment of the optical axis can be predicted and controlled via external stimulus.

2.5 A closer look at mesogens

As mentioned in the introductory section, a mesogen is an organic molecule, which displays LC properties. For this reason, mesogens may be regarded as disordered solids or ordered liquids, as each mesogen consists of rigid and flexible parts, which determine the order and mobility of the structure [78]. At the macroscopic length scale, a collection of compatible mesogens forms the LC system of a typical device. As might be anticipated from the traits expressed above, mesogens directly determine the properties of the LC device. The rigid parts align mesogen moieties along one direction and prompt distinctive shapes, which may resemble rods or disks. The flexible parts, which are usually alkyl chains, impart flexibility to the mesogen and hinder crystallization to a certain degree. It follows that a balanced combination of

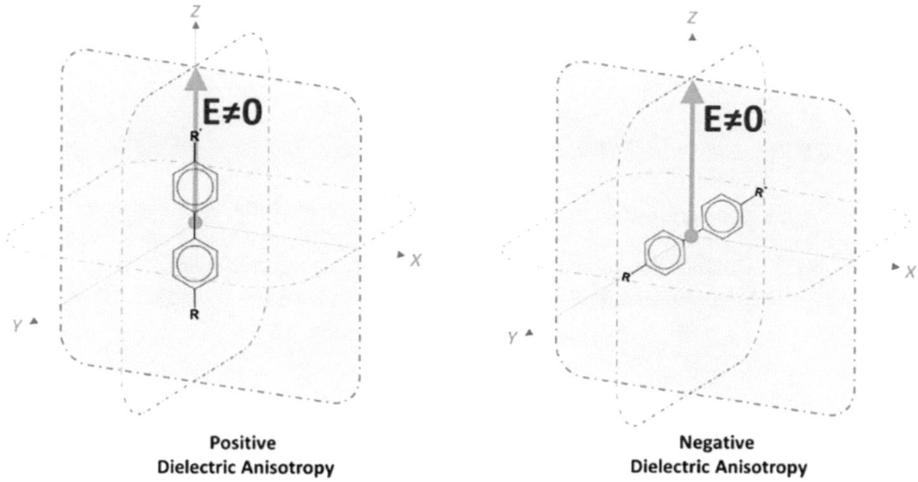

(a) (b)

Figure 4.
Orientation of LC displaying (a) positive and (b) negative dielectric anisotropy in an externally applied electric field.

appropriately rigid and flexible moieties in the mesogen should result in structural alignment and fluidity between individual LC molecules and a functional device formed by the collection of these LC molecules. On a lighter note, there is a tendency to interchangeably use the terms mesogen and LC, and for the most part, this practice is acceptable. To be absolutely accurate, however, one must bear in mind that the mesogen is truly the smallest molecular building block of an LC system, whereas the true LC is a semi-ordered collection or ensemble of many associated mesogens at a size scale much greater than that of a mesogen. That being said, it is always true that an LC molecule is a mesogen. It should also be noted that mesogens have evolved from their classic role as LC molecules to additional roles as "reactive" mesogens wherein they partake in polymer network formation as chemically reactive monomers.

In the context of PDLC technologies, the most important LC compounds fall into the thermotropic class. The three LC types of the thermotropic class are rod-like [79], discotic [80], and conic (bowlic) mesogens (**Figure 5**) [81]. The geometry of rod-shaped mesogens is long and anisotropic, allowing for preferential alignment in one spatial direction. Discotic mesogens are disc-like with a flat core and feature adjacent aromatic rings. Conic LC mesogens display cores, which are not flat, but rather rice bowl-like in appearance. Both discotic and conic LCs can engage in two-dimensional columnar ordering. The proper selection of LC candidates requires various considerations, which are typically based on physicochemical structure. For instance, low-temperature mesomorphic behavior, promoted by alkyl terminal groups, is generally sought as it would avoid metastable, monotropic liquid crystalline phases and prove technologically useful under typical operating conditions. In making a selection among the various available mesogens, a low melting point is usually preferable. Aromatic rings are also very much sought, as liquid crystalline behavior is strongly promoted by extended, structurally rigid, highly anisotropic shapes. Hence, many liquid crystalline materials are benzene derivatives.

The assortment of available mesogens and resultant LC systems is virtually endless. Complex compounds with integrated LCs also form a subject of active interest. These are designed and synthesized so that the mesogenic moiety of the molecule is

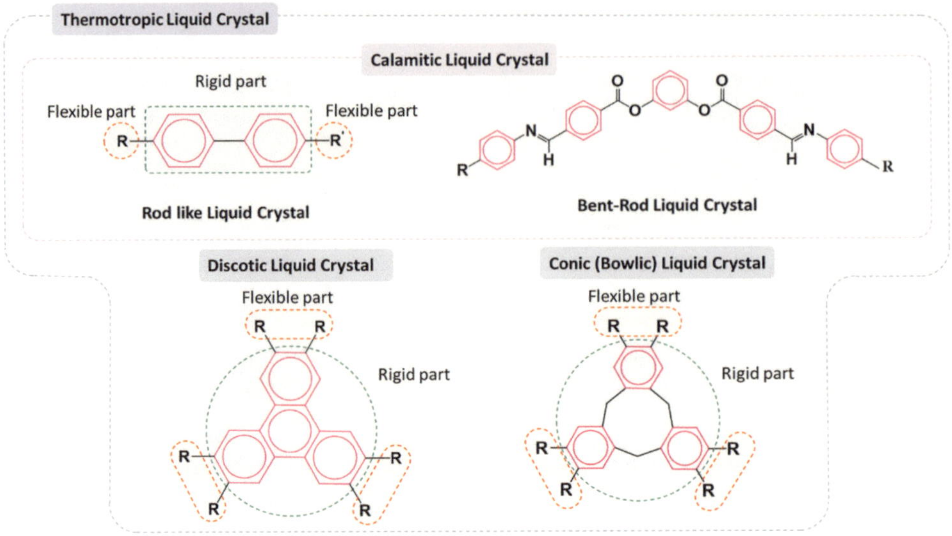

Figure 5.
Three thermotropic LC classes as exemplified by their characteristic mesogen types, which are demarcated into rigid and flexible moieties.

physically distant from the remaining non-mesogenic flexible linker moiety [82]. Such "reactive mesogen" molecules may be grafted, for instance, to the reactive side groups of a polymer chain utilizing the flexible spacer. Due to the limited scope of this chapter, only selected LC examples will be extensively discussed. To begin this discussion, a very commonly used commercial example of the nematic (rod-like) LC type is shown below. Termed E7, this LC formulated by Merck is the most frequently studied and used nematic LC in normal-mode PDLC-based smart windows [83]. E7 is composed of 4 hematogenic compounds depicted in **Figure 6**. Pure E7 has two transition temperatures, namely, a low-temperature glass transition at T_g = −62°C and a high-temperature nematic–isotropic transition at T_{NI} = +61°C.

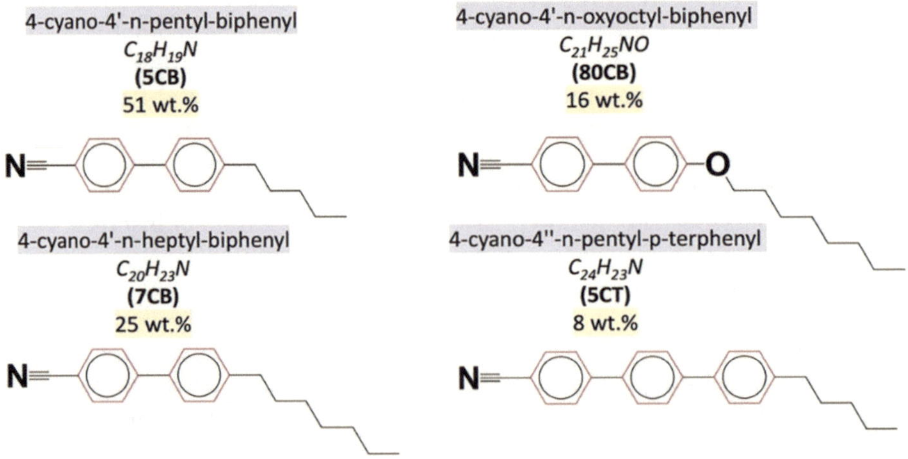

Figure 6.
Components of the Merck E7 LC formulation with empirical formula, chemical structures, and relative weight percentages shown.

2.5.1 The LC mesophase

As the temperature drops, the isotropic phase, which displays neither preferential positional nor orientational order, forms a nematic phase. Within a specific temperature range, the LC assumes a low-viscosity liquid of either a transparent or translucent appearance. In general, the isotropic phase displays short-range order and highest thermal mobility. Lowering the temperature yields long-range positional and orientational order. The LC mesophase lies within these extremes and thus reflects traits intermediate to those of long-range orientational and short-range positional order. Depending on the temperature, orientation, and position of the LCs, several mesophases may potentially form. Nematic, chiral nematic (i.e., cholesteric or twisted nematic), and smectic mesophases are the most commonly studied types. These are also utilized in smart windows manufacturing. The nematic phase is the most common phase among the calamatic or rod-shaped LCs. The orientation and positioning of LCs with respect to the local director ($\overline{n}$) is what distinguishes one mesophase from another. For instance, in the nematic mesophase, the LC molecule is oriented along the local director. In comparison, in any of the possible smectic phases, a different alignment angle (smectic C) and/or positional ordering (smectic A, B) is observed. In the chiral nematic mesophase, the general orientation of an ensemble of LCs mimics the helical structure of a DNA molecule. **Figure 7** summarizes these attributes of the various mesophases.

When these mesophases form layers with an inter-layer distance (d), they produce structures with the highest packing density and greatest long-range orientational and positional order.

Chiral dopants are optically active LC materials used to create helical structures in a host nematic LC mixture [35]. In keeping with this theme, cholesteric mesophases can be obtained by doping a nematic LC with the appropriate chiral agents (**Figure 8** [84]. The most distinguishing and desirable outcome of obtaining a cholesteric mesophase is the ability to selectively reflect light. Here, the reflected wavelength, λ_{max}, is determined by n, the average refractive index, and P, the helix pitch (Eq. (5)):

$$\lambda_{max} = nP \tag{5}$$

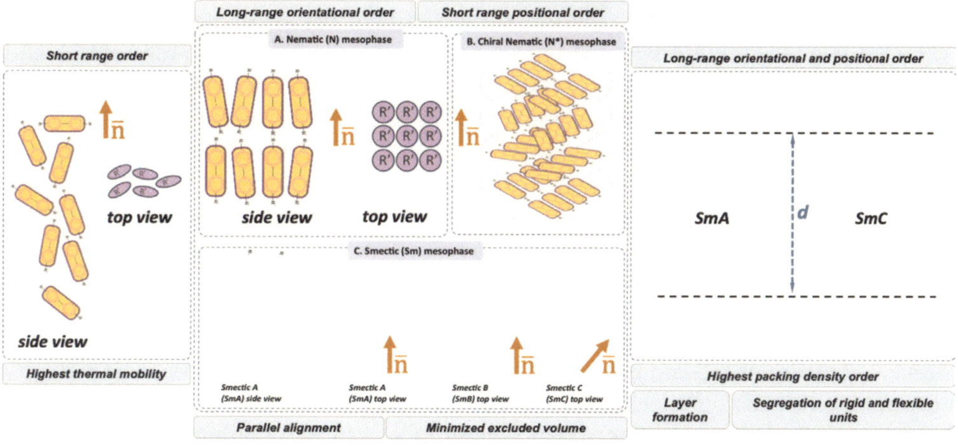

Figure 7.
Summary of orientational and positional ordering of LCs with respect to optical direction ($\overline{n}$). Thermal mobility descreases in proceeding from left to right.

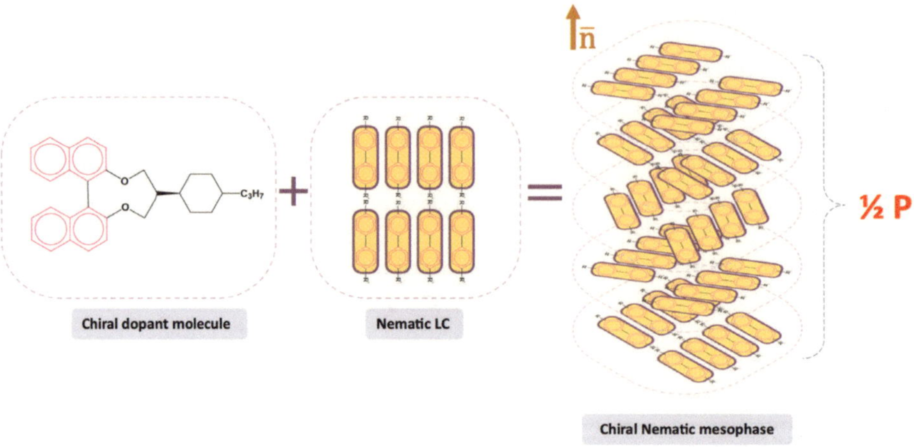

Figure 8.
Illustration of the formation of a chiral nematic mesophase via the addition of chiral dopant to a nematic LC.

The size of the cholesteric mesophase formed is half of the helix pitch. The helical pitch obtained will depend on the molecular structure and concentration of the chiral agent.

2.6 LC based film compositions

To permit the industry-scale production and processing of LC-based smart device films, a prepolymer component with appropriate physicochemical traits is selected to (i) physically disperse/support the LC component prior to curing; (ii) impart solid state LC immiscibility and liquid state LC miscibility, as well as film compliancy during curing; and (iii) yield acceptable transparency, amorphicity, and mechanical traits in the cured product [83]. The principle role of the resultant polymer matrix is to (i) encapsulate the LC as nanodroplets, (ii) direct the orientation of LC molecules in contact with the alignment layer (generally reverse-mode PDLC (R-PDLC)), and (iii) physically stabilize the LC mixture (i.e., PSLC). In the case of direct-mode PDLC (D-PDLC), the most conventional means of polymer matrix formation is photopolymerization. In this approach, LC-compatible monomers and photo-initiators are used, forming a polymer matrix. The influence of various matrix monomers and photoinitiators on LC device performance has been characterized using the example of commercially available Merck E7, a well-known mixture of four mesogens [85–87]. Such work on E7 served as a useful reference, encouraging many research groups to conduct similar investigations on industrial or academically significant alternative systems.

In this section, we will focus on the LC response to electric field stimuli. Conventional PDLCs (also known as direct-mode, D-PDLC, or normal-mode, N-PDLC) are the most popular compounds used in smart windows production. Overall, the formulation is a mixture of LC and polymer, the latter typically being greater than >40 wt%. In producing N-PDLC devices, mesogens are initially blended into a polymer matrix. The insolubility of these low molecular weight LCs within the polymer matrix leads to a phase separation, yielding nanosized LC droplets embedded along the matrix. The mixture obtained initially appears milky white, given the mismatch between the refractive indices of LC and polymeric matrix [88]. Said mixture is then sandwiched between two transparent conductive electrodes (with PET films typically used as backing). Further cross-linking (curing) of reactive groups within the mixture is achieved via UV exposure or the application of heat.

In N-PDLC, electric fields applied across the electrodes are used to toggle the device between the opaque (Voltage OFF) and transparent (Voltage ON) states. Typically, AC voltages (50–100 V_{rms}) at frequencies of less than 100 Hz are employed [89]. When a sufficiently high voltage is applied so as to align the LC director in the direction of electric field, the refractive index difference between LC and polymer matrix minimizes, allowing incident light to enter the system so as to yield an optically transparent mixture. This minimum refractive index value is highly dependent on the chemical structure of the LC and polymer. Work is ongoing to further diminish this value.

2.6.1 Reverse-mode PDLC (R-PDLC)

Several reverse-mode PDLC techniques have been reported previously. As might be surmised by the term "reverse", reverse-mode PDLC (R-PDLC) is transparent during the voltage OFF state and opaque during the voltage ON state [90]. The starting materials used to prepare the R-PDLC largely depend on the fabrication method. Various fabrication methods combined with different LCs have yielded satisfactorily performing R-PDLC devices [91]. To date, various strategies have formed the basis of the performance of these devices. Among the list of notable approaches, some of the most successful include the use of large liquid crystals and a polymer matrix formed out of reactive mesogenic network precursors [92], the surface energy modification of polymer droplets [93], the utilization of dual-frequency responsive liquid crystals [65], the technique of nematic emulsion polymerization [94], the use of rough and supportive alignment layers [95], and, finally, the introduction of an anisotropic polymer matrix [96]. From a technical viewpoint, the underlying rationale of these approaches has been to alter the surface energy of positive/negative dielectric anisotropy-type LCs using active monomers, to disperse double frequency-responsive LCs in a polymer matrix, to vertically align LC molecules along rough glass substrates, to disperse LCs into an anisotropic polymer matrix, and to replace passive polymeric matrix elements with low amounts of mesogenic networks (PSLC based systems). Among these approaches, the alignment method [95] is the most conventional and practical strategy to mass produce LC-based smart windows. This method is also conveniently general in that LCs with either positive [92] or negative anisotropic [97] permittivity can be used to produce the R-PDLC devices [90]. PSLC is the only strategy that has shown commercialization potential in the smart glass industry. The performance of some of the other strategies fell short of meeting commercial expectations in terms of haze, contrast ratio, and driver voltage, whereas the remaining ones required complex power electronics or were unsuitable for roll-to-roll film production methods. Given the above, the remainder of this section will focus on PSLC devices, which utilize an alignment layer on transparent conductive electrodes and feature a relatively low weight percentage content of polymer compared to other PDLCs.

2.6.2 Polymer stabilized liquid crystal (PSLC)

In normal-mode PDLC, an unfortunate consequence of utilizing high polymer loadings (>40 wt%) during preparation is that high voltages are required to toggle the resulting device (48–65 V_{rms}). In this respect, normal as well as reverse-mode PSLC provides a viable and convenient alternative to the requirement and dangers of using high voltages. In PSLC, the prepolymer content is reduced to such an extent (<5 wt%) that the device transitions from transparent (Voltage OFF) to opaque (voltage ON) states at only $\sim 30\ V_{rms}$. In preparing a typical PSLC, a small amount (2–10 wt%) of monomer (typically a reactive mesogen) and photoinitiator are dissolved in a low molecular weight liquid crystal. The resultant precursor mixture is injected between ITO-glass or ITO-PET substrates and cured under UV illumination or via heat. Subsequent cross-linking of monomers affords a myriad of LC ensembles enclosed inside a polymer network. During polymerization, the uncured mixture is made to assume the "voltage ON" alignment between two conductive electrodes using an alignment layer to prompt reorientation. The so-called alignment layer, from which this method was coined, refers to a thin layer of material, typically a commodity plastic, which is pre-coated/pre-melted along the electrode surfaces and subsequently imparted a particular micron-scale surface topology using a brush affixed to a specialized "rubbing" machine. It is this regular micron-scale topology along the plastic layer that associates and enforces appropriate orientational traits to that very thin fraction of liquid crystals directly layered along the electrode surface. Collections of disoriented liquid crystals in the uncured mixture contact this topologically oriented layer and follow suit, orienting themselves in the same direction as the initially aligned body of liquid crystals via what may be loosely described as a templating effect. This effect is transmitted layer-by-layer in a direction normal to the electrode until the entire glaze layer assumes the same alignment. Finally, a small amount of reactive mesogens present within the uncured mixture are chemically cured via UV irradiation or heat, affording a cross-linked polymer network extending from one plastic surface to the other. Like a flexible glue, this network serves to hold all substances in place and provides much needed mechanical support against delamination. As might be surmised by this description of events, the correct initial orientation of liquid crystals and hence the correct performance of the PSLC device critically depend on the quality of the alignment layer [98]. The alignment layer influences device performance directly, via surface interaction with the LCs, and indirectly, via setting up the circumstances by which a proper polymer network comprising cured reactive mesogens is established. The most compatible and generally utilized plastic material for the alignment layer in both industry and academia is polyimide (PI) [99, 100]. Just like normal mesogens, which form the LCs, a most distinguishing aspect of reactive mesogen monomers compared to polyacrylate, polymethacrylate, or epoxy groups used in classical PDLC systems is that reactive mesogens can be oriented or aligned by their interaction with an electric field, other liquid crystals, or alignment layers such as polyimide due to their liquid crystal structure. An oriented polymer network formed from reacted mesogens can be achieved by applying bias to the substrates during UV curing or by incorporating brush-rubbed polyimide or nylon alignment layers along the ITO surfaces. Once oriented to assume a specific morphology, the reactive mesogens can be polymerized, affording a desired PSLC structure and imparting a targeted function. The desired structure is typically a polymer network arranged in parallel or perpendicularly to the substrate surface. Just as in the case of the alignment layer itself, the presence and quality of a mesogenic polymer network on the local alignment of LC molecules also determines the operating mode and performance of the PSLC system.

Compared to conventional N-PDLC, the view angle transparency of PSLC is higher due to the relative scarcity of the polymer network, which greatly limits the influence of any refractive index mismatch existing between the polymer and LC. A lower polymer content also limits the total area delineated by the polymer-LC interface. Since the interfacial area is limited, the influence of stereo-electronic "anchors" or energy terms, which otherwise would resist the reorientation of LCs, is also limited. With a low number of anchors in the collective whole, the device can operate at substantially lower working voltages compared to conventional PDLCs. Monomers used in today's PSLCs are often based on terminally functionalized acrylate or methacrylate groups. These moieties are chemically attached to both sides of the mesogenic core, typically via a flexible alkyl chain, which provides a measure of distance and isolation. To exemplify the above, the chemical structure of a monomer (and necessary photoinitiator) used in the fabrication of a PSLC device is given in **Figure 9** [101].

2.7 Doped LC devices

The addition of organic and inorganic substances to a mixture of LC and polymer is of one the well-established routes to modify the device characteristics. The main approaches used to tailor various features of the LC-polymer mixture include dye-doping (guest-host) [102], nanostructured materials-doping [103], and cholesteric (chiral) molecules-doping [84] of liquid crystals.

Guest-host LCs (GH-PDLC) may be viewed as a composite of dye dissolved in a host LC [102]. Generally, a dichroic dye molecule, which is a geometrically anisotropic (rod-like), is used to absorb light of a specific wavelength or range. Being anisotropic, the operative feature of the dye is that it absorbs incident light better from one orientation compared to the others. Given the intimate guest-host-type interaction between dye and LC molecules, it follows that both will reorient together upon the application of and external electric field. Dyes rotating together with their host LC will

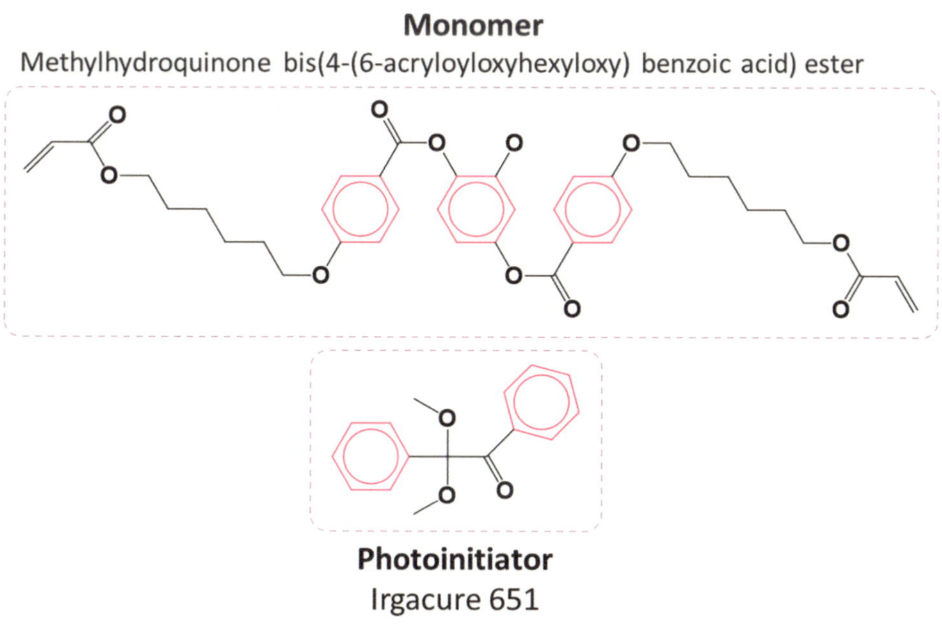

Figure 9.
Example monomer and photoinitiator used to produce commercial PSLC.

absorb light so as to favorably modulate the transmission of light. That being said, many traits are to be considered in choosing a dye. The dyes selected should absorb light best along the molecular chain. Potential candidates should also display a high dichroic ratio, a high-order parameter, a high chemical stability, and good solubility in the LC molecule (but not in traces of unreacted monomer or in the polymer matrix). In general, dyes possessing a positive anisotropic permittivity value are preferred. In such dyes, the absorption transition dipole lies along the long molecular axis. For this reason, incident light polarized perpendicularly to the long axis of the dye is weakly absorbed. In contrast, when light polarization lies parallel, light absorption is fastest. As an example, a dichroic azo dye possessing the N=N functional moiety absorbs specific light wavelengths more along one direction than along the others (**Figure 10**).

The modification of LCs using various nanostructured dopants has been vastly cited in scientific literature. For LCs used in smart windows, the general aim of LC doping has been to enhance the EO properties. Nanostructures including nanoparticles [104], nanorods [105], metallic nanoparticles (MNP) [106], Carbon nanotubes (CNT) [107], magnetic nanoparticles [108], and quantum dots (QDs) [70] were all reported to impart a positive effect on the performance of the LC devices.

Alongside modification methods using doping with nanostructured materials, other alternative methods were reported to effectively impact the EO properties of various LCs. In addressing the vulnerability of LCs in harsh environments, the notion of stabilizing physical traits and/or improving optical traits by tailoring the chemical structure of the parent compound proved promising. For instance, microencapsulation of LCs with a protective layer of compatible polymer to yield a core-shell structure has been widely employed as a protective strategy to preserve the essential traits of LC materials. Also, Reza et al. [109] EO properties of novel side-chain liquid crystalline polymers (SCLCPs) in which LC polymers featuring mesogenic coumarin-based side-groups were fixed along the PMMA backbone via flexible bridging alkyl spacers. A similar strategy utilizing highly compatible Polyurethane (PU) was reported to protect cholesteric LCs and to improve their EO properties, yielding functionality at lower fabrication expenses [110].

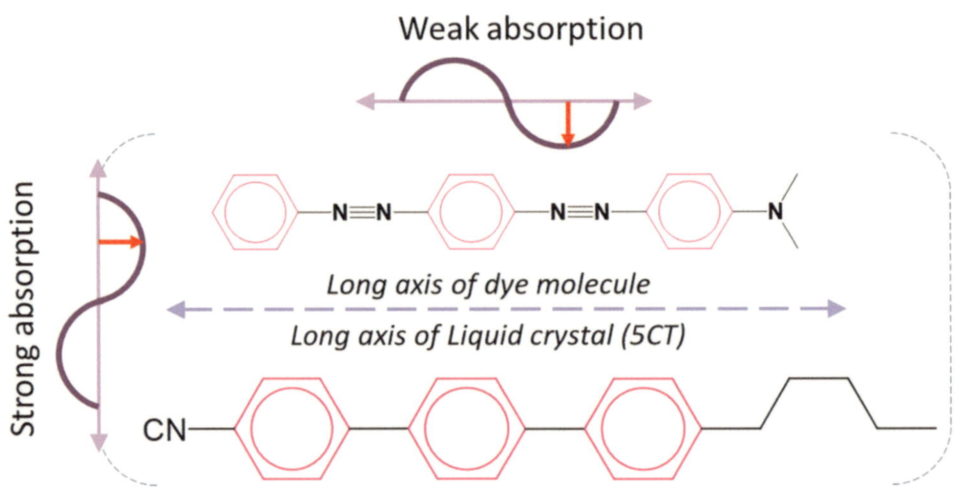

Figure 10.
Orientation of Duest dye along the LC long axis and depiction of orientation-dependent absorption.

3. PDLC displays

3.1 General remarks

The first example of a PDLC system was reported in 1986 by Doane et al. [111]. The classic PDLC structure may be described as a film comprising nematic liquid crystals dispersed in a solid polymer matrix as spherical or elliptical-shaped droplets [112–115]. Again, an externally applied electric field controls the optical transmittance of the film. The potential difference to create the electric field is typically established by sandwiching the PDLC film between glass/PET substrates coated with a transparent conductive surface such as ITO [116]. It should not be surprising that seminal contributions along these lines prompted the emergence of the smart glass industry [14, 15]. Significant refinements notwithstanding, the conventional PDLC systems still operate by essentially the same principles as those of the original design. To ameliorate haze, for instance, modern designs strive to match the extraordinary refractive indices (n_e) of the LCs and the refractive index (n_m) of the isotropic matrix polymer to high accuracy [112–115]. In the absence of any electrical potential across the PDLC film, the alignment of the LC directors is set by the LC-matrix interface. This alignment differs for each LC droplet because the droplets and surrounding matrix structure are heterogeneous [117, 118]. To a first approximation, the directors of the droplets are randomly oriented. As such, incident light, traversing from the polymer matrix to the LC, generally experiences scattering at the matrix-LC interface, as the LC in question has a refractive index value between n_o and n_e [116, 119]. The specific value, which is subject to change on the basis of crystal orientation and interfacial integrity, is in all cases marginally different from the refractive index of the matrix [120]. The light scattering mechanism is shown in **Figure 11a**. Scattering occurs at all visible light wavelengths. Scattering is random due to the spherical shape of the LC droplets, which, again, are randomly dispersed in the heterogeneously structured polymer matrix. As such, the PDLC film appears hazy when switched off [66]. When a

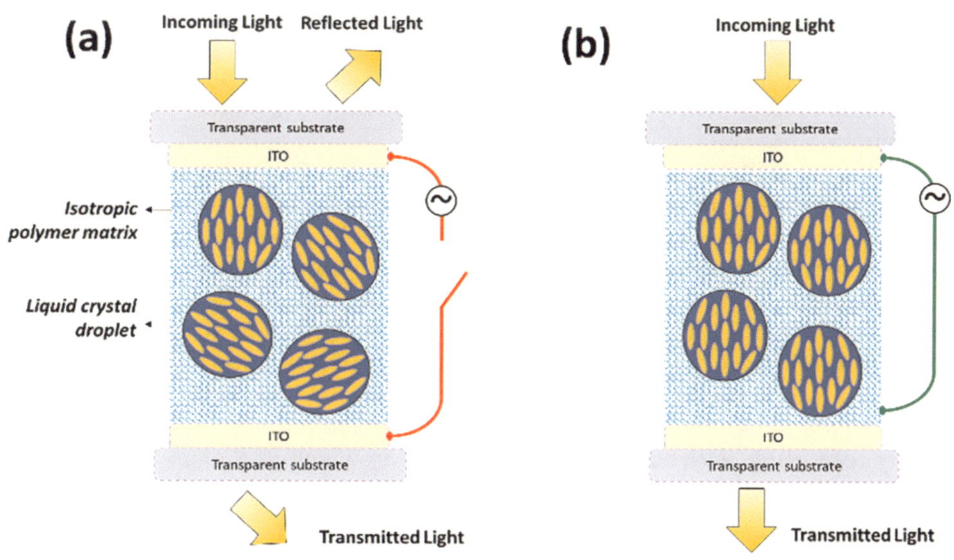

Figure 11.
Operating principle of a PDLC composite fixed between ITO electrodes.

potential is applied to the PDLC film, directors having positive dielectric anisotropy in the LC droplets align in the direction of the electric field (**Figure 11b**). In this well-known EO phenomenon, the refractive indices of the matrix polymer and LC droplets match, so the PDLC film looks transparent [66, 111, 119, 120]. This is to say that light striking the outer surface transmits through the PDLC film without scattering [75].

When a potential is applied, the ideal PDLC film should display high transparency and clarity (i.e., low haze) from a broad range of viewing angles, a millisecond respond time, and a low threshold voltage [15, 66, 121]. These ideals reflect current commercial concerns and demands. Happily, modern PDLC films are not far from these ideals in that commercial PDLC products typically exhibit 80% total optical transmittance and an acceptable base 3.5% haze value when switched on [122–125]. While haze increases with the viewing angle, this angular dependency is one technical problem, which has slowed the more widespread use of classical PDLC films [126–128]. Work is ongoing in our laboratories to further ameliorate these technical aspects, and new-generation marketable materials are forecast in the near future. Currently, PDLC films are being used to a limited extent in industrial sectors such as construction, energy, and transport, to name a few [129–131]. Although the automotive sector is a particularly important market for the smart glass industry, this haze limitation coupled with very strict consumer desires has thus far limited the use of PDLC films to sunroofs [124, 132]. It should be emphasized that the discussion above relates to "normal-mode" PDLC, whereof the device is transparent when powered up and hazy when switched off. However, new-generation reverse-mode PDLC versions are currently being developed in-house to overcome the continuous power consumption issue, and the fruits of these efforts in the auto industry are forecast. Academic reports on reverse-mode PDLC began during the 1990s, and commercial applications as smart glass products began to emerge in 2021 [93, 133, 134].

Even now, basic reverse mode devices have been developed and marketed by a small number of manufacturers. The current industry standard for smart glass, that is, electrochromic devices, is routinely touted for being more energy efficient, and to date, this fact has led to their more widespread adaptation in motor vehicles. Still, projections strongly support the notion that reverse-mode PDLC systems can be refined to such an extent in the near future so as to rehash the shortcomings inherently related to electrochromic devices and thereby negate favorable perceptions of ED technologies in this sector.

Reverse-mode PDLC was primarily developed to address energy concerns in sectors requiring long-term transparency of normal-mode PDLC devices. Indeed, the use of normal-mode PDLC has also been restricted by the high potential and thus high power consumption requirements needed to maintain transparency [48, 70, 131, 135]. For classical (i.e., normal-mode) PDLC films to exhibit maximum optical transmittance, an AC potential of 48 V – 75 V_{rms} must be applied [135–138]. AC is the preferred signal, as DC is known to prompt electrolysis of LCs and degradation of ITO coatings. In fact, first-generation PDLC products can show potential requirements as high as 110 V_{rms} [137–140]. Understandably, AC signals and high voltage ratings are not very compatible with automotive applications in which 12 V or 24 V DC power supplies have become standard. Moreover, all automotive devices requiring $<50\ V_{rms}$ become subjected to directives demanding the use of complex power electronics, prompting a substantial cost increase. The problem of incompatibility is not as severe in the building sector, as high voltage 50–60 Hz AC power lines can be adopted to standard building needs including architectural glass using simple inexpensive step-down transformers [141]. Still, the architectural glass industry is becoming more energy conscience by the day, so the high voltage requirements of PDLC remains a formidable impasse to rectify [57, 142, 143]. From a commercial viewpoint, normal mode PDLC devices are inherently limited due to the EO nature of the composite films and high power requirements [141, 142]. These EO properties are influenced by the LC-matrix interface, the anchoring effect at this interface, and the size and shape of the LC droplets [144–147]. The most challenging problem of normal-mode PDLC is the unwanted presence of haze. While haze is minimal in viewing the device straight-on (i.e., normal to the plane of the device), haze becomes increasingly pronounced in viewing the device at increasingly oblique angles compared to the ideal straight-on perspective. The root problem of haze and its angular dependency relates to light scattering. Scattering originates from 2 issues. Firstly, some LC molecules may inadvertently become polymerized with prepolymer molecules. This side reaction slightly alters the refractive index of the isotropic polymer matrix, so haze will occur at every viewing angle due to refractive index mismatches involving no and np. The second issue relates to the isotropic refractive index of the polymer matrix. The refractive index of the polymer matrix is constant and equal in all directions. In an ideal PDLC system, the ordinary refractive index of the LC will match the refractive index of the polymer matrix. However, LCs display birefringence, and more importantly, the ordinary and extraordinary refractive indices differ from one another. As a result, when an electric potential is applied to the PDLC film, the two refractive indices of the LC will only match the matrix refractive index in the direction of the electric field. This results in substantial haze and haze variability at all "off-normal" viewing angles. While strategies exist to increase the miscibility of the LC in the matrix, thus ameliorating the first issue, no strategies founded in theory exist to address the second issue, which originates from material properties. That being said, practical solutions are reported. For example, haze was reduced to an optimal level by adjusting the size of the LC droplets and by using external polarizer filters. Still, these approaches negatively affect the total transmittance and threshold voltage. Optimization has met with partial success; however, a need to perform beyond the intrinsic limitations of normal mode PDLC is what prompted the development of LC-matrix polymer composites, which operate in reverse mode [66, 148, 149]. Such reverse-mode composites naturally consume less power. More importantly, selected variants also exhibit very low haze when transparent, regardless of the viewing angle. Among these, the PSLC device, functioning in reverse mode (R-PSLC), achieved commercialization in the smart glass industry [67, 94, 150–152]. Indeed, PSLC technology is a candidate, which not only can replace PDLC films in that it displays none of the standard technical shortcomings known to PDLC; moreover, PSLC also exhibits technical advantages, which permit additional applications outside the scope of PDLC and desirable features compared to many other commercial smart glass alternatives [92, 153–155].

3.2 Production methods of conventional PDLC films and technical facts

In 1984, Fergason et al. patented a microemulsion approach to fabricate an EO film system consisting of LC droplets dispersed in a polyvinyl alcohol (PVA) matrix [6, 156]. Such LC systems predated PDLC and were classified as the nematic curvilinear aligned phase (NCAP) [157, 158]. The NCAP technology lies outside the scope of this review and has been mentioned just as a point of comparison and historical reference. While NCAP films may look and performed similarly to PDLC, the two systems differ from each other in terms of microstructure [159, 160].

The first academic report of a conventional "true" PDLC system appeared 2 years later in 1986 (J. W. Doane et al.) [111]. Many reports followed this seminar contribution in which nematic LCs with positive dielectric anisotropy were dispersed in a polymer matrix either in interconnected pores or as separate droplets [66, 161–163].

Thermally induced, solvent-induced, or polymerization-induced phase separation techniques are used to produce PDLC films. For practical and economic reasons, commercial PDLC films are preferably manufactured using roll-to-roll (R2R) processes. In R2R processing, polymerization-induced phase separation (PIPS) is understandably advantageous and preferred over solvent or heat-induced phase separation techniques for reasons of simplicity, convenience, and limited processing steps [15, 164, 165]. In this technique, LCs and miscible prepolymer molecules are combined, yielding a PDLC precursor mixture. For nematic type LCs, low molecular weight monomers or oligomers are typically preferred [136, 149, 166, 167]. The precursor mixture obtained is injected between ITO-coated PET substrates, and the sandwiched precursor material is polymerized to yield a flexible film. The LC and prepolymer combination is so chosen that polymerization gradually prompts immiscibility of the two components. As polymerization continues, a gel-like film is obtained [168]. During gelation, the LC undergoes phase separation, precipitating out as droplets dispersed in the polymer matrix [169, 170]. Polymerization is continued until all low molecular weight components are cured and incorporated into the polymer, effectively separating them from the LC phase. The overall polymerization event may thus be arbitrarily divided into an initial gelation stage, in which the film consolidates and phase separation occurs, and a secondary "purification" stage, in which the LCs become purified of any residual reactive monomers [166]. In conventional PDLC systems, the weight ratio of LC to the polymer matrix is a maximum of 80/20 and typically 60/40 [66, 162, 171, 172]. The first PIPS studies utilized epoxide-based curing. In these studies, precursors were prepared by mixing epoxy monomers, a curing agent, and nematic LCs. Polymerization was carried out at temperatures moderately above room temperature [66, 166, 173]. As with the expected temperature dependence of many chemical reactions, the curing temperature was found to affect the size of the LC droplets [174]. Currently, free radical polymerization and UV curing is the most commonly exploited chemical basis of the PIPS approach in producing commercial PDLC films [66, 148, 162, 172]. Compared to epoxide curing, this approach benefits from less shrinkage, zero heat input, and precise control of the UV dose and intensity, which permits fine adjustment of the LC droplet size [117, 175, 176].

3.3 Modern PSLC systems

PSLC composite film systems address and resolve much of the classical PDLC haze problem by deliberately utilizing relatively small amounts of polymer matrix in the composite film. A negligible LC-polymer interfacial surface area is thus obtained. The refractive index mismatch experienced at such a limited interface affords a virtually imperceptible total haze. Early scientific reports on the development of PSLC structures were published in 1991 by R.A.M. Hikmet et al. [153]. In this groundbreaking work, the polymer content of the total film was 10 wt.% or less [150]. Improvements followed such that modern-day PSLC systems utilize approximately 5 wt% of polymer. In spite of the low loading of polymer, the LCs nonetheless still align themselves by interacting with the polymer network [151]. While polyacrylate, polymethacrylate, polyacrylate-thiol, or epoxy systems form standard matrices in PDLC, bifunctional photo-reactive mesogenic monomers are typically used in PSLC to yield the polymeric network [66, 150]. As might be surmised, photo-reactive mesogen monomers are highly advantageous compared to acrylate, methacrylate, or epoxy prepolymers used to construct PDLC systems in the sense that reactive mesogens have an LC structure, and thus, they can be oriented or made to align by their interaction with an electric field, other LCs, and alignment layers such as polyimide (PI) [64, 67]. The reactive mesogens typically utilized are large, rod-shaped molecules terminated at both ends by acrylate or methacrylate groups [177]. As monomeric mesogens exhibit LC properties, it follows to correctly reason that their incorporation into a polymer matrix will also bestow substantial LC traits to the matrix [178–180].

Normal and reverse mode PSLC structures with different polymer network alignments have been previously reported. Both nematic and cholesteric (i.e., chiral nematic) LC phases can be used to prepare PSLC structures [64, 150, 181, 182]. Those phases based on cholesteric LCs have been subclassified under the category of polymer-stabilized cholesteric texture (PSCT) [150, 183]. Devices operating in normal and reverse modes are available as both phase types. Although many different PSLC-based systems have been reported, three basic structural models have been explained below.

Model 1 - Reverse-mode PSLC: The working principle of a film prepared using a positive nematic LC is shown in **Figure 11**. A PSLC precursor mixture is prepared by dissolving reactive mesogen monomers in a nematic LC, and a trace of photoinitiator is added to the mixture. A polyimide layer is coated along the conductive surface of the sandwiching substrates, and the polyimide is subsequently rubbed, affording a regularly patterned, unidirectional alignment layer. Now functionally competent, the alignment layer ensures that the LC molecules making up the bulk of the PSLC precursor mixture will align themselves parallel to the substrate surface (**Figure 12a**) [184]. Interestingly, this alignment of LC molecules also prompts the mesogens, which are dispersed throughout the LC molecules, to similarly align. UV curing follows, and an appropriate film thickness maintains the alignment during this period. Once the film is UV-cured, a polymer network will have formed with an orientation parallel to the substrate surface (**Figure 12b**) [154]. The LC molecules remain aligned parallel to the surface of substrates without an external electric field (**Figure 12a,b**) [64, 67]. In this case, the system is transparent because undesired light scattering from the PSLC cell is very low at all angles. The small amount of polymer network in the

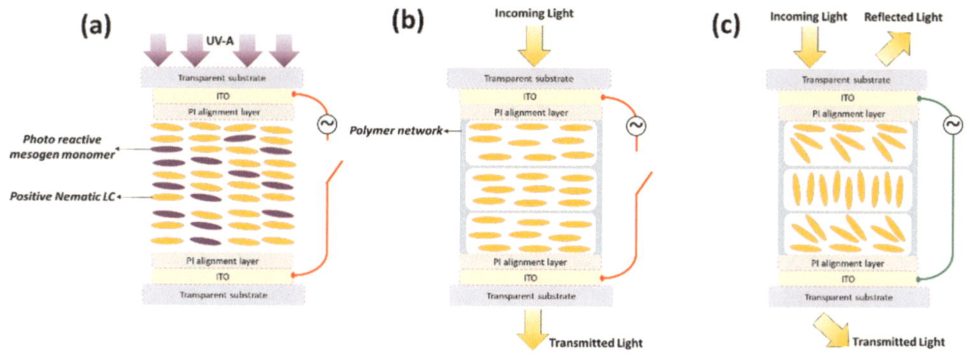

Figure 12.
Reverse-mode operation of a PSLC device showing: (a) UV-curable precursor sandwiched between alignment layer-coated electrodes; (b) UV-cured film and LC alignment in the absence of an external electric field; and (c) UV-cured film subjected to an externally applied electric field showing perturbed LC alignment.

PSLC cell reaps a refractive index mismatch between polymer and LC exhibiting negligible haze. When a small electric potential is applied to the film, LC molecules in direct contact with the alignment layer and polymer network remain parallel to the substrate surface because adsorptive forces predominate. In contrast, LC molecules within the interior of the film tend to align in the direction of the electric field. The difference in refractive index between perpendicular and parallel LCs imparts a substantial haze to the film, causing it to become opaque (**Figure 12c**) [185].

Model - 2 Normal-mode PSCT: A PSLC precursor mixture is prepared by dissolving reactive mesogen monomers in a chiral nematic LC, and a trace of photoinitiator is added to the mixture. The precursor mixture is injected between two conductive transparent substrates and cured under UV light (**Figure 13a**). An alignment layer is not utilized. Instead, the film curing is conducted under AC potential [64]. In this case, the reactive mesogen monomers adopt a perpendicular alignment with respect to the substrate surface. Polymerization yields perpendicular polymer growth, and the resultant polymer network separates the chiral nematic LCs into distinct regions (**Figure 13b**) [64, 150]. The phase of these regions is sometimes termed a focal conic texture. In the absence of an electric current, chiral nematic LCs are randomly dispersed within the focal conic texture, causing incident light to scatter. The non-energized state of the normal mode will thus be opaque. When a sufficiently strong electric field is applied to the device, the LCs assume a homeotropic phase (**Figure 13c**). As such, incident light perpendicular to the cell passes through without scattering [186, 187]. Light scattering (i.e., haze) at all other viewing angles is very low, thanks to the negligible optical mismatch between the LC and small amount of polymer network.

Model 3 - Reverse-mode PSCT structure: A PSCT precursor mixture is injected in between conductive transparent substrates, which have been furbished with a competent alignment layer. Again, reactive mesogen monomers directly contacting the alignment layers are aligned in parallel to the substrate. However, this time, the alignment layer also arranges the chiral nematic LC molecules perpendicularly to the substrate surface. In so doing, some to the mesogens intimately contacting the LC molecules follow suit. The dynamics of alignment between alignment layer, mesogens, and LCs is such that when the PSCT mixture is UV-cured, a polymer network oriented parallel and perpendicular to the substrate surfaces is obtained, imparting a multicellular appearance to the film cross section (**Figure 14a**) [91, 150]. In the absence of an electric potential, the film is transparent because focal conic structural units are aligned in the same direction [64]. When some electrical potential

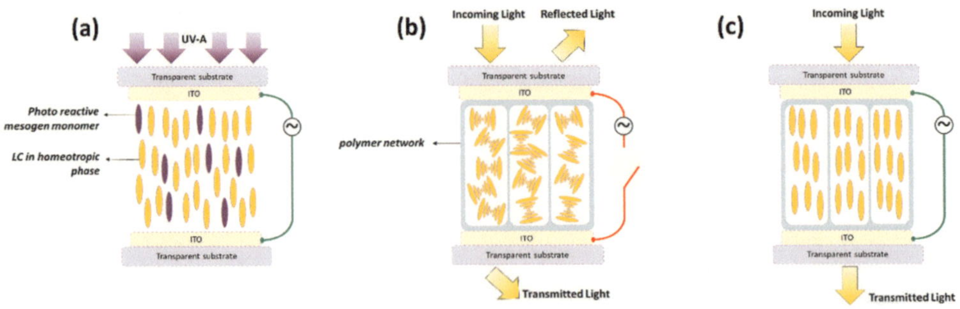

Figure 13.
Normal-mode operation of a PSLC device showing: (a) UV curing of a PSCT precursor mixture under the influence of an external electric field as opposed to alignment layers; (b) LC orientation of cured film in the absence of an electric field; and (c) LC orientation of cured film under the influence of an applied potential.

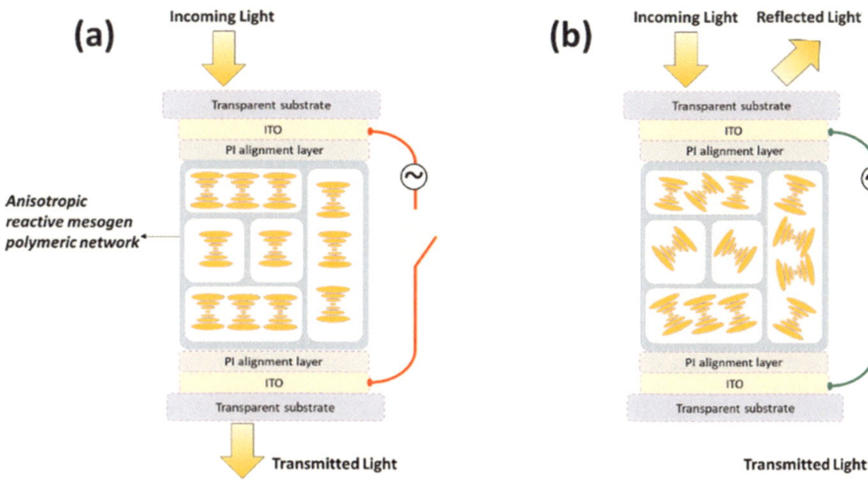

Figure 14.
Reverse-mode operation of a PSCT film illustrating (a) LC orientation in (a) the absence and (b) the presence of an externally applied electric field.

is applied to the device (**Figure 14b**), focal conical structures directly contacting the alignment layer retain some element of their perpendicular alignment to the surface of the substrates. Focal conical structures not directly contacting the alignment layers align more randomly due to their interaction with the electrically responsive mesogenic polymer network [67]. Hence, the net orientation of LCs becomes a weight-averaged contribution of many energy terms working together. In summary, orientational differences of LCs near the substrate surface and those in more inner regions of the film prompt the complete scattering of transmitted light, creating a strong haze in the film [64, 188].

Selected examples of normal and reverse mode PSLC and PSCT systems as well constraints underlying their preparation are given in **Table 3**. LCs and reactive mesogens well established in the LCD sector are listed along with their commercial codes and abbreviations [195]. As some reactive mesogens have been identified using more than one commercial code, synonyms are also provided where appropriate. Reactive mesogen content, film thickness, and cure conditions have also been provided. The data listed is typical and provides a useful reference from where to begin, but these starting materials and parameter values are in no way written in stone, and they are certainly subject to change on the basis of need. Given that temperature and cure conditions (i.e., UV intensity and dosing time) alone can markedly affect the microstructure, morphology, and EO performance of a PSLC system [63], the readership is encouraged to experiment further. For instance, the polymerization cure time has been empirically found to relate to the fraction of prepolymer in the curable mixture. The polymerization time of a PSLC film with a 6 wt.% prepolymer loading typically ranges between 30 and 60 min [64]. Other reports have documented that droplet sizes increase at higher UV intensities and diminish when longer cure times are applied at weaker light intensities. In general, PSLC and PSCT films are fabricated thinner than conventional PDLC films in order to extend the stability of the LCs. As a side-benefit to this added longevity, the thinner film and limited interaction between LCs and a small amount of polymer network effectively lowers the threshold voltage requirements [67, 90, 91].

Model	Liquid Crystal (LC)	Reactive mesogen (RM)	RM (%)	Film thickness (μ)	Curing Cond.	Ref.
Reverse mode PSLC	E7 (PN)	C6H (C6/RM1)	3–10	6	UV curing @ 25°C	[92]
Normal- and reverse-mode PSCT	E31 (PN) CB15 (CA)	RM206	2	8	UV curing 4 mW/ cm^2 - 1 h @ 25°C	[189]
PSLC	5CB (PN)	R6M (RM82/ HCM009/C6M)	10–90	2–7	UV curing @ 25°C	[153]
Normal-mode PSCT	E48 (PN) CB15 (CA)	BMBB6	2	10	UV curing @ 25° C + 100 V BIAS	[150]
Normal-mode PSLC and PSCT	ZLI4389 (PN) E48 (PN) R1011 (CA)	BAB6 BMBB6	4.5–7	15	UV curing @ 25° C + 100 V BIAS	[154]
Normal-mode PSCT	E48 (PN) S6 (SA) ZLI-4572 (CA)	R6M (RM82/ HCM009/C6M)	2	14	UV curing - 20 min	[190]
Normal-mode PSCT	SLC-1717 (PN) S811 (CA)	R6M (RM82/ HCM009/C6M)		20	UV curing 1.1 mW/ cm^2–30 min @ 29°C	[191]
Reverse-mode PSLC	ZLI 4788 (NN)	PAHB AHBB	14	25	UV curing 0.5 mW/ cm^2–10 min @ 45°C	[152]
Reverse-mode PSCT	E48 (PN) CB15 (CA) CPCA-1	RM257 (HCM008)	1–6	5–10	UV curing	[192]
Reverse-mode PSLC	MLC-2136 (PN) MLC-2053 (PN)	ARLM-002	5	5–25	UV curing @ 25°C	[193]
Reverse-mode PSLC	ZLI-4792 (PN) MLC-2136 (PN) E7 (PN) RDP-86483 (PN)	RM257 (HCM008) ARLM002 UCL-017	5	10	UV curing 20 mW/cm^2–5 min	[194]

PN: Positive dielectric anisotropy; NN: Negative dielectric anisotropy; SA: Smectic A; CA: Chiral agent; CPCA-1: Chiral photochromic dopant.

Table 3.
Some PSLC and PSCT contributions.

Lastly, the nature of PSLC technologies provides flexibility and a diversity of functional tools to researchers striving to further develop smart-glass systems. In the simplest sense, new EO devices can be derived from the strategies utilized in the basic PSLC models listed above. For instance, vertical alignment layers can be used to replace electrical bias during polymerization, or particle-substituted nematic or chiral nematic phase LCs can be selected to prompt an unusual trait or performance attribute [196]. Even dichroic or non-dichroic rod-like dyes have been mixed with nematic or chiral nematic LCs [197–199]. The dyes become oriented via electrostatic forces created within neighboring LCs molecules and mesogens. Once cured and energized, films containing dyes will display interesting traits because the dyes will remain oriented along the direction of the LCs even when the system is toggled by an external electric field. Indeed, there are several reports of dye-doped PSLC and PSCT cells [199–201]. Shown in **Figure 14**, a film containing dye-doped chiral nematic LCs

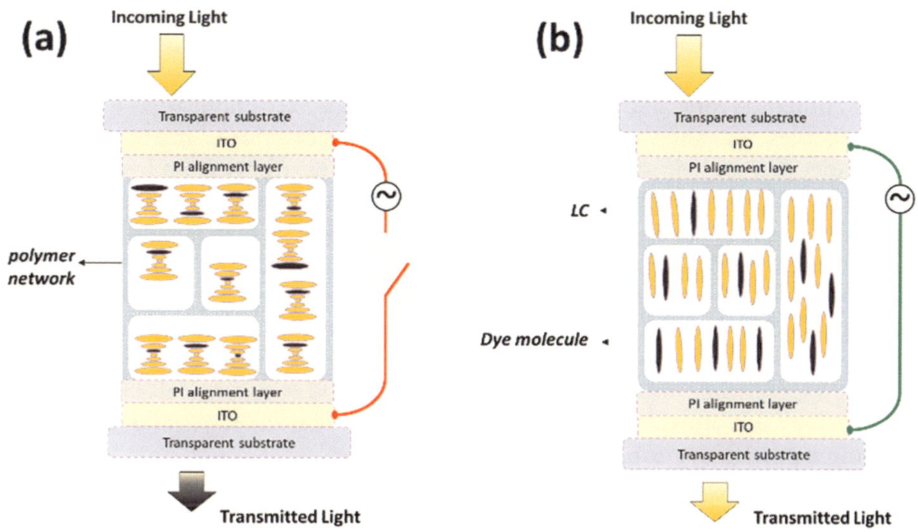

Figure 15.
A dye-doped PSCT showing (a) the absorption of a particular wavelength of light in the absence of an electric field and (b) the full & free transmission of light under an applied electric field.

perpendicularly aligned to the substrate depicts a good example of an advanced normal mode PSCT with potentially new applications. In the absence of an electric field (**Figure 15a**), dye particles, LC molecules, and the alignment layer are parallel. Incident light travels without scattering, but a certain wavelength is nonetheless absorbed by the dye molecules. For this reason, the PSCT film is clear but tinted [202]. When a sufficiently strong electric field is applied, the LCs reorganize into the homeotropic phase, and the rod-like dye particles retain their parallel alignment to the LC molecules. Now, the PSCT cell exhibits marginally higher clarity and not tinted in this state [203].

This tinting attribute of dye-doped PSLC and PSCT structures is significant and substantial in the sense that it enables LC-based smart glass technologies to compete against SPD and EC approaches in the highly lucrative facade and automotive glass markets. Moreover, future PSLC technological advances can potentially spark the manufacture of even more original smart glass products displaying desirable attributes such as selective transmittance and adjustable IR reflectance [204–207].

4. Concluding remarks and prospects

Clearly, PDLC appears destined to become a most instrumental and unavoidable construction material of window technologies. This unwavering interest in PDLC is a principle factor driving efforts to continually develop better films. Despite the aforementioned progress and highly encouraging developments, current LC technologies still require subtle yet important improvements to meet client and energy demands. With the basic theory of operation being long established, efforts have focused by and large on resolving more practical problems such as robustness, power consumption, haze, longevity (i.e., stability), and tinting. Still, the basis to prepare impressive next-generation variants nonetheless appears to rest upon better understanding and addressing microscopic factors such as the physicochemical and optoelectronic nature of the various individual components making up the film; the state of the various interfaces separating immiscible components; the surface microstructure and topology of alignment layers; the adsorption, desorption, and reorganization energy terms of LC molecules; and the thermodynamic and kinetic behavior of the system at macroscopic, mesoscopic, and microscopic length scales. Apart from lowering production and operating costs, a most foreseeably general yet crucial technical target might be to optimize the interaction energetics of molecular participants making up the film, thereby facilitating the kinetics of optical change at all ambient temperatures. Such an undertaking will require a precise knowledge of the thermodynamic and kinetic behavior and packing interactions between various LC molecules, matrix polymers, and additives. It follows to reason that many impressive improvements can potentially be realized within the next decade or two by systematically identifying, characterizing, and optimizing all physicochemical and electrooptical contributions underlying a particular problem.

Author details

Canhan Sen[1]*, Berk Alkan[1], Omid Mohammadmoradi[1] and Alpay Taralp[2]

1 Teta Glass Technologies Ltd., Istanbul, Turkey

2 Prosyn Science LLC, Cheyenne, WY, USA

*Address all correspondence to: canhansen@tetacam.com

References

[1] Reinitzer F. Beiträge zur Kenntniss des Cholesterins. Monatshefte für Chemie und verwandte Teile anderer Wissenschaften. 1888;**9**:421-441

[2] Mitov M. Liquid-crystal science from 1888 to 1922: Building a revolution. ChemPhysChem. 2014;**15**(7):1245-1250

[3] GmbH LP. LiqCryst Datenbank für Flüssigkristalle [Internet]. LCI SYSTEMS GmbH. Available from: https://www.lci-systems.com//liqcryst/ [Accessed: May 16, 2023]

[4] Castellano JA. Liquid Gold - the Story of Liquid Crystal Displays and the Creation of an Industry. Singapore: World Scientific Publishing Co. Pte. Ltd; 2005. 315 p

[5] Koide N, editor. The Liquid Crystal Display Story-50 Years of Liquid Crystal R&D That Lead the Way to the Future. Japan: Springer; 2014. 426 p

[6] Fergason JL. Encapsulated Liquid Crystal and Method. United States of America: US4435047A; 1984. p. 15

[7] Doane JW, Chidichimo G, Vaz NA. Light Modulating Material Comprising a Liquid Crystal Dispersion in a Plastic Matrix. United States of America: US4688900A; 1985. p. 19

[8] Wu B-G, Doane JW. Liquid Crystalline-Plastic Material Having Submillisecond Switch Times and Extended Memory. United States of America: US4671618A; 1987. p. 12

[9] Geelhaar T, Griesar K, Reckmann B. 125 years of liquid crystals—A scientific revolution in the home. Angewandte Chemie International Edition. 2013; **52**(34):8798-8809

[10] Coutts TJ, Mason TO, Perkins J, Ginley DS. Transparent conducting oxides: Status and opportunities in basic research. Proceedings of the Electrochemical Society. 1999;**1999**(99): 274-288

[11] Minami T. Transparent conducting oxide semiconductors for transparent electrodes. Semiconductor Science and Technology. 2005;**20**(4):S35

[12] Ellmer K. Past achievements and future challenges in the development of optically transparent electrodes. Nature Photonics. 2012;**6**(12):809-817

[13] Hakemi H. Polymer-dispersed liquid crystal technology 'industrial evolution and current market situation. Liquid Crystals Today. 2017;**26**(3):70-73

[14] Hakemi H. Polymer dispersed liquid crystal (PDLC) "industrial technology and development in Europe.". Molecular Crystals and Liquid Crystals. 2019; **684**(1):7-14

[15] Hakemi H. Industrial development of plastic PDLC: Is there a future? Liquid Crystals Today. 1998;**8**(3):7-12

[16] Bechinger C, Ferrere S, Zaban A, Sprague J, Gregg BA. Photoelectrochromic windows and displays. Nature. 1996; **383**(6601):608-610

[17] Lampert CM. The world of large-area glazing and displays. In: Switchable Materials and Flat Panel Displays. Vol. 3788. SPIE; 8 Oct 1999. pp. 2-11

[18] Rosseinsky DR, Mortimer RJ. Electrochromic systems and the prospects for devices. Advanced Materials. 2001;**13**(11):783-793

[19] Somani PR, Radhakrishnan S. Electrochromic materials and devices:

Present and future. Materials Chemistry and Physics. 2003;**77**(1):117-133

[20] Slovak SM, Chen XP, Saxe RL. SPD Light Valves Incorporating Films Comprising Improved Matrix Polymers and Methods for Making Such Matrix Polymers. USA: US7791788B2; 2006

[21] Vergaz R, Sanchez-Pena JM, Barrios D, Vázquez C, Contreras-Lallana P. Modelling and electro-optical testing of suspended particle devices. Solar Energy Materials and Solar Cells. 2008; **92**(11):1483-1487

[22] Leinberg S, Kisand V, Šutka A, Saal K, Lõhmus R, Joost U, et al. Switchable optical transmittance of TiO2 submicron-diameter wire suspension-based "smart window" device. Optical Materials. 2015;**46**:418-422

[23] United Nations. World Population Prospects 2022: Summary of Results [Internet]. United Nations; 2022. (Statistical Papers - United Nations (Ser. A), Population and Vital Statistics Report). Available from: https://www.un-ilibrary.org/content/books/9789210014380 [Accessed: June 8, 2023]

[24] Venables AJ. Urbanisation in developing economies: Building cities that work. The Region. 2018;**5**(1):91-100

[25] Towards a more urban world (GMT 2) — European Environment Agency [Internet]. Available from: https://www.eea.europa.eu/soer/2015/global/urban-world [Accessed: June 8, 2023]

[26] 2018 Global Status Report Towards a zero-emission, efficient and resilient buildings and construction sector [Internet]. International Energy Agency (IEA). Available from: https://wedocs.unep.org/bitstream/handle/20.500.11822/27140/Global_Status_2018.pdf?sequence=1&isAllowed=y [Accessed: June 8, 2023]

[27] Pérez-Lombard L, Ortiz J, Pout C. A review on buildings energy consumption information. Energy and Buildings. 2008;**40**(3):394-398

[28] Isaac M, van Vuuren DP. Modeling global residential sector energy demand for heating and air conditioning in the context of climate change. Energy Policy. 2009;**37**(2):507-521

[29] International Energy Agency (IEA). Global Alliance for Buildings and Construction, International Energy Agency and the United Nations Environment Programme (2019): 2019 Global Status Report for Buildings and Construction: Towards a Zero-Emission, Efficient and Resilient Buildings and Construction Sector. International Energy Agency (IEA); 2019. p. 41

[30] Cao X, Dai X, Liu J. Building energy-consumption status worldwide and the state-of-the-art technologies for zero-energy buildings during the past decade. Energy and Buildings. 2016;**128**:198-213

[31] Fetting C. The European Green Deal. Vienna: ESDN Office; 2020. p. 22. ESDN Report

[32] UNFCCC. Conference of the Parties (COP). Warsaw International Mechanism for Loss and Damage associated with climate change impacts. [Internet]. 2016. Available from: https://unfccc.int/documents/311124 [Accessed: June, 2023]

[33] IPCC. Summary for policymakers. In: Climate Change 2021: The Physical Science Basis. Contribution of Working Group I to the Sixth Assessment Report of the Intergovernmental Panel on Climate Change. Cambridge, United Kingdom and New York, NY, USA: Cambridge University Press; 2021. pp. 3-32

[34] Proposal for a directive of the European parliament and of the council amending Directive (EU) 2018/2001 of the European Parliament and of the Council, Regulation (EU) 2018/1999 of the European Parliament and of the Council and Directive 98/70/EC of the European Parliament and of the Council as regards the promotion of energy from renewable sources, and repealing Council Directive (EU) 2015/652 [Internet]. 2021. Available from: https://eur-lex.europa.eu/legal-content/EN/TXT/?uri=CELEX:52021PC0557 [Accessed: May 19, 2023]

[35] Li W, Koo C, Hong T, Oh J, Cha SH, Wang S. A novel operation approach for the energy efficiency improvement of the HVAC system in office spaces through real-time big data analytics. Renewable and Sustainable Energy Reviews. 2020;**127**:109885

[36] Ascione F, Bianco N, Iovane T, Mastellone M, Mauro GM. The evolution of building energy retrofit via double-skin and responsive façades: A review. Solar Energy. 2021;**224**: 703-717

[37] Frontini F, Bonomo P, Moser D, Maturi L. 8 - building integrated photovoltaic facades: Challenges, opportunities and innovations. In: Gasparri E, Brambilla A, Lobaccaro G, Goia F, Andaloro A, Sangiorgio A, editors. Rethinking Building Skins. United Kingdom: Woodhead Publishing; 2022. pp. 201-229

[38] Kougias I, Taylor N, Kakoulaki G, Jäger-Waldau A. The role of photovoltaics for the European green Deal and the recovery plan. Renewable and Sustainable Energy Reviews. 2021; **144**:111017

[39] Lacal Arantegui R, Jäger-Waldau A. Photovoltaics and wind status in the European Union after the Paris agreement. Renewable and Sustainable Energy Reviews. 2018;**81**:2460-2471

[40] Bodart M, De Herde A. Global energy savings in offices buildings by the use of daylighting. Energy and Buildings. 2002;**34**(5):421-429

[41] Krarti M. Energy performance of control strategies for smart glazed windows applied to office buildings. Journal of Building Engineering. 2022; **45**:103462

[42] Al-Obaidi KM, Ismail M, Rahman AMA. A review of skylight glazing materials in architectural designs for a better indoor environment. Modern Applied Science. 2014;**8**(1):68

[43] Gago EJ, Muneer T, Knez M, Knez M, Köster H. Natural light controls and guides in buildings. Energy saving for electrical lighting, reduction of cooling load. Renewable & Sustainable Energy Reviews. 2015;**41**:1-13

[44] Boyce P, Hunter C, Howlett O. The Benefits of Daylight through Windows. Troy, New York: Rensselaer Polytechnic Institute; 2003

[45] Vine E, Lee E, Clear R, DiBartolomeo D, Selkowitz S. Office worker response to an automated venetian blind and electric lighting system: A pilot study. Energy and Buildings. 1998;**28**(2):205-218

[46] European Commision. Directive 2010/31/EU of the European Parliament and of the Council of 19 May 2010 on the energy performance of buildings [Internet]. Available from: https://eur-lex.europa.eu/legal-content/EN/TXT/?qid=1583922805643&uri=CELEX:02010L0031-20181224 [Accessed: May 27, 2023]

[47] International Energy Agency (IEA. IEA (2021), Net Zero by 2050, IEA, Paris, License: CC BY 4.0 [Internet]. Available from: https://www.iea.org/reports/net-zero-by-2050 [Accessed: June 7, 2023]

[48] Casini M. Active dynamic windows for buildings: A review. Renewable Energy. 2018;**119**:923-934

[49] Rezaei SD, Shannigrahi S, Ramakrishna S. A review of conventional, advanced, and smart glazing technologies and materials for improving indoor environment. Solar Energy Materials and Solar Cells. 2017; **159**:26-51

[50] Kuhn TE. Calorimetric determination of the solar heat gain coefficient g with steady-state laboratory measurements. Energy and Buildings. 2014;**84**:388-402

[51] Oh M, Park J. Evaluation of building energy and daylight performance according to applying electrochromic and pdcl (polymer dispersed liquid crystal) to office building in South Korea. International Journal of Sustainable Building Tech. 2019;**10**(4): 227-240

[52] Oh M, Sung LC, Park JS, Lee K, Tae S. Evaluation of energy and daylight performance of old office buildings in South Korea with curtain walls remodeled using polymer dispersed liquid crystal (PDLC) films. Energies. 2019;**12**(19):3679

[53] Hemaida A, Ghosh A, Sundaram S, Mallick TK. Evaluation of thermal performance for a smart switchable adaptive polymer dispersed liquid crystal (PDLC) glazing. Solar Energy. 2020;**195**:185-193

[54] Shaik S, Gorantla K, Venkata Ramanan M, Mishra S, Kulkarni KS. Thermal and cost assessment of various polymer-dispersed liquid crystal film smart windows for energy efficient buildings. Construction and Building Materials. 2020;**263**:120155

[55] Lami M, Al-naemi F, Alrashidi H, Issa W. Quantifying of vision through polymer dispersed liquid crystal double-glazed window. Energies. 2022;**15**(9): 3196

[56] Hemaida A, Hemaida A, Ghosh A, Sundaram S, Mallick TK. Simulation study for a switchable adaptive polymer dispersed liquid crystal smart window for two climate zones (Riyadh and London). Energy and Buildings. 2021; **251**:111381

[57] Iluyemi DC, Nundy S, Shaik S, Tahir A, Ghosh A. Building energy analysis using EC and PDLC based smart switchable window in Oman. Solar Energy. 2022;**237**:301-312

[58] Mesloub A, Ghosh A, Kolsi L, Alshenaifi M. Polymer-dispersed liquid crystal (PDLC) smart switchable windows for less-energy hungry buildings and visual comfort in hot desert climate. Journal of Building Engineering. 2022;**59**:105101-105101

[59] Hasan AH, Ismail S. A study of the practical use of green engineering-based Technology for Building Sustainability: Users' perspective. IOP Conference Series: Materials Science and Engineering. 2021;**1200**(1):012029

[60] Frattolillo A, Loddo G, Mastino CC, Baccoli R. Heating and cooling loads with electrochromic glazing in Mediterranean climate. Energy and Buildings. 2019;**201**: 174-182

[61] Zeng Z, Chen J, Augenbroe G. Movable window insulation as an instantiation of the adaptive building

envelope: An investigation of its cost-effectiveness in the U.S. Energy and Buildings. 2021;**247**:111138

[62] Elkhayat YO, Ibrahim MG, Tokimatsu K, Ali AAM. Life cycle cost analysis on three high-performance glazing systems for an office building in new Cairo, Egypt. Architectural Engineering and Design Management. 2021;**17**(1–2):131-145

[63] Dierking I, Kosbar LL, Afzali-Ardakani A, Lowe AC, Held GA. Network morphology of polymer stabilized liquid crystals. Applied Physics Letters. 27 Oct 1997;**71**(17):2454-2456

[64] Dierking I. Polymer network–stabilized liquid crystals. Advanced Materials. 1 Feb 2000;**12**(3):167-181

[65] Gotoh T, Murai H. Preparation and characteristics of new reverse mode film of polymer dispersed liquid crystal type. Applied Physics Letters. 1992;**60**(3):392-394

[66] Coates D. Normal and reverse mode polymer dispersed liquid crystal devices. Displays. 1993;**14**(2):94-103

[67] Chen J, Hu X, Zhao W, Broer DJ, Zhou G. Reverse mode polymer dispersed liquid crystal-based smart windows: A Progress report. Recent Progress in Materials. 2021;**3**(4):1-1

[68] Liu S, Li Y, Zhou P, Chen Q, Su Y. Reverse-mode PSLC multi-plane optical see-through display for AR applications. Optics Express. 2018;**26**(3):3394-3403

[69] Mishra KK, Dubey SK, Mani SA. Optical characterization of inorganic nanoparticles doped in polymer dispersed liquid crystal. Molecular Crystals and Liquid Crystals. 2017;**647**(1):244-252

[70] Hsu CC, Chen YX, Li HW, Shan HJ. Low switching voltage ZnO quantum dots doped polymer-dispersed liquid crystal film. Optics Express. 2016;**24**(7):7063-7068

[71] He T, Yang B, Zhang L, Shi Z, Gong X, Geng P, et al. A study on electro-optical properties of polymer dispersed liquid crystal films doped with barium titanate nanoparticles prepared by nucleophile-initiated thiol-ene click reaction. Liquid Crystals. 2020;**47**(7):1004-1018

[72] Sun H, Xie Z, Ju C, Hu X, Yuan D, Zhao W, et al. Dye-doped electrically smart windows based on polymer-stabilized liquid crystal. Polymers. 2019;**11**(4):694

[73] Goda K, Miyauchi H, Hattori R, Fujinaga Y, Yoshidome S. A sunlight-driven smart window using guest-host liquid crystals based on opto-thermal conversion of near-infrared dye. Molecular Crystals and Liquid Crystals. 2023;**756**(1):38-45

[74] Sharma V, Kumar P. Studies of absorption coefficient cum electro-optic performance of polymer dispersed liquid crystal doped with CNT and dichroic dye. Physica B: Condensed Matter. 2017;**524**:118-122

[75] Saeed MH, Zhang S, Cao Y, Zhou L, Hu J, Muhammad I, et al. Recent advances in the polymer dispersed liquid crystal composite and its applications. Molecules. 2020;**25**(23):5510

[76] Fergason JL. Cholesteric Structure-1 optical properties. Molecular Crystals. 1966;**1**(2):293-307

[77] Haller I. Elastic constants of the Nematic liquid crystalline phase of p-Methoxybenzylidene-p-n-Butylaniline

(MBBA). The Journal of Chemical Physics. 1972;**57**(4):1400-1405

[78] Bigham KJ. LCP Introduction to Liquid Crystal Polymers. SC, USA: Zeus Industrial Products, Inc.; 2018

[79] Mandle RJ, Cowling SJ, Goodby JW. A nematic to nematic transformation exhibited by a rod-like liquid crystal. Physical Chemistry Chemical Physics. 2017;**19**(18):11429-11435

[80] Bushby RJ, Lozman OR. Discotic liquid crystals 25 years on. Interface Science. 2002

[81] Wang L, Huang D, Lam L, Cheng Z. Bowlics: History, advances and applications. Liquid Crystals Today. 2017;**26**(4):85-111

[82] Zentel R. Polymers with side-chain Mesogenic units. In: Allen G, Bevington JC, editors. Comprehensive Polymer Science and Supplements. Amsterdam: Pergamon; 1989. pp. 723-732

[83] Mucha M. Polymer as an important component of blends and composites with liquid crystals. Progress in Polymer Science. 2003;**28**(5):837-873

[84] Ryabchun A, Bobrovsky A. Cholesteric liquid crystal materials for tunable diffractive optics. Advanced Optical Materials. 2018;**6**(15):1800335

[85] Allouchery V, Roussel F, Buisine JM. Thermodynamic and electro-optic characteristics of UV-cured Monofunctional acrylate/Nematic liquid crystal mixtures. Molecular Crystals and Liquid Crystals Science and Technology Section A Molecular Crystals and Liquid Crystals. 1999;**329**(1):227-237

[86] Smith GW, Vaz NA. The relationship between formation kinetics and microdroplet size of epoxy-based polymer-dispersed liquid crystals. Liquid Crystals. 1988;**3**(5):543-571

[87] Hsin LP, Chin WK. TSC study and electro-optical properties of epoxy/acrylic polymer-dispersed liquid-crystal film in DICY thermal cure. Journal of Polymer Science Part B: Polymer Physics. 2001;**39**(5):507-514

[88] Luo L, Liang Y, Feng Y, Mo D, Zhang Y, Chen J. Recent Progress on preparation strategies of liquid crystal smart windows. Crystals. 2022;**12**(10):1426

[89] Meng X, Li J, Lin Y, Liu X, Zhao J, Li D, et al. Periodic electro-optical characteristics of PDLC film driven by a low-frequency square wave voltage. Crystals. 2022;**12**(2):163

[90] Ahmad F, Jamil M, Jeon YJ. Current trends in studies on reverse-mode polymer dispersed liquid-crystal films — A review. Electronic Materials Letters. Jul 2014;**10**(4):679-692

[91] Ahmad F, Jamil M, Jeon YJ. Reverse mode polymer stabilized cholesteric texture (PSCT) light shutter display – A short review. Journal of Molecular Liquids. May 2017;**233**:187-196

[92] Hikmet RAM. Electrically induced light scattering from anisotropic gels. Journal of Applied Physics. 1990;**68**(9):4406-4412

[93] Ma YD, Wu BG, Xu G. Reverse-mode microdroplet liquid crystal display. In: Liquid Crystal Displays and Applications. Santa Clara, CA, United States: SPIE; 1990. pp. 46-57

[94] Nicoletta FP, De Filpo G, Lanzo J, Chidichimo G. A method to produce reverse-mode polymer-dispersed liquid-

crystal shutters. Applied Physics Letters. 1999;**74**(26):3945-3947

[95] Macchione M, Cupelli D, Filpo GD, Nicoletta FP, Chidichimo G. Rough surfaces for orientation control in reverse mode polymer dispersed liquid crystal films. Liquid Crystals. 2000; **27**(7):917-920

[96] Yamaguchi RYR, Waki YWY, Sato SSS. Reverse mode and wide viewing angle properties in polymer dispersed liquid crystal cells prepared using a UV curable liquid crystal. Japanese Journal of Applied Physics. 1997;**36**(5R):2771

[97] Romani A, Chidichimo G, Formoso P, Manfredi S, Favaro G, Mazzucato U. Photochromic behavior of a Spiro-indolino-oxazine in reverse-mode polymer-dispersed liquid crystal films. The Journal of Physical Chemistry. B. 2002;**106**(37):9490-9495

[98] Rajaram CV, Hudson SD, Chien LC. Morphology of polymer-stabilized liquid crystals. Chemistry of Materials. 1995; 7(12):2300-2308

[99] Epure EL, Stoica I, Albu RM, Hulubei C, Barzic AI. New strategy for inducing surface anisotropy in polyimide films for Nematics orientation in display applications. Nanomaterials. 2021; **11**(11):3107

[100] Lee TR, Kim JH, Lee SH, Jun MC, Baik HK. Investigation on newly designed low resistivity polyimide-type alignment layer for reducing DC image sticking of in-plane switching liquid crystal display. Liquid Crystals. 2017; **44**(4):738-747

[101] Zhang Y, Wang C, Zhao W, Li M, Wang X, Yang X, et al. Polymer stabilized liquid crystal smart window with flexible substrates based on low-temperature treatment of polyamide acid technology. Polymers. 2019;**11**(11):1869

[102] Lee J, Nam S, Choi SS. Design of chiral guest-host liquid crystals for a transmittance-tunable smart window. Optical Materials Express. 2022;**12**(7):2568

[103] Nimmy John V, Varanakkottu SN, Varghese S. Flexible, ferroelectric nanoparticle doped polymer dispersed liquid crystal devices for lower switching voltage and nanoenergy generation. Optical Materials. 2018;**80**:233-240

[104] Kumari A, Sinha A. Role of $BaTiO_3$ nanoparticles on electro-optic performance of epoxy-resin-based PDLC devices. Liquid Crystals. 2021;**48**(1): 23-34

[105] Mishra S, Manjuladevi V, Gupta RK, Kumar S. Investigation on physical properties of silver nanorod doped nematic liquid crystal. AIP Conference Proceedings. 2021;**2369**: 020065

[106] Ahmad F, Luqman M, Jamil M. Advances in the metal nanoparticles (MNPs) doped liquid crystals and polymer dispersed liquid crystal (PDLC) composites and their applications - a review. Molecular Crystals and Liquid Crystals. 2021;**731**(1):1-33

[107] Manaila Maximean D, Cîrcu V, Bărar A, Konkolovich A, Miskevich A, Staicu T, et al. Polymer dispersed liquid crystals films doped with carbon nanotubes: Preparation methods. In: Cristea I, Vladescu M, Tamas RD, editors. Advanced Topics in Optoelectronics, Microelectronics, and Nanotechnologies IX. Constanta, Romania: SPIE; 2018. p. 169

[108] Kopcanský P. Morphology and dielectric properties of polymer dispersed liquid crystal with magnetic

nanoparticles. Semicond Phys Quantum Electron Optoelectron. 2010;**13**(4): 343-347

[109] Najjar R, Bigdeli E. Synthesis of novel core-shells of PMMA with coumarin based liquid crystalline side chains and PMMA shell as electro-optical materials. European Polymer Journal. 2018;**104**:136-146

[110] Najjar R, Bigdeli E, Asadpour-Zeynali K, Zaker-Hamidi MS. Liquid crystalline nanomaterials extracted from egg yolk: Encapsulation and characterization of their electro-optical activity as potential materials for flexible LCD displays. Journal of Electronic Materials. 2018;**47**(12):7143-7150

[111] Doane JW, Vaz NA, Wu BG, Žumer S. Field controlled light scattering from nematic microdroplets. Applied Physics Letters. 1986;**48**(4):269-271

[112] Barbero G, Oldano C. Derivative-dependent surface-energy terms in nematic liquid crystals. Il Nuovo Cimento D. 1985;**6**(5):479-493

[113] Vaz NA, Smith GW, Montgomery GP Jr. A light control film composed of liquid crystal droplets dispersed in an epoxy matrix. Molecular Crystals and Liquid Crystals. 1987;**146**(1):17-34

[114] Montgomery GP, Vaz NA. Contrast ratios of polymer-dispersed liquid crystal films. Applied Optics. 1987;**26**(4):738-743

[115] Broer DJ, Finkelmann H, Kondo K. In-situ photopolymerization of an oriented liquid-crystalline acrylate. Makromolekulare Chemie. 1988;**189**:185

[116] Jain SC, Rout DK. Electro-optic response of polymer dispersed liquid-crystal films. Journal of Applied Physics. 1991;**70**(11):6988-6992

[117] Vaz NA, Smith GW, Montgomery GP. A light control film composed of liquid crystal droplets dispersed in a UV-curable polymer. Molecular Crystals and Liquid Crystals. 1987;**146**(1):1-15

[118] Kelly JR, Palffy-Muhoray P. The optical response of polymer dispersed liquid crystals. Molecular Crystals and Liquid Crystals Science and Technology Section A Molecular Crystals and Liquid Crystals. 1994;**243**(1):11-29

[119] Coates D. Polymer-dispersed liquid crystals. Journal of Materials Chemistry. 1995;**5**(12):2063-2072

[120] West JL, Jewell K, Francl JJ, Ji Y, Kelly JR. Surface anchoring, polymer glass transition, and polymer-dispersed liquid crystal electro-optics. In: Liquid Crystal Materials, Devices, and Applications. San Jose, CA, United States: SPIE; 1992. pp. 8-12

[121] Chaudhary A, Shukla RK, Rani P, Choudhary A, Malik P, Mehra R, et al. Polymer dispersed liquid crystals devices: Role of photopolymerisation to control defect orientation, optical and electro-optical properties. Liquid Crystals. 2023;**50**:1-12

[122] Ghosh A, Mallick TK. Evaluation of optical properties and protection factors of a PDLC switchable glazing for low energy building integration. Solar Energy Materials and Solar Cells. 2018; **176**:391-396

[123] Park S, Hong JW. Polymer dispersed liquid crystal film for variable-transparency glazing. Thin Solid Films. 2009;**517**(10):3183-3186

[124] Manfre G. PDLC building and automotive glazing applications. Molecular Crystals and Liquid Crystals Science and Technology Section A

Molecular Crystals and Liquid Crystals. 2001;**360**(1):41-60

[125] TETA Cam PDLC Films & Research [Internet]. 2023. Available from: https://www.tetacam.com/ [Accessed: May 23, 2023]

[126] Kim BK, Hong EY, Ok YS. Wide view angle polymer/liquid crystal composite films. Korea Polymer Journal. 1997;**5**(2):77-83

[127] Wu BG, West JL, Doane JW. Angular discrimination of light transmission through polymer-dispersed liquid-crystal films. Journal of Applied Physics. 1987;**62**(9):3925-3931

[128] De Filpo G, Mashin AI, Beneduci A, Chidichimo G. Reduced OFF-axis haze in polymer-dispersed liquid crystals. Liquid Crystals. 2017;**44**(9): 1477-1483

[129] West J. Polymer-dispersed liquid crystals. In: Progress in Liquid Crystal Science and Technology. Series on Liquid Crystals. Vol. 4. Singapore: World Scientific; 2012. pp. 575-595

[130] West JL, Fredley DS, Carrell JC. Haze-free polymer dispersed liquid crystals utilizing linear polarizers. Applied Physics Letters. 1992;**61**(17): 2004-2005

[131] Chan CH, Wu TY, Yen MH, Lin CE, Cheng KT, Chen CC. Low power consumption and high-contrast light scattering based on polymer-dispersed liquid crystals doped with silver-coated polystyrene microspheres. Optics Express. 2016;**24**(26):29963-29971

[132] Cupelli D, Nicoletta FP, De Filpo G, Formoso P, Chidichimo G. Reverse mode operation polymer dispersed liquid crystal with a positive dielectric anisotropy liquid crystal. Journal of Polymer Science Part B: Polymer Physics. 2011;**49**(4):257-262

[133] eyrise® Dynamic Liquid Crystal Glazing [Internet]. 2023. Available from: https://www.eyrise.com/ [Accessed: June 26, 2023]

[134] PNLC Smart Film - FILMBASE [Internet]. 2023. Available from: https://www.smartfilmbase.com/smart-film/pnlc-smart-film.html [Accessed: June 26, 2023]

[135] Lampert CM. Optical switching technology for glazings. Thin Solid Films. 1993;**236**(1):6-13

[136] Kitzerow HS. Polymer-dispersed liquid crystals from the nematic curvilinear aligned phase to ferroelectric films. Liquid Crystals. 1994;**16**(1):1-31

[137] Uranus HP. Characterization and driving of polymer-dispersed liquid-crystal sheet for smarthome applications. Journal of Physics: Conference Series. 2022;**2274**(1):012010

[138] Nundy S, Mesloub A, Alsolami B, Ghosh A. Electrically actuated visible and near-infrared regulating switchable smart window for energy positive building: A review. Journal of Cleaner Production. 2021;**301**:126854

[139] Lee J, Mok HS, Choi JH, Lee JW. Power electronic converter topology with regulable transmittance of PDLC applications. In: 2018 21st International Conference on Electrical Machines and Systems (ICEMS). Jeju, Korea (South): IEEE; 2018. pp. 2347-2350

[140] Macrelli G. Optical characterization of commercial large area liquid crystal devices. Solar Energy Materials and Solar Cells. 1995;**39**(2):123-131

[141] Korner W, Scheller H, Beck A, Fricke J. PDLC films for control of light

transmission. Journal of Physics D: Applied Physics. 1994;**27**(10):2145

[142] Baetens R, Jelle BP, Gustavsen A. Properties, requirements and possibilities of smart windows for dynamic daylight and solar energy control in buildings: A state-of-the-art review. Solar Energy Materials and Solar Cells. 2010;**94**(2):87-105

[143] Mesloub A, Ghosh A, Touahmia M, Albaqawy GA, Alsolami BM, Ahriz A. Assessment of the overall energy performance of an SPD smart window in a hot desert climate. Energy. 2022;**252**: 124073

[144] Pane S, Caporusso M, Hakemi H. Haze and opacity control in polymer dispersed liquid crystal (PDLC) films with phase separation method. Liquid Crystals. 1997;**23**(6):861-867

[145] Pane S, Caporusso M. Effect of liquid crystal birefringence on the opacity and off-axis haze of PDLC films. In: Liquid Crystals: Physics, Technology, and Applications. Vol. 3318. Zakopane, Poland: SPIE; 1998. pp. 431-434

[146] Drzaic PS. Droplet size and shape effects in nematic droplet/polymer films. In: Liquid Crystal Displays and Applications. Vol. 1257. Santa Clara, CA, United States: SPIE; 1990. pp. 29-36

[147] Montgomery GP Jr, West JL, Tamura-Lis W. Light scattering from polymer-dispersed liquid crystal films: Droplet size effects. Journal of Applied Physics. 1991;**69**(3): 1605-1612

[148] Nolan P, Tillin M, Coates D. High on-state clarity polymer dispersed liquid crystal films. Liquid Crystals. 1993; **14**(2):339-344

[149] Bouteiller L, Barny PL. Polymer-dispersed liquid crystals: Preparation, operation and application. Liquid Crystals. 1996;**21**(2):157-174

[150] Yang D-K, Chien L-C, Doane JW. Cholesteric liquid crystal/polymer dispersion for haze-free light shutters. Applied Physics Letters. 1992;**60**(25): 3102-3104

[151] Sun J, Wu ST, Haseba Y. A low voltage submillisecond-response polymer network liquid crystal spatial light modulator. Applied Physics Letters. 2014;**104**(2):023305

[152] Chidichimo G, De Filpo G, Manfredi S, Mormile S, Tortora L, Gallucci C, et al. High contrast reverse mode PDLC films: A morphologic and electro-optical analysis. Molecular Crystals and Liquid Crystals. 2009; **500**(1):10-22

[153] Hikmet RAM. Anisotropic gels and plasticized networks formed by liquid crystal molecules. Liquid Crystals. 1991; **9**(3):405-416

[154] Fung YK, Yang DK, Ying S, Chien LC, Zumer S, Doane JW. Polymer networks formed in liquid crystals. Liquid Crystals. 1995;**19**(6): 797-801

[155] Shen W, Li G. Recent Progress in liquid crystal-based smart windows: Materials, structures, and design. Laser & Photonics Reviews. 2023;**17**(1): 2200207

[156] Fergason JL. Polymer encapsulated nematic liquid crystals for display and light control applications. In: SID International Symposium Digest. 1985. pp. 268-270

[157] Reamey RH, Montoya W, Wartenberg M. Switching speeds in NCAP displays: Dependence on collection angle and wavelength.

Proceedings of the SPIE. 1991;**1455**: 39-44

[158] Drzaic PS, Muller A. Droplet shape and reorientation fields in nematic droplet/polymer films. Liquid Crystals. 1989;**5**(5):1467-1475

[159] Konynenburg P, Wipfler R, Smith JL. Optical and environmental properties of NCAP glazing products. Proceedings of SPIE. 1989;**1080**:62-71

[160] Kinugasa N, Yano Y, Takigawa A, Kawahara H. Characterization of light scattering in nematic droplet-polymer films. Proceedings of the SPIE. 1992; **1665**:57-63

[161] Marin L, Ailincai D, Paslaru E. Monodisperse PDLC composites generated by use of polyvinyl alcohol boric acid as matrix. RSC Advances. 2014;**4**(72):38397-38404

[162] Hou SL, Choi WK, Su GD. High efficient polymer dispersed liquid crystal for ultra-bright projected image. In: Liquid Crystals XVII 2013 Sep 12. Vol. 8828. San Diego, California, United States: SPIE; 2013. pp. 165-172

[163] Chen G, Hu W, Zhou L, Zhang H, Wang L, Zhang C, et al. The electro-optical properties and adhesion strength of epoxy-Polymercaptan-based polymer dispersed liquid crystal films. Crystals. 2021;**11**(6):576

[164] Ma H, Zhou L, Han C, Zhang C, Zhang L. The fabrication of novel optical diffusers based on UV-cured polymer dispersed liquid crystals. Liquid Crystals. 2019;**46**(1):138-144

[165] Yamagishi FG, Miller LJ, Van Ast CI. Morphological control in polymer-dispersed liquid crystal film matrices. In: Liquid Crystal Chemistry, Physics, and Applications. Vol. 1080. Los Angeles, CA, United States: SPIE; 1989. pp. 24-31

[166] West JL. Phase separation of liquid crystals in polymers. Molecular Crystals and Liquid Crystals Incorporating Nonlinear Optics. 1988; **157**(1):427-441

[167] Bronnikov S, Kostromin S, Zuev V. Polymer-dispersed liquid crystals: Progress in preparation, investigation, and application. Journal of Macromolecular Science, Part B. 2013; **52**(12):1718-1735

[168] Lee JC. Polymerization-induced phase separation. Physical Review E. 1999;**60**(2):1930-1935

[169] White TJ, Natarajan LV, Tondiglia VP, Bunning TJ, Guymon CA. Polymerization kinetics and monomer functionality effects in thiol–Ene polymer dispersed liquid crystals. Macromolecules. 2007;**40**(4):1112-1120

[170] Nakazawa H, Fujinami S, Motoyama M, Ohta T, Araki T, Tanaka H, et al. Phase separation and gelation of polymer-dispersed liquid crystals. Computational and Theoretical Polymer Science. 2001;**11**(6):445-458

[171] Abbate M, Mormile P, Musto P, Petti L, Ragosta G, Villano P. Thermosetting PDLCs: Cure, morphology and Thermo-optical response. Molecular Crystals and Liquid Crystals Science and Technology Section A Molecular Crystals and Liquid Crystals. 1999;**336**(1):61-81

[172] Noh CH, Jung JE, Kim JY, Sakong DS, Choi KS. A study on the morphology and electro-optic properties of liquid crystal-polymer composite films. Molecular Crystals and Liquid Crystals Science and Technology Section

A Molecular Crystals and Liquid Crystals. 1993;**237**(1):299-309

[173] Kim JY, Palffy-muhoray P. Phase separation kinetics of a liquid crystal-polymer mixture. Molecular Crystals and Liquid Crystals. 1991;**203**(1):93-100

[174] Liu YJ, Sun XW. Holographic polymer-dispersed liquid crystals: Materials, formation, and applications. Advances in OptoElectronics. 2008; **2008**:1-52

[175] Sutherland RL, Tondiglia VP, Natarajan LV, Bunning TJ, Adams WW. Electrically switchable volume gratings in polymer-dispersed liquid crystals. Applied Physics Letters. 1994;**64**(9): 1074-1076

[176] Ren H, Wu ST. Inhomogeneous nanoscale polymer-dispersed liquid crystals with gradient refractive index. Applied Physics Letters. 2002;**81**(19): 3537-3539

[177] Galli G, Laus M, Angeloni AS. Synthesis and thermotropic properties of new mesogenic diacrylate monomers. Die Makromolekulare Chemie. 1986; **187**(2):289-296

[178] Dierking I. Crystallisation of a bent-core liquid crystal mesogen. Physica B: Condensed Matter. 2001;**304** (1–4):51-59

[179] Broer DJ, Mol GN, Challa G. In situ photopolymerization of an oriented liquid-crystalline acrylate, 2. Die Makromolekulare Chemie: Macromolecular Chemistry and Physics. 1989;**190**(1):19-30

[180] Broer DJ, Boven J, Mol GN, Challa G. *In-situ* photopolymerization of oriented liquid-crystalline acrylates, 3. Oriented polymer networks from a mesogenic diacrylate. Die Makromolekulare Chemie: Macromolecular Chemistry and Physics. 1989;**190**(9):2255-2268

[181] Bacchiocchi C, Foschi G, Miglioli I, Shoarinejad S, Arcioni A, Zannoni C. Nematic director configuration, local order and microviscosity in a PSLC cell. Molecular Crystals and Liquid Crystals. 2015;**614**(1):2-10

[182] Li H, Xu J, Ren Y, Han R, Song H, Huang R, et al. Preparation of highly durable reverse-mode polymer-stabilized liquid crystal films with polymer walls. ACS Applied Materials & Interfaces. 2023;**15**(1):2228-2236

[183] Dierking I, Kosbar LL, Afzali-Ardakani A, Lowe AC, Held GA. Two-stage switching behavior of polymer stabilized cholesteric textures. Journal of Applied Physics. 1997;**81**(7):3007-3014

[184] Hikmet RAM. Anisotropic gels in liquid crystal devices. Advanced Materials. 1992;**4**(10):679-683

[185] Ren H, Fan YH, Wu ST. Prism grating using polymer stabilized nematic liquid crystal. Applied Physics Letters. 2003;**82**(19):3168-3170

[186] Ma J, Shi L, Yang DK. Bistable polymer stabilized cholesteric texture light shutter. Applied Physics Express. 2010;**3**(2):021702

[187] Lin YH, Ren H, Fan YH, Wu YH, Wu ST. Polarization-independent and fast-response phase modulation using a normal-mode polymer-stabilized cholesteric texture. Journal of Applied Physics. 2005;**98**(4):043112

[188] Ren H, Wu ST. Reflective reversed-mode polymer stabilized cholesteric texture light switches. Journal of Applied Physics. 2002;**92**(2):797-800

[189] Doane JW, Yang DK, Chien LC. Current trends in polymer dispersed liquid crystals. In: Conference Record of the 1991 International Display Research Conference. San Diego, CA, USA: IEEE; 1991. pp. 175-178

[190] Yang H, Kikuchi H, Kajiyama T. Temperature dependent light transmission-light scattering switching of (homeotropic liquid crystalline polymer network/liquid crystals/chiral dopant) composite film. Liquid Crystals. 2000;**27**(12):1695-1699

[191] Pan G, Yu L, Zhang H, Guo J, Guo R, Cao H, et al. Effects on thermo-optical properties of the composition of a polymer-stabilised liquid crystal with a smectic A–chiral nematic phase transition. Liquid Crystals. 2008;**35**(9): 1151-1160

[192] Bobrovsky A, Shibaev V. Novel type of combined photopatternable and electro-switchable polymer-stabilized cholesteric materials. Journal of Materials Chemistry. 2009;**19**(3): 366-372

[193] Yamaguchi R, Sakurai S. Cell thickness dependence on electric optical property of reverse mode liquid crystal display. Journal of Photopolymer Science and Technology. 2014;**27**(3):287-290

[194] Yamaguchi R, Inoue K, Kurosawa R. Effect of liquid crystal material on polymer network structure in polymer stabilized liquid crystal cell. Journal of Photopolymer Science and Technology. 2016;**29**(2):289-292

[195] Yun CJ, Song JK. Functional films using reactive mesogens for display applications. Journal of Information Display. 2017;**18**(3):119-129

[196] Wang MH, Choi WK, Su GD. Transparent image generator by using vertically aligned polymer-stabilized liquid crystal (VA-PSLC) for see-through display applications. In: Twelfth International Conference on Solid State Lighting and Fourth International Conference on White LEDs and Solid State Lighting. Vol. 8484. SPIE; 15 Oct 2012. pp. 125-131

[197] Yadav SP, KKR P, AKR M, Manohar R. Electro-optical behavior of dye doped Nematic liquid crystal. Acta Physica Polonica A. 2011;**119**(6):824-828

[198] West JL, Ondris-Crawford R, Erdmann MA. Dichroic dye containing polymer-dispersed liquid crystal films. In: Proc. SPIE 1257, Liquid Crystal Displays and Applications. SPIE; 1 April 1990

[199] Wu JJ, Wang CM, Li WY, Chen SH. Electrooptical properties of aligned dye-doped polymer dispersed liquid crystal films. Japanese Journal of Applied Physics. 1998;**37**(Part 1, 12A):6434-6439

[200] Hurley S, Ma J, Yang DK. P-108: A flexible display using dye-doped PSCT. SID Symposium Digest. 2009; **40**(1):1520

[201] Lin YH, Yang JM, Lin YR, Jeng SC, Liao CC. A polarizer-free flexible and reflective electro-optical switch using dye-doped liquid crystal gels. Optics Express. 2008;**16**(3):1777

[202] Yu BH, Huh JW, Heo J, Yoon TH. Simultaneous control of haze and transmittance using a dye-doped cholesteric liquid crystal cell. Liquid Crystals. 2015;**42**(10):1460-1464

[203] Huh JW, Yu BH, Heo J, Yoon TH. Double-layered light shutter using long-pitch cholesteric liquid crystal cells. Applied Optics. 2015; **54**(12):3792

[204] Lee CS, Kumar TA, Kim JH, Lee JH, Gwag JS, Lee GD, et al. An electrically switchable visible to infra-red dual frequency cholesteric liquid crystal light shutter. Journal of Materials Chemistry C. 2018;**6**(15):4243-4249

[205] Binet C, Mitov M, Mauzac M. Switchable broadband light reflection in polymer-stabilized cholesteric liquid crystals. Journal of Applied Physics. 2001;**90**(4):1730-1734

[206] Li F, Wang L, Sun W, Liu H, Liu X, Liu Y, et al. Dye induced great enhancement of broadband reflection from polymer stabilized cholesteric liquid crystals: Polymer stabilized cholesteric liquid crystals. Polymers for Advanced Technologies. 2012;**23**(2): 143-148

[207] Min Lee K, Tondiglia VP, Godman NP, Middleton CM, White TJ. Blue-shifting tuning of the selective reflection of polymer stabilized cholesteric liquid crystals. Soft Matter. 2017;**13**(35):5842-5848

Chapter 7

Sustaining Sub-Bandgap Photons *via* Upconversion for Solar Splitting Cells

Amit Kumar Verma, Prerna Tripathi, Akhoury Sudhir Kumar Sinha and Shikha Singh

Abstract

Conservation of the entire spectrum of the sun is crucial to raising the efficiency of solar splitting cells or any photochemical conversion. With the aid of upconversion nanomaterials, it could potentially be achievable. In general, solar splitting technologies are associated with numerous losses. Remarkably, inadequate utilization of the light spectrum is the primary cause of losses in photophysical processes. This is usually caused by a particular band gap in semiconductor materials, where higher-energy photons dissipate as energy and lower-energy photons, or sub-bandgap photons, are unable to be absorbed. The process of absorbing two or more photons and then emitting one photon with more energy than the sum of the individual energies of the previously absorbed ones is known as upconversion. Introducing an appropriate upconverter can significantly improve the photoconversion process's efficiency. Efforts have been made in the past few years to enhance the efficiency, broad-range sensitivity, and activity of semiconductors by integrating upconversion systems. This chapter provides a detailed discussion of the upconversion strategies that have been used thus far to increase the efficiency of solar splitting cells. It will undoubtedly assist the researchers in advancing in this area.

Keywords: photophysical, solar splitting cells, photochemical conversion, semiconductors, upconversion materials

1. Introduction

1.1 Sub-bandgap photons

When photons strike the surface of a semiconductor, they can be absorbed, reflected, or transmitted through the material. The energy of a photon is a key factor in determining whether it is transmitted or absorbed during photonic interaction. Photon interaction with a semiconductor depends entirely on the energy of the photon in relation to the semiconductor's band gap. There are two pathways: one when the photon energy (E_{photon}) is less than the band gap energy (E_{bg}), resulting in weak interactions with the semiconductor, that photon is known as sub-bandgap photons, and the other, when E_{photon} is greater than or equal to E_{bg}, leading to efficient absorption and the formation of an electron-hole pair. Since photons with energy below the bandgap are not absorbed, cells often lose a large portion of the energy that is incident upon them from the sun. Sub-bandgap light photon upconversion (UC) offers a practical solution to this core problem while preserving the advantages of semiconductor devices.

1.2 Photon upconversion

In a process known as photon upconversion (UC), light at a wavelength lower than the excitation wavelength is released because of the consecutive absorption of two or more photons. Multiple mechanisms can lead to upconversion in both organic and inorganic materials. **Figure 1** highlights the distinctions from the traditional photoluminescence process and illustrates the general idea of the upconversion luminescence method. In upconversion luminescence, an excitation photon can be used by a luminous center in the ground state 1 to absorb energy and move toward the excited state 2. Following that, the luminous center will be promoted to the excited state by an additional excitation photon 3. A higher-energy photon emission is the outcome of a radiative transition from this excited state 3 back to the ground state or another lower-energy state [1]. In actuality, the metastable intermediate excited state is responsible for the upconversion luminescence process. Two primary strategies have been established for achieving upconversion luminescence emission with low-energy stimulation. These are triplet–triplet annihilation (TTA)-based upconversion [2] and the upconversion luminescence emission of lanthanide ions [3]. Both upconversion luminescence methods rely on intermediate excitation states [2, 3].

1.3 Upconversion material

Upconversion can occur in both organic and inorganic materials *via* various methods.

Polycyclic aromatic hydrocarbons (PAHs) are capable of achieving photon upconversion by triplet-triplet annihilation [4–10]. Triplet-triplet annihilation (TTA) through photon upconversion (UC) often produces high-energy (specifically, UV)

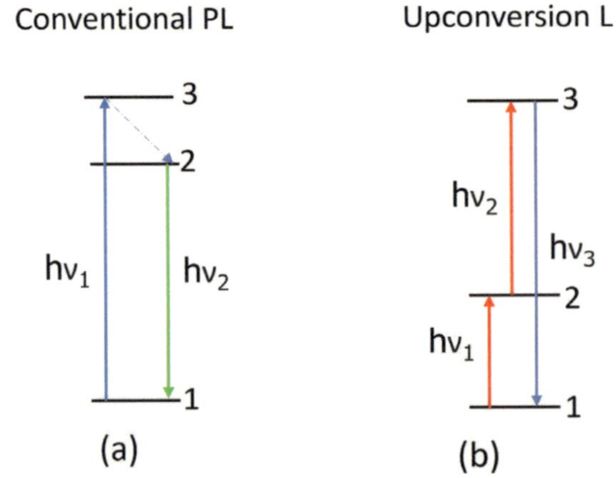

Figure 1.
Figure representing (a) conventional photoluminescence and (b) the upconversion luminescence process.

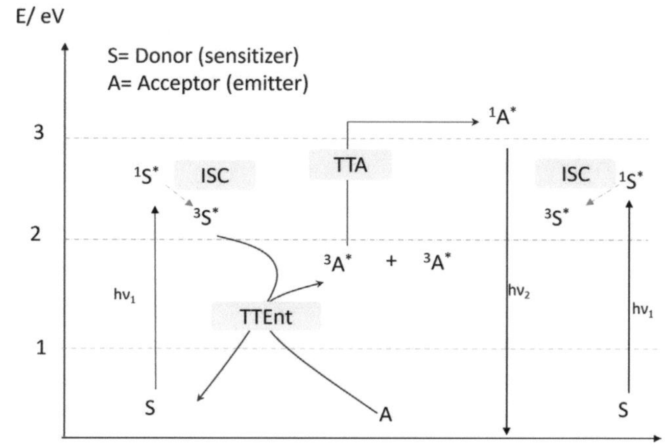

Figure 2.
Presentation of photochemical events linked to the triplet-triplet annihilation (TTA) through photon upconversion (UC) approach; ISC = intersystem crossing; TTEnt = triplet-triplet energy transfer; and TTA = triplet-triplet annihilation.

light from lower-energy (visible) light. The TTA-based upconversion system relies on a combination of a sensitizer and an acceptor (**Figure 2**).

The fundamental properties of these two components play a crucial role in determining TTA-UC properties. The sensitizer must possess two primary properties. (i) The capacity to capture light in the visible to near-infrared range of the electromagnetic spectrum, facilitating low-energy stimulation. (ii) The triplet excited state can last several microseconds or more, allowing for efficient and progressive energy dissipation. Later, this quenching process occurs when the energy of the acceptor's triplet state is smaller than that of the sensitizer. A higher energy differential between the triplet sensitizer and triplet acceptor leads to a stronger driving force for the TTA, resulting in a more favorable triplet energy transfer mechanism. The conditions for observing upconverted fluorescence from the sample are adequate if the energy criteria are satisfied and the combined triplet energy from two acceptor molecules is larger than or equal to the acceptor singlet state energy. **Table 1** summarizes the available annihilators and sensitizers for the TTA-based upconversion process.

Currently, the most widely recognized upconversion sensitizer is (2, 3, 7, 8, 12, 13, 17, 18-octaethylporphyrinato) Pt(II), which exhibits a substantial phosphorescence quantum yield (QY) of 50%. Phthalo cyanine has a strong ability to absorb light in the red to near-infrared (NIR) ranges. It also has a low level of energy in its triplet state, which makes it excellent for enhancing long wavelength emissions in upconversion systems based on TTA (triplet-triplet annihilation). The introduction of a heavy-metal atom, such as Pd, into phthalocyanine significantly increases its capacity for intersystem crossing. Ruthenium(II) polyimine complexes demonstrate highly effective intersystem crossover. The singlet to triplet transition in Ru(II) complexes is highly efficient, approaching unity. This, together with the extended duration of the triplet excited state, makes these complexes advantageous as sensitizers. Platinum(II) polyimine acetylide complexes are appealing as triplet sensitizers due to their ability to adjust their photophysical properties by employing various acetylide ligands. Rhenium(I) diimine tricarbonyl complexes are typically constrained in their ability to function as triplet sensitizers due to their low absorption in the visible spectrum. Cyclometalated iridium(III) complexes have durable excited states and exhibit

Sensitizer with core properties	(2,3,7,8,12,13,17,18-octaethylporphyrinato) Pt(II), exhibits a substantial PL quantum yield (QY) of 50% [4].
	Palladium(II) octaethylporphyrin employed as an upconversion sensitizer, similar to the above one [5].
	Phthalo cyanine (Pc) has a strong ability to absorb light in the red to near-infrared (NIR) ranges [6].
	Ruthenium(II) polyimine complexes demonstrate highly effective intersystem crossover [7].
	Rhenium(I) diimine tricarbonyl complexes are typically constrained in their ability to function as triplet sensitizers due to their low absorption in the visible spectrum [8].
	Cyclometalated iridium(III) complexes have durable excited states and absorption bands extend into the visible spectrum [9].
Activators	Anthracene, pyrene, tetracene, perylene, fluorene, and its derivatives are used [10]. 9,10-diphenylanthracene, 9,10-bis(phenylethynyl)anthracene are most extensively utilized.

Table 1.
Lists the common available annihilators and sensitizers for organic-based UC material.

absorption bands that extend into the visible spectrum. This enables these complexes to function as sensitizers for TTA-based upconversion.

Regarding acceptor or annihilators combination, polycyclic aromatic hydrocarbons, such as anthracene, tetracene, pyrene, perylene, fluorene, and their derivatives have been widely used as the primary annihilators in TTA-based upconversion systems documented in scientific literature. Despite being colorless, anthracene emits a blue fluorescence of approximately 420 nm with an 18% quantum yield.

Inorganic materials that contain ions of d- or f-block elements have the ability to upconvert photons. Inorganic-based upconverting particles are usually made of a host material doped with optically active activator and sensitizer ions [11, 12]. Among all common host materials, including oxides (ZrO_2, Y_2O_3), vanadates (YVO_4, $GdVO_4$), and phosphates ($LuPO_4$, YPO_4), it has been demonstrated that the $NaREF_4$ family of host matrices is one of the most successful due to its higher chemical stability and lower phonon energy [1, 3]. The purpose of the sensitizers in the UC materials is to increase the sensitivity of the activators. The properties of an ideal sensitizer include a significant absorption cross section at the targeted excitation wavelength, a sufficiently long lifespan of the excited state, and resonant energy levels that are compatible with the activators' appropriate excited energy states. For the reasons described above, Yb^{3+} and Nd^{3+} have been extensively studied as sensitizers. All things considered, Yb^{3+} is the best candidate for a sensitizer because its energy level diagram only has one excited state ($^2F_{5/2}$) that roughly matches the f-f transitions of many RE activators, including Er^{3+}, Ho^{3+}, and Tm^{3+} [13]. Ions that are employed as activators have a to some degree rich energy level structure along with a comparatively extended duration of UCL states. Following the stimulation of ground-state electrons to the metastable state, activators are likely to absorb energy from nearby sensitizers to become more excited at higher levels. Luminescent Pr^{3+}, Nd^{3+}, Sm^{3+}, Eu^{3+}, Tb^{3+}, Dy^{3+}, Ho^{3+}, Er^{3+}, and Tm^{3+} are popular activator ions [1, 3]. The six primaries proposed upconversion mechanisms in lanthanide include cross relaxation (CR), photon avalanche (PA), cooperative upconversion (CU), cooperative luminescence (CL), ground state and excited state absorption (GSA/ESA), energy transfer upconversion (ETU), and cooperative upconversion (CU) [14].

1.4 Solar splitting cells

Solar splitting cells include the utilization of semiconductor material where chemical reactions occur upon the interaction of a light source with the material surface. In this process, it is important for a minimum of two processes to take place simultaneously: the oxidation reaction, involving holes generated by light and the reduction reaction, involving electrons generated by light. The primary obstacle in the development of efficient photocatalytic cells for water splitting, degradation, or conversions is the search for cost-effective materials that meet the criteria of an ideal photoelectrode. The criteria for finding an ideal photoelectrode in this context include [15] (i) the capacity to capture all types of solar radiation, (ii) strong chemical stability in water-based electrolyte solutions in both dark and illuminated conditions, (iii) appropriate alignment of energy levels for specific oxidation and reduction reactions, (iv) minimal excess energy required for these reactions to occur, and finally (v) the ability to selectively transfer charges at the interface between the semiconductor and electrolyte for the purpose of splitting or converting substances. The book chapter offers a thorough examination of an upconversion integrated semiconductor nanosystem (Section 2) that possesses the capability to efficiently capture sub-bandgap photons for photo reactions.

2. Literature outlook of UC integrated semiconductors systems

2.1 Organic upconverter integrated semiconductor system (TTA-based upconversion systems)

Barawi et al. [16] presented a theoretical model of a photoelectrochemical (PEC) device utilizing photon upconversion, particularly in the visible-to-ultraviolet spectrum. The system used titanium dioxide (TiO_2) as the photoanode to generate hydrogen gas (H2). Photoelectrochemical tests were carried out using a standard three-electrode PEC cell that included the TTA-UC system (**Figure 3**). The photoanode employed was a semiconductor made of TiO_2 with a bandgap energy (E_g) of 3.2 eV. A laser pointer with a wavelength (λ_{exc}) of 445 ± 10 nm, power of 2 W, and intensity of 1.64 mW cm^{-2} was used as the irradiation source. The group selected 2,3-butanedione (biacetyl, BA) and 2,5-diphenyloxazole (PPO) as an appropriate TTA-UC pair system. BA absorbed visible light with an excitation wavelength of 430-450 nm. The UV band at 370 nm appeared by the delayed emission of PPO (1 PPO*), which has an excited singlet energy (ES) of 3.6 eV in dimethylformamide.

The TiO_2 photoanode displayed a photocurrent density of approximately 4.5 μA cm^{-2} and demonstrated good stability throughout the experiment. When exposed to visible light, the upconversion (TTA-UC) system emits UV photons that activate the TiO_2 photoanode, resulted in the formation of charge separation or hole and electron pairs. Sulfite ions trap the empty spaces in the valence band, facilitated the formation of sulfate. Concurrently, the electrons in the conduction band had moved to the platinum counter electrode. At the counter electrode, they participated in water reduction, leading to the production of hydrogen gas. The conversion of visible light into UV radiation was accomplished by the interaction of two basic organic chromophores using the triplet-triplet annihilation mechanism. Felix N. Castellano and his colleagues [17] introduced a system that utilizes upconversion to enhance photoelectrochemical processes. The system involved the combination of

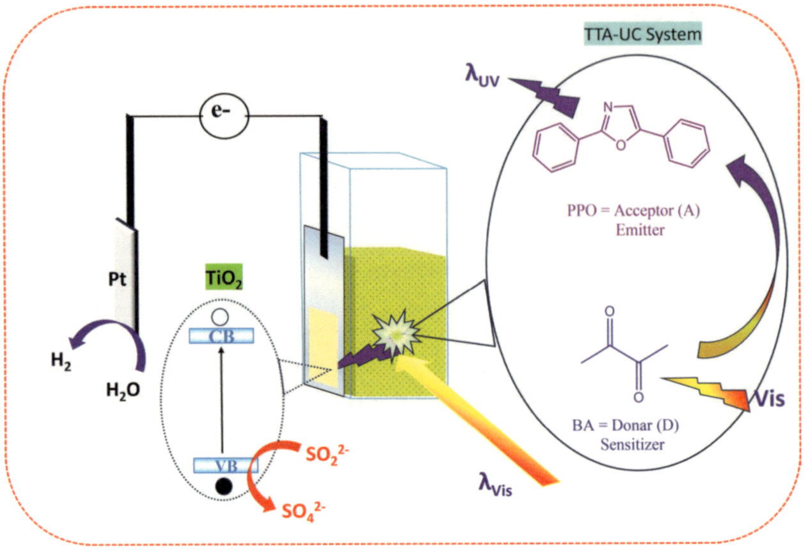

Figure 3.
Diagram depicting the 2,3-butanedione (biacetyl, BA) and 2,5-diphenyloxazole (PPO) (TTA-UC) powered PEC cell for hydrogen generation via sulfite to sulfate conversion.

palladium(II) octaethylporphyrin (PdOEP) and 9,10-diphenylanthracene (DPA) in toluene. The aforementioned combination, consisting of noncoherent green photons, was discovered to efficiently enhance the sensitivity of nanostructured WO_3 photoanodes with an energy gap of 2.7 eV. The experiment represented in **Figure 4** utilized thin WO_3 colloidal photoanodes, with a thickness of 10 mm, that were coated onto FTO glass. These photoanodes of active area 1 cm^2 were employed as the working electrode in the experiment, using a 1.0 M H_2SO_4 electrolyte. The electrode was set to a potential of +0.9 V compared to the Ag/AgCl reference electrode. The upconversion pumping commenced using a noncoherent lamp that emitted light with wavelengths over 500 nm.

The blue photons (with a maximum wavelength of 432 nm) produced by the upconversion process, manifesting as singlet DPA fluorescence, were haphazardly absorbed by the WO_3 anode, leading to detectable photocurrents under relatively moderate noncoherent excitation conditions. The photo responses obtained in these proof-of-concept studies were low because of the imperfect capture of photons at the WO_3 photoanode. The experiment showed that the WO_3 electrode could not sense anything below the bandgap level because there was no TTA-UC in the electrolyte. Nevertheless, the authors believe that there is considerable scope for enhancement if there is a change in geometric configuration, light condition, and photon retrieval. D. Choi et al. [18] showed that incorporating a luminescent back reflector capable of upconverting photons (UC LBR) improves the efficiency of water splitting in $BiVO_4$. The LBR was created by dispersing a mixture of meso-tetraphenyltetrabenzoporphine palladium (PdTPBP) and perylene organic fluorophore. PdTPBP absorbs light between 600 and 650 nm and emits light at 470 nm with higher photon energy *via* Dexter energy transfer and triplet-triplet annihilation (TTA) (**Figure 5**). Perylene was discovered to absorb light at 350-450 nm and emit light at 470 nm by photoluminescence. The $BiVO_4$ photoelectrode absorbs the emissions, improving the water splitting process. Using the light bending reflector (LBR) increased the photocurrent

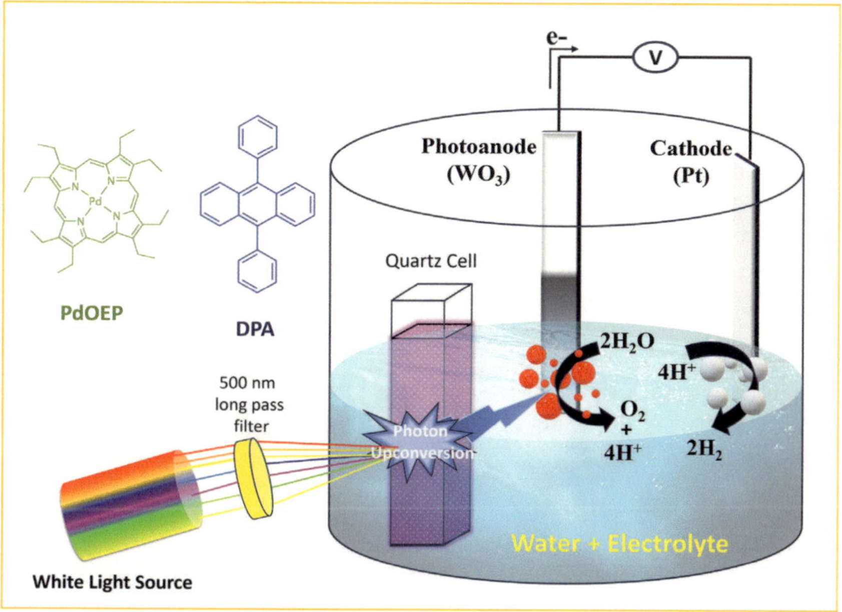

Figure 4.
Schematic representation of PdOEP and DPA (TTA-UC) powered PEC cells for hydrogen generation via acidic water spitting.

for water splitting in $BiVO_4$ electrodes by 17%, achieving photocurrent densities of up to 5.25 mA/cm^2 at 1.23 V vs. RHE. **Figure 5** displays the attachment of the LBR film to the rear of the Mo:$BiVO_4$ photoelectrode. The Mo:$BiVO_4$ film was found very transparent for wavelengths above 500 nm but possessed transmittance of around 10% for wavelengths below 500 nm due to photon absorption.

The J-V characteristics of Mo:$BiVO_4$ with PdTPBP LBR films were changed by changing the concentrations to 0.10, 0.15, and 0.19 mM and the film thicknesses to 1, 2, and 3 mm. A direct association was found between the thickness of the LBR coating and the increase in emission intensity. A maximum water splitting photocurrent of 5.34 mA/cm^2 was achieved when a Mo:$BiVO_4$ photoelectrode was paired with a 3 mm LBR film containing 0.15 mM PdTPBP, primarily due to the TTA-UC effect. The value was approximately 16% more than the photocurrent detected using only the Mo:$BiVO_4$ photoelectrode, under the same conditions.

2.2 Inorganic upconverter integrated semiconductor system (lanthanide-based upconversion systems)

Yun-Mo Sung et al. [19] described infrared (IR)-powered photoelectrochemical (PEC) cells with upconversion glass-ceramics as bases. UC glass-ceramics as substrates for photocatalysts can address the chemical instability of fluoride nanoparticles, prevent incident light blocking, and enhance the exposure of photocatalysts to liquid electrolytes. Oxyfluoride glass-ceramics containing nanocrystals doped with (Yb,Er)-doped YF_3 and (Yb,Tm)-doped YF_3 have generated green and ultraviolet/blue emissions, respectively, when exposed to 980 nm light. High-density ZnO nanorods were grown on upconversion glass-ceramic substrates using the hydrothermal technique. They were then covered with CdSe nanocrystals to create

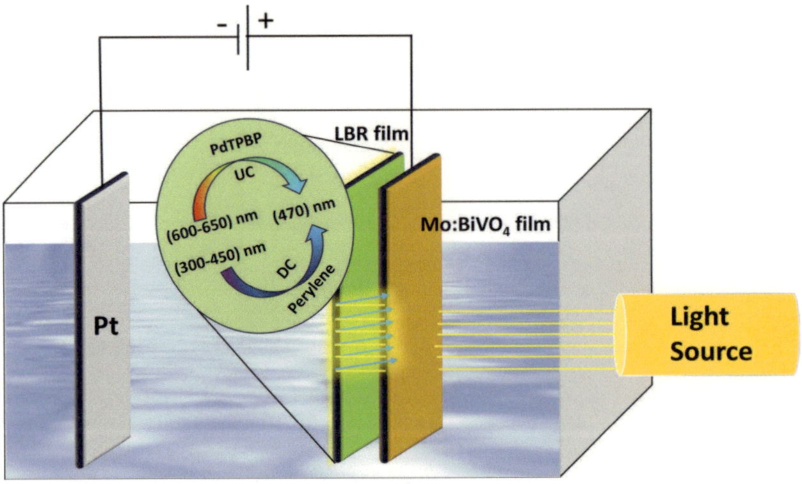

Figure 5.
Luminescent rear reflector with the ability to upconvert photons fabricated using combination of meso-tetraphenyltetrabenzoporphine palladium (PdTPBP) and perylene organic fluorophore (TTA-UC system) powered PEC cells for water splitting application.

CdSe/ZnO heterostructures using chemical bath deposition. CdSe nanoparticles were stimulated by the UC UV emission from Tm and the visible emission from Er and Tm, whereas ZnO nanorods were mostly stimulated by the UC UV emission from Tm. The photocurrent densities of CdSe/ZnO-Er and CdSe/ZnO-Tm were around 0.65 $\mu A\ cm^{-2}$ and 1.4 $\mu A\ cm^{-2}$, respectively. The PEC performances of CdSe/ZnO-Er and CdSe/ZnO-Tm were enhanced compared to ZnO-Er (0.03 $\mu A\ cm^{-2}$) and ZnO-Tm (0.2 $\mu A\ cm^{-2}$). Neither ZnO nor CdSe/ZnO on the ITO substrate exhibited any photocurrent densities when exposed to 980 nm irradiation. The photocurrent density of CdSe/ZnO-Tm is higher than that of CdSe/ZnO-Er. The CdSe/ZnO material created on the glass-ceramics with (Yb,Tm)-doped YF_3 had higher photocurrent density than that created on the glass-ceramics with (Yb,Er)-doped YF_3 attributed to charge separation induced by the type-II cascade structure. Yoongu Lim et al. [20] developed a Core@shell@shell UCNPs ($NaYF_4$:Yb,Nd,Tm [19/1/0.5 mol%] @$NaYF_4$:Yb,Nd [10/20 mol%]@$NaYF_4$) -coated $ZnFe_2O_4/TiO_2$ photocatalyst (UCNPs-ZFO/TiO_2) by layering ZFO nanoparticles and UCNPs onto TiO_2 nanopillars on a transparent fluorine-doped tin oxide (FTO) substrate. The UCNP-ZFO/TiO_2 showed a markedly greater photocurrent density of 0.795 $mA\ cm^{-2}$ in comparison to the pristine TiO_2, which had a value of 0.260 $mA\ cm^{-2}$. The UCNP-ZFO/TiO_2 had a photon-to-current efficiency that was 4.1 times greater than TiO_2. Additionally, its photoanode exhibited the lowest charge transfer resistance and the largest charge density compared to the other photoelectrodes. Trang Nguyen Thi Thuy et al. [21] detailed the development of TiO_2 thin films enhanced with an upconversion (UC) phosphor for solar water splitting. The study investigated how varying the UC phosphor concentration affected the morphology and photoelectrochemical (PEC) performance of the samples. $YF_3:Yb^{3+}$, Tm^{3+} showed notable upconversion emits UV photons when exposed to IR light. Adding UC phosphor to a TiO_2, thin layer improved the performance of the photoelectrochemical system. A photoelectrode with 7 wt.% UC phosphors showed the maximum photocurrent of 0.073 $mA\ cm^{-2}$ at 1.23 V compared to other samples.

Qianfan Jian et al. [22] had reported displayed plasmonic-enhanced water splitting photoanode using hematite-lanthanide. $NaYF_4$:Er,Yb upconversion nanoparticles nanocomposites that captured photons below the bandgap of hematite. Hematite, a conventional semiconductor oxide photoelectrode, absorbs only UV and visible light from the sun spectrum, resulting in a 40% loss of infrared energy. The $NaYF_4$:Er,Yb upconversion process converts photons from 980 nm to (510–570) nm, falling under the bandgap of hematite. The sample containing only hematite and UCNPs showed minimal photocurrent signal due to the UCNPs' very poor upconversion efficiency. Conversely, the sample containing hematite and AuNDs but lacking UCNPs exhibited a modest yet detectable photocurrent at the identical potential. Plasmonic materials found interact with semiconductors by light scattering, hot electron injection (HEI), and plasmonic resonance-induced energy transfer (PRIET). The author suggested two-pathway method, involving plasmonic-enhanced upconversion process and hot electron injection procedure. Yuxiang Zhu et al. [23] reported the photocatalytic conversion of N_2 to NH_3 using $NaYF_4$:Yb,Tm (NYF) upconversion nanoparticles (NPs) attached to carbon nitride nanotubes with nitrogen vacancies (NYF/NV-CNNTs) in water under near-infrared (NIR) light exposure. Nanoflowers with a particle size of approximately 20 nanometers were evenly spread throughout the surface of nitrogen-vacancy carbon nanotubes. The NYF/NV-CNNTs containing 15 wt % NYF showed the greatest NH_3 synthesis yield of 1.72 mmol L^{-1} $gcat^{-1}$ with an apparent quantum efficiency of 0.50% under NIR light. This activity was almost three times more than that of the naked CNNTs under UV-filtered solar light. Hannah Kwon et al. [24] reported a tri-doped β-$NaYF_4$:Yb^{3+},Tm^{3+},Gd^{3+} upconversion (UC) nanorods, which were incorporated into a carbon-doped mesostructured TiO_2 hybrid film with triblock copolymer P123, serving as a mesoporous template and carbon source. The novel material's photoactivity was demonstrated by the breakdown of nitrobenzene, a typical organic waste. UC nanorod-embedded C-doped TiO_2 showed enhanced absorption throughout a wide spectrum from UV to NIR, resulting in a significant improvement in nitrobenzene degradation (83%) within 3 hours under light exposure, compared to pristine TiO_2 (50%). This study demonstrates a synergistic connection between the impact of an innovative photon-trapping TiO_2 structure, enhanced NIR light-capturing effectiveness with UC nanorod integration, and a concurrent reduction in band gap energy and heightened absorption of visible light through C-doping of the oxide lattice. Anushree A. Chilkalwar et al. [25] demonstrated that the combined effect of surface plasmon resonance of Au nanoparticles and upconversion properties of (Yb,Er) NYF greatly improves broadband absorption and the formation and transmission of photogenerated charges. Gold acts as a plasmonic nanoparticle on (Yb,Er) NYF-TiO_2, serving the dual purpose of absorbing solar energy and demonstrating exceptional catalytic properties. The (Yb,Er) NYF-TiO_2/Au heterostructure exhibited a consistent and notably high hydrogen generation rate of 350 μmol h^{-1} under solar AM 1.5, roughly 3.1 times more than Au/TiO_2 (Degussa P-25) in donor-assisted photocatalytic hydrogen generation testing. Ramireddy Boppella et al. [26] showed 3D composite photoanode containing near-infrared responsive upconversion nanocrystals (UCNs) such as NaYF4:Yb^{3+}-based UCNs doped with Er^{3+} or Tm^{3+} ions, and visible-responsive plasmonic gold nanoparticles (NPs) within 3D titanium dioxide inverse opal nanostructures (Au/UCN/TiO_2). It showed enhanced solar energy absorption across the ultraviolet-visible-near-infrared spectrum. Their hybrid nanostructure, consisting of Au/Er-UCN/TiO_2, showed a tenfold enhancement in photocurrent density when exposed to UV-vis-NIR light compared to a pure TiO_2 sample. Tm-doped upconverting nanoparticles (UCN) generated a higher photocurrent compared to the Er-doped UCN. This result supports the alignment between the UV emission bands of Tm-UCN (345 and 361 nm) and the absorption band of TiO_2 (360 nm). Kuang Feng et al. [27] have developed core-shell or core-shell-shell shaped upconversion nanoparticles (UCNPs) to act as "nano-transducers" that transform NIR light into visible light, enabling NIR light to be indirectly absorbed by visible-responsive photocathodes. Upon introducing $NaYF_4$:$Yb_{20\%}$/$Er_{2\%}$ @$NaYF_4$, $NaYF_4$@ $NaYbF_4$:$Tm_{1\%}$, and $NaYbF_4$:$Er_{60\%}$@$NaYF_4$ core-shell UCNPs into the buffer solution, the Cu_2ZnSnS_4 (CZTS)-based photocathode demonstrated a significant photocurrent density of -4 mA cm^{-2} at 0 V_{RHE} when exposed to 980 nm irradiation. A new experimental approach was demonstrated to enhance the use of long or short-wavelength lights for CZTS photocathode. The report detailed the utilization of a CZTS-based photocathode with a buffer solution containing UCNPs to facilitate water splitting processes under NIR light exposure. The device generated 16 micromoles of hydrogen over a 3-hour period. The concentration of H_2 increased steadily at a fixed rate of 0.09 μmol min^{-1}. Wei Gao et al. [28] demonstrated an effective photocatalyst consisting of RGO, CdS, and an upconversion component ($NaYF_4$-Yb^{3+}/Er^{3+}) for efficient water splitting to produce hydrogen. A cooperative catalyst was developed by merging an upconversion component, $NaYF_4$-Yb^{3+}/Er^{3+}, with a visible light catalyst for 980 nm NIR-driven water splitting. An optimal hydrogen evolution rate of 4.54 μmol g^{-1} h^{-1} was attained when exposed to 980 nm NIR irradiation. The 980 nm laser energy-to-hydrogen conversion efficiency (LTH) was 0.011%. Wei Gao et al. [29] displayed the catalytic production of hydrogen using a photocatalyst made of semiconductor CdS and an upconversion component $NaYF_4$-Yb^{3+}/Er^{3+} (NYF) under infrared light. The 980 nm laser irradiation turned the input infrared light into visible light, which was then used to stimulate CdS for the production of hydrogen and oxygen. The maximum rate of hydrogen evaporation achieved is 3.38 μmol g^{-1} h^{-1}. An attempt has been made to achieve 0.008% AQE under NIR light exposure. Amit Kumar Verma et al. [30] demonstrated hydrogen production from aqueous methanol using a TiO_2 upconversion (CeF_3:Ho^{3+}) nanosystem (CHT) under visible light exposure. CeF_3:Ho^{3+} nanoparticles were incorporated into TiO_2 *in situ* by a polyol reduction process with triethanolamine as the reducing agent. The CeF_3:Ho^{3+}-incorporated TiO_2 nanosystem (CHT) demonstrated increased photocatalytic activity and showed a peak hydrogen evolution rate of 79.85 μmol h^{-1} under visible light. The solar-to-hydrogen conversion efficiency of 1.37% and the apparent quantum efficiency of 4.00% were reported for the photocatalyst system. Wei Gao et al. [31] demonstrated visible light-induced water splitting by combining reduced graphene oxide (RGO) with $CaTiO_3$ and Pr^{3+}-Y_2SiO_5, leading to enhanced charge transfer inside the $CaTiO_3$/Pr^{3+}-Y_2SiO_5/RGO catalyst. UV-responsive nano-$CaTiO_3$ was activated by upconverting visible light into UV light using the upconversion luminescent material Pr^{3+}-Y_2SiO_5 (Pr^{3+}-YSO) to generate charge pairs. The efficient transfer of charge to the hydrogen-producing site Pt by RGO was successfully accomplished. The reverse recombination reaction of hydrogen and oxygen was notably hindered by removing newly produced oxygen from the reaction mixture with an artificial gill. The anodic quantum efficiency (AQE) of reduced graphene oxide (RGO)/Praseodymium (III) ion-coordinated thioglycolic acid (CTYS)/Platinum (Pt) for hydrogen evolution was recorded 0.003% at a wavelength of 400 nm. In **Table 2**, we summarize the UC photoluminescence properties of a few upconversion materials available for splitting purposes.

2.3 Carbon or graphene-based upconverter integrated semiconductor system

Graphene and carbon quantum dots exhibit impressive optical and electrical properties. Several groups have utilized them as upconversion materials. While the exact mechanism is challenging to determine, these materials demonstrate exceptional upconversion capabilities. We have summarized a few of these reports (**Table 3**). Ensafi et al. [32] developed a novel form of graphene quantum dots known as nitrogen-doped graphene quantum dots (N-GQDs). The N-GQDs exhibited a strong tendency to generate light by photoluminescence and upconversion emission. It was observed

Sr. No.	Up-conversion material	Up-conversion photoluminescence property	Ref.
1	Yb,Er)-doped YF_3 and (Yb,Tm)-doped YF_3 upconversion glass-ceramics as substrate	Standard photoluminescence (PL) methods with a 980 nm laser were used to study the UC light emission properties of GC-Er and GC-Tm. GC-Tm had emitted blue light at around 455 nm and 480 nm, as well as UV light at around 348 nm and 365 nm. On the other hand, GC-Er mostly emitted green light at around 550 nm.	[19]
2	YF_3:Yb^{3+}, Tm^{3+}	The YF3:Yb^{3+}, Tm^{3+} microrods exhibited upconversion emission spectra when excited at a wavelength of 980 nm, ranging from 320 to 440 nm. The YF_3:Yb^{3+}, Tm^{3+} upconversion emission peaks were identified as resulting from the $^1I_6 \rightarrow {}^3F_4$ and $^1D_2 \rightarrow {}^3H_6$ transitions in Tm^{3+}.	[21]
3	UCNPs $NaYF_4$: Er, Yb UCNPs	The process of upconversion converted photons from a wavelength of 980 nm to a range of 510 nm–570 nm, which falls inside the bandgap of hematite.	[22]
4	$NaYF_4$:Yb^{3+},Tm^{3+},Gd^{3+} Nanorods	The tri-doped UCN system emitted intense blue light visible to the naked eye when stimulated by a 980 nm laser. The photoluminescence spectra of UCN showed peaks at 450, 474, 645, 695, and 802 nm, corresponding to Tm^{3+} transitions of $^1D_2 \rightarrow {}^3F_4$, $^1G_4 \rightarrow {}^3H_6$, $^1G_4 \rightarrow {}^3F_4$, $^3F_3 \rightarrow {}^3H_6$, and $^3H_4 \rightarrow {}^3H_6$.	[24]
5	(Yb,Er)NaYF4	Under excitation at 980 nm (NIR), (Yb,Er)NYF produced four emission bands in the upconversion emission spectra. The UV region at 408 nm exhibited blue emission, corresponding to the transition from $^2H_{9/2}$ to $^4I_{15/2}$. Green emission in the visible area was observed (520 and 540 nm), which corresponds to the transitions from $^2H_{11/2}$ to $I_{15/2}$ and from $^4S_{3/2}$ to $^4I_{15/2}$. The red emission was produced by the transition of erbium ions from the $^4F_{9/2}$ to $^4I_{15/2}$ state, resulting in light at 654 nm.	[25]
6	Hexagonal phase $NaYF_4$:Yb^{3+},Er^{3+}/Tm^{3+} (80:18:2)	Tm-UCN, when exposed to radiation, produced two peaks in the blue spectrum at 451 and 476 nm, and three peaks in the red to near-infrared spectrum at 646, 697, and 743 nm. The emissions listed were attributed to electronic transitions from $^1D_2$ to $^3F_4$, $^1G_4$ to $^3H_6$, $^1G_4$ to $^3F_4$, $^3F_2$ to $^3H_6$, and $^3F_3$ to $^3H_6$. The Er-UCN displayed three distinct peaks at 525, 541, and 655 nm corresponding to transitions from $^4S_{3/2}$, $^2H_{11/2}$, and $^4F_{9/2}$ to the $^4I_{15/2}$ level, respectively.	[26]
7	$NaYF_4$–Yb^{3+}/Er^{3+}	When exposed to 980 nm irradiation, the green light emission was observed from the excited states $^2H_{11/2}$ and $^4S_{3/2}$to the ground state $^4I_{15/2}$ of Er^{3+}, whereas the blue emission was found associated with the excitation route from $^2H_{9/2}$ to $^4I_{15/2}$ of Er3+.	[28]

Sr. No.	Up-conversion material	Up-conversion photoluminescence property	Ref.
8	$NaYF_4$–Yb^{3+}-Er^{3+}	When excited at 975 nm, the NYF exhibited a prominent emission peak at 650 nm and a less intense peak at 450 nm.	[29]
9	CeF_3: Ho^{3+}	The photoluminescence upconversion spectra were recorded by applying wavelengths from 400 to 700 nm, resulting in observed upconversion responses within the visible range. The upconversion reactions were caused by the energy levels of 460 nm ($^5I_8$–$^5F_1$), 480 nm ($^5I_8$–$^5F_3$), 535 nm ($^5I_8$–$^5F_4$), and 640 nm ($^5I_8$–$^5F_5$) present in rare earth Ho(III) ions.	[30]
10	Pr^{3+}-Y_2SiO_5 (Pr^{3+}-YSO)	Pr^{3+}-YSO, Pr^{3+}- CTYS/Pt, and RGO/Pr^{3+}-CTYS/Pt effectively converted visible light wavelengths of 447 nm, 488 nm, 515 nm, and 540 nm into UV emission wavelengths ranging from 253 nm to 371 nm.	[31]

Table 2.
The up-conversion material integrated with semiconductors with their up-conversion photoluminescence property for splitting applications is tabulated.

Sr. No.	UC material/ semiconductor	Synthesis conditions	Activity (photocurrent density)	Reference
1	N-GQDs/TiO_2	produced by carbonizing precursors of tetraethylenepentamine (TEPA) and citric acid (CA) hydrothermally under pressure and at a high temperature of 180°C.	3.0 $mA.cm^{-2}$ (at 1.23 V vs. RHE in 1 M KOH solution); eight times better than unmodified TiO_2	[32]
2	CQDs/TiO_2, CQDs-H/TiO_2	Ascorbic acid and ethanol precursor materials are used in hydrothermal synthesis at 160°C.	3.0 $mA.cm^{-2}$ (at 1.23 V vs. RHE in 1 M KOH solution), 3.3 $mA.cm^{-2}$ (at 0.8 V vs. Ag/AgCl)	[33]
3	NCD-gC_3N_4/α-Fe_3O_4	Using ethylenediamine and citric acid as precursors, NCD was produced in an oil bath at 120°C. The NCD@CNFO nanocomposite was developed by using the EDC/NHS process to conjugate the carboxylic groups of NCD with the amine groups of the 3-aminopropyl)-triethoxysilane (APTES) coated CNFO.	95% TMP degradation (at 0.1 M N_2SO_4), 3.07 $mA.cm^{-2}$ (at 1.6 V vs. NHE in 0.1 M N_2SO_4) 550 μmol cm^{-2} h^{-1} Hydrogen generation	[34]
4	CQDs/TiO_2	Precursor materials L-glutamic acid (L-Glu) and m-phenylenediamine (MPD) were prepared hydrothermally at 200°C.	70.56% of MO (40 ppm) degradation in 6 h at 600 nm light source, 78.75% under 365 nm.	[35]

Table 3.
Summarizes the graphene-based up-conversion materials available for photo splitting cells.

that the addition of nitrogen-doped graphene quantum dots (N-GQDs) enhanced the photoelectrochemical (PEC) performance of the TiO_2 photoanode. The photo electrocatalytic (PEC) efficiency of TiO_2 was evaluated with and without the inclusion of upconversion N-GQDs. The upconversion N-GQDs displayed a high quantum yield of 51.5% and, when paired with TiO_2 nanoarrays, showed an enhanced photocurrent density of 3.0 mA.cm^{-2} at 1.23 V vs. RHE. Zhao Liang et al. introduced hydrogenated TiO_2 nanorod arrays adorned with carbon quantum dots to improve the effectiveness of photoelectrochemical water splitting, as detailed in their publication [33]. The limited ability to absorb light and collect charges due to its large energy gap is acknowledged as a significant obstacle in developing highly efficient TiO_2 photoanodes. To overcome these intrinsic constraints, the group has developed a photoanode made of TiO_2 nanoarrays that underwent hydrogenation treatment and were embellished with carbon quantum dots (CQDs). The study showed that hydrogenation treatment leads to the formation of oxygen vacancies, which effectively prevents the recombination of photoinduced carriers. The decorated CQDs acted as electron reservoirs to absorb photoinduced electrons and enhanced the absorption of solar light through their upconversion action. The newly made photoanodes showed a significant water splitting photocurrent density of approximately 3.0 mA.cm^{-2} at 1.23 V compared to the reversible hydrogen electrode under simulated sunshine. This value was almost six times higher than that of the original TiO_2 photoanodes. Van Dien Dang et al. [34] described the photoelectrochemical (PEC) breakdown of trimethoprim and hydrogen evolution. The accomplishment was realized utilizing a photoanode composed of N-doped carbon dots (NCD) mixed with g-C_3N_4/αFe_2O_3 (CNFO) shell/core nanocomposite. The NCD@CNFO photoanode exhibited a photocurrent density of 3.07 mA.cm^{-2} at 1.6 V relative to the normal hydrogen electrode (NHE). This value was found four times more than CNFO and fifteen times greater than intact αFe_2O_3. The PEC system demonstrated the concurrent creation of hydrogen and the decomposition of TMP, leading to a hydrogen generation rate of 550 μmol cm^{-2} h^{-1}. The high efficiency labeled a photoelectrochemical (PEC) system using carbon dots/semiconductor hybrid catalysts to break down antibiotics and generate hydrogen through photocatalysis in wastewater. The article provided a different approach to dealing with issues with environmental pollution and the energy problem. NCD efficiently absorbed longer wavelengths (600-800 nm) from the light source and released shorter wavelengths (300-450 nm) because of its upconversion capabilities. As a result, the absorption efficiency of the NCD@CNFO nanocomposite was found improved. Yiqing Deng et al. [35] produced upconversion carbon quantum dots (CQDs) by a hydrothermal method involving L-glutamic acid (L-Glu) and m-phenylenediamine (MPD). The CQDs were combined with commercial nano-TiO_2 to produce CQDs/TiO_2 composites for photocatalytic destruction of methyl orange. The fluorescence spectra showed that the generated CQDs effectively converted visible light at approximately 600 nm into UV light at 350 nm. In photocatalysis experiments, a composite of carbon quantum dots (CQDs) and titanium dioxide (TiO_2) with a molar ratio of L-Glu to TiO_2 of 1:1 showed the best performance in breaking down methyl orange (MO). When exposed to a 600 nm light source, CQDs/TiO2 exhibited 70.56% degradation of MO (40 ppm) within 6 hours, a degradation rate equal to its 78.75% under 365 nm UV light. The degradation process was discretely associated with the presence of H^+ and $^{\bullet}OH$ under upconverted visible light, proposed by free radical scavenging studies and electron spin resonance (ESR) testing.

3. Conclusion

The book chapter offered a succinct and lucid rationale for the importance of sub-bandgap harvesting. Upconversion integration is a proposed strategy for collecting all sub-bandgap photons. However, limited research has been conducted in the field of solar splitting or fuel generation by upconversion. This book chapter offers a systematic examination of upconversion integrated semiconductor systems. These systems include graphene-based, inorganic lanthanide-based, and organic triptycene annihilation-based systems developed for solar cell splitting. Graphene-based upconversion materials are highly prospective candidates for efficient solar energy conversion due to their considerable specific surface area, high electrical conductivity, and tunable properties. Inorganic upconversion materials composed of rare earth elements are known for converting low-energy photons from the solar spectrum into higher-energy photons due to their exceptional photostability and efficient upconversion. In contrast, organic upconversion materials provide several benefits, including cost-effectiveness, straightforward processing, and the capacity to customize optical characteristics *via* molecular engineering. While organic materials may have a lower upconversion efficiency than their inorganic counterparts, ongoing research endeavors are focused on surmounting these constraints in order to fully exploit the capabilities of organic materials in the area of solar energy conversion. Through ongoing research and development, scientists anticipate uncovering novel materials that exhibit superior upconversion properties and demonstrate enhanced performance in practical applications. Upconversion materials are crucial for advancing toward a sustainable energy future. This technology holds great promise in addressing the pressing global issues surrounding energy and the environment. To increasingly capture sub-bandgap photons, we require a substantial amount of active participation in order to develop a significantly more effective upconversion semiconductor integrated system. This chapter provides guidance to those who wish to do research on sub-bandgap photon harvesting for solar splitting applications.

Acknowledgements

The authors would like to thank the Department of Science and Technology (DST/TMD-EWO/AHFC-2021/2021/111). The authors are grateful for the invaluable help provided by Rajiv Gandhi Institute of Petroleum Technology.

Conflict of interest

The authors declare no conflict of interest.

Author details

Amit Kumar Verma[1], Prerna Tripathi[1], Akhoury Sudhir Kumar Sinha[2*] and Shikha Singh[1*]

1 Department of Sciences and Humanities, Rajiv Gandhi Institute of Petroleum Technology, Jais, Amethi, India

2 Department of Chemical Engineering and Biochemical Engineering, Rajiv Gandhi Institute of Petroleum Technology, Jais, Amethi, India

*Address all correspondence to: asksinha@rgipt.ac.in and shikhas@rgipt.ac.in

References

[1] Zhou J, Liu Q, Feng W, Sun Y, Li F. Upconversion luminescent materials: Advances and applications. Chemical Reviews. 2015;**115**(1):395-465. DOI: 10.1021/cr400478f

[2] Singh-Rachford TN, Castellano FN. Photon upconversion based on sensitized triplet–triplet annihilation. Coordination Chemistry Reviews. 2010;**254**(21):2560-2573. DOI: 10.1016/j.ccr.2010.01.003

[3] Gai S, Li C, Yang P, Lin J. Recent Progress in rare earth micro/nanocrystals: Soft chemical synthesis, luminescent properties, and biomedical applications. Chemical Reviews. 2014;**114**(4):2343-2389. DOI: 10.1021/cr4001594

[4] Dienel T, Proehl H, Fritz T, Leo K. Novel near-infrared photoluminescence from platinum(II)-porphyrin (PtOEP) aggregates. Journal of Luminescence. 2004;**110**(4):253-257. DOI: 10.1016/j.jlumin.2004.08.017

[5] Singh A, Johnson LW. Phosphorescence spectra and triplet state lifetimes of palladium octaethylporphyrin, palladium octaethylchlorin and palladium 2,3-dimethyloctaethylisobacteriochlorin at 77 K. Spectrochimica Acta Part A: Molecular and Biomolecular Spectroscopy. 2003;**59**(5):905-908. DOI: 10.1016/S1386-1425(02)00258-5

[6] Urbani M, Ragoussi M-E, Nazeeruddin MK, Torres T. Phthalocyanines for dye-sensitized solar cells. Coordination Chemistry Reviews. 2019;**381**:1-64. DOI: 10.1016/j.ccr.2018.10.007

[7] Wu W, Ji S, Wu W, Shao J, Guo H, James TD, et al. Ruthenium(II)–Polyimine–Coumarin light-harvesting molecular arrays: Design rationale and application for triplet–triplet-annihilation-based Upconversion. Chemistry – A. European Journal. 2012;**18**(16):4953-4964. DOI: 10.1002/chem.201101377

[8] Yi X, Zhao J, Wu W, Huang D, Ji S, Sun J. Rhenium(i) tricarbonyl polypyridine complexes showing strong absorption of visible light and long-lived triplet excited states as a triplet photosensitizer for triplet–triplet annihilation upconversion. Dalton Transactions. 2012;**41**(29):8931-8940. DOI: 10.1039/C2DT30804E

[9] Yi X, Yang P, Huang D, Zhao J. Visible light-harvesting cyclometalated Ir(III) complexes with pyreno[4,5-d]imidazole C^N ligands as triplet photosensitizers for triplet–triplet annihilation upconversion. Dyes and Pigments. 2013;**96**(1):104-115. DOI: 10.1016/j.dyepig.2012.07.020

[10] Gao C. Organic Chromophore Aggerates for Solid-State Photon Upconversion. The University of Melbourne; 2019. Available from: http://hdl.handle.net/11343/230661

[11] Wang X, Zhuang J, Peng Q, Li Y. Hydrothermal synthesis of rare-earth fluoride nanocrystals. Inorganic Chemistry. 2006;**45**(17):6661-6665. DOI: 10.1021/ic051683s

[12] Li C, Yang J, Quan Z, Yang P, Kong D, Lin J. Different microstructures of β-NaYF4 fabricated by hydrothermal process: Effects of pH values and fluoride sources. Chemistry of Materials. 2007;**19**(20):4933-4942. DOI: 10.1021/cm071668g

[13] Vetrone F, Naccache R, Mahalingam V, Morgan CG,

Capobianco JA. The active-core/active-Shell approach: A strategy to enhance the upconversion luminescence in lanthanide-doped nanoparticles. Advanced Functional Materials. 2009;**19**(18):2924-2929. DOI: 10.1002/adfm.200900234

[14] Manoj Kumar M, Hans Christian H, Ulrich V. Photon-Upconverting materials: Advances and prospects for various emerging applications. In: Jagannathan T, editor. Luminescence. Rijeka: IntechOpen; 2016. p. Ch. 6. DOI: 10.5772/65118

[15] Bie C, Wang L, Yu J. Challenges for photocatalytic overall water splitting. Chem. 2022;**8**(6):1567-1574. DOI: 10.1016/j.chempr.2022.04.013

[16] Barawi M, Fresno F, Pérez-Ruiz R, de la Peña O'Shea VA. Photoelectrochemical hydrogen evolution driven by visible-to-ultraviolet photon Upconversion. ACS Applied Energy Materials. 2019;**2**(1):207-211. DOI: 10.1021/acsaem.8b01916

[17] Khnayzer RS, Blumhoff J, Harrington JA, Haefele A, Deng F, Castellano FN. Upconversion-powered photoelectrochemistry. Chemical Communications. 2012;**48**(2):209-211. DOI: 10.1039/c1cc16015j

[18] Choi D, Nam SK, Kim K, Moon JH. Enhanced photoelectrochemical water splitting through bismuth vanadate with a photon upconversion luminescent reflector. Angewandte Chemie International Edition. 2019;**58**(21):6891-6895. DOI: 10.1002/anie.201813440

[19] Lee J-W, Cho K-H, Yoon J-S, Sung Y-M. Enhanced IR-driven photoelectrochemical responses of CdSe/ZnO heterostructures by up-conversion UV/visible light irradiation. Nanoscale. 2020;**12**(15):8525-8535. DOI: 10.1039/d0nr00477d

[20] Lim Y, Lee SY, Kim D, Han M-K, Han HS, Kang SH, et al. Expanded solar absorption spectrum to improve photoelectrochemical oxygen evolution reaction: Synergistic effect of upconversion nanoparticles and ZnFe2O4/TiO2. Chemical Engineering Journal. 2022;**438**:135503. DOI: 10.1016/j.cej.2022.135503

[21] Thuy TNT, Atabaev TS, Vu HHT, Lee D, Kim HK, Hwang Y-H. TiO2 thin films sensitized with upconversion phosphor for efficient solar water splitting. Journal of Nanoscience and Nanotechnology. 2017;**17**(10):7647-7650. DOI: 10.1166/jnn.2017.14772

[22] Jiang Q, Xie X, Riley DJ, Xie F. Harvesting the lost photon by plasmonic enhanced hematite-upconversion nanocomposite for water splitting. The Journal of Chemical Physics. 2020;**153**(1):011102. DOI: 10.1063/5.0013060

[23] Zhu Y, Zheng X, Zhang W, Kheradmand A, Gu S, Kobielusz M, et al. Near-infrared-triggered nitrogen fixation over upconversion nanoparticles assembled carbon nitride nanotubes with nitrogen vacancies. ACS Applied Materials & Interfaces. 2021;**13**(28):32937-32947. DOI: 10.1021/acsami.1c05683

[24] Kwon H, Marques Mota F, Chung K, Jang YJ, Hyun JK, Lee J, et al. Enhancing solar light-driven photocatalytic activity of mesoporous carbon–TiO2 hybrid films via upconversion coupling. ACS Sustainable Chemistry & Engineering. 2018;**6**(1):1310-1317. DOI: 10.1021/acssuschemeng.7b03658

[25] Chilkalwar AA, Rayalu SS. Synergistic plasmonic and upconversion effect of the (Yb,Er)NYF-TiO2/Au composite for photocatalytic hydrogen generation. The Journal of Physical

Chemistry C. 2018;**122**(46):26307-26314. DOI: 10.1021/acs.jpcc.8b05480

[26] Boppella R, Marques Mota F, Lim JW, Kochuveedu ST, Ahn S, Lee J, et al. Plasmon and upconversion mediated broadband spectral response in tio2 inverse opal photocatalysts for enhanced photoelectrochemical water splitting. ACS Applied Energy Materials. 2019;**2**(5):3780-3790. DOI: 10.1021/acsaem.9b00469

[27] Feng K, Cai Z, Huang D, Li L, Wang K, Li Y, et al. Near-infrared-driven water splitting for hydrogen evolution using a Cu2ZnSnS4-based photocathode by the application of upconversion nanoparticles. Sustainable Energy & Fuels. 2020;**4**(6):2669-2674. DOI: 10.1039/D0SE00152J

[28] Gao W, Wu Y, Lu G. 980 nm NIR light driven overall water splitting over a combined CdS–RGO–NaYF4–Yb3+/Er3+ photocatalyst. Catalysis Science & Technology. 2020;**10**(8):2389-2397. DOI: 10.1039/D0CY00256A

[29] Gao W, Tian B, Zhang W, Zhang X, Wu Y, Lu G. NIR light driven catalytic hydrogen generation over semiconductor photocatalyst coupling up-conversion component. Applied Catalysis B: Environmental. 2019;**257**:117908. DOI: 10.1016/j.apcatb.2019.117908

[30] Verma AK, Tripathi P, Dubey A, Vishwakarma NK, Sinha ASK, Singh S. Visible light-promoted enhanced photocatalytic hydrogen generation by the CeF3:Ho3+-incorporated TiO2 Nanosystem. ACS Applied Energy Materials. 2023;**6**(11):5739-5752. DOI: 10.1021/acsaem.3c00113

[31] Gao W, Zhang W, Tian B, Zhen W, Wu Y, Zhang X, et al. Visible light driven water splitting over CaTiO3/Pr3+-Y2SiO5/RGO catalyst in reactor equipped artificial gill. Applied Catalysis B: Environmental. 2018;**224**:553-562. DOI: 10.1016/j.apcatb.2017.10.072

[32] Saeidi S, Rezaei B, Ensafi AA. Fabrication and characterization of upconversion N-doped graphene quantum dots for improving photoelectrocatalytic performance of rutile hierarchical TiO_2 nanowires under visible and near-infrared light irradiations. Materials Today Chemistry. 2022;**23**:100742. DOI: 10.1016/j.mtchem.2021.100742

[33] Liang Z, Hou H, Fang Z, Gao F, Wang L, Chen D, et al. Hydrogenated TiO2 Nanorod arrays decorated with carbon quantum dots toward efficient photoelectrochemical water splitting. ACS Applied Materials & Interfaces. 2019;**11**(21):19167-19175. DOI: 10.1021/acsami.9b04059

[34] Dang VD, Annadurai T, Khedulkar AP, Lin J-Y, Adorna J, Yu W-J, et al. S-scheme N-doped carbon dots anchored g-C3N4/Fe2O3 shell/core composite for photoelectrocatalytic trimethoprim degradation and water splitting. Applied Catalysis B: Environmental. 2023;**320**:121928. DOI: 10.1016/j.apcatb.2022.121928

[35] Deng Y, Chen M, Chen G, Zou W, Zhao Y, Zhang H, et al. Visible–ultraviolet upconversion carbon quantum dots for enhancement of the photocatalytic activity of titanium dioxide. ACS Omega. 2021;**6**(6):4247-4254. DOI: 10.1021/acsomega.0c05182

Chapter 8

The Two-Step Innovative Smart Energy Storage: Photoelectrochemical Materials for the Fabrication of High-End, High-Efficiency Smart Energy Storage Devices

Farai Dziike

Abstract

The topic explores advances in innovative high-end technological developments that revolutionize energy loading schemes through high-energy storage capacity. A highly efficient energy conversion mechanism for photoelectron charging and discharging systems is engineered. The result is a smart energy storage design that is sustainable and conforms to a smart energy distribution with zero energy losses through the transmission infrastructure. The topic unpacks the choice of chalcogenide materials previously known to have exceptional photoelectrochemical properties and their innovative morphological manipulation into few-layered thin films of metal chalcogenides such as In_xSe_y, Mo_xS_y, $In_xTi_ySe_z$, Mo_xSe_y, and many other photoelectrochemical materials. These materials have been used to fabricate supercapacitors, solar cells, sensors, batteries, and other superior smart energy conversion and storage devices. These latest innovative smart storage devices composed of photoelectrochemical materials have paved the attainment of high-end, highly efficient smart storage devices that have translated into the advancement of artificial intelligence and remote technologies including robotic devices, drones, satellite equipment. The two-step innovative smart energy devices are characterized by advanced mechanisms of high quantum energy packing and then smart discharge and energy deployment with minimum of zero losses during transmission.

Keywords: supercapacitors, dye-sensitized solar cells, chalcogenides, electropositive, sensors

1. Introduction

Recent technological developments in "smart energy storage" enable a system wherein energy storage devices such as batteries, supercapacitors, and fuel cells charge and reserve energy at and release electrical discharge through transmission systems of the energy transfer channels. Smart storage technologies have found a wide range of applications, including micro-scales in miniaturized surveillance cameras, codeless audio devices, and micro-chip detectors in security and medical applications. The macro-levels include power banks such as industrial-scale batteries, supercapacitors, and giant solar electropower stations. In recent years, advancements in smart energy storage have facilitated the innovative development of robotics, which has unlocked the advent of unmanned high-tech machines that perform sophisticated tasks including drones and self-driving motor vehicles that are electrically powered (EV). These high-tech machines rely on two-step innovative smart energy storage of photoelectrochemical materials of high-end, high-efficiency smart energy storage devices, as shown in **Figure 1**.

Military technology performs sophisticated operations on a backdrop of high-end, high-efficiency smart energy storage devices and has also embraced smart energy systems including laser power for scanning and targeted precision destruction points. Military equipment is now powered using smart storage devices such as supercapacitors and highly efficient batteries, including, more recently, hydrogen power from a photoelectrochemical catalyzed water splitting fuel tank with a fixed bed catalytic bed. The solar-powered public EV charging stations (**Figure 1**), and the military laser weapon equipment (**Figure 2**), are backed by a powerful energy storage device that has high efficiency in both energy harnessing and energy dissipation to sustain the performance operations. Similarly, photoelectrochemical materials used to fabricate innovative smart storage devices have extended uses by companies and households that may generate their own electricity under stipulated regulations and install smart energy storage systems.

Power grids provide constant power supply by optimally harmonizing supply and demand on the backdrop of the installed power capacity. Thus, the two-step innovative smart energy storage photoelectrochemical materials for the fabrication of high-end, high-efficiency smart energy storage devices guarantee sustainable energy supply. However, the use of solar power and other renewable energy sources have

Figure 1.
Solar-powered public EV charging station.

Figure 2.
High energy laser military—Grade weapon. Fitted with high-efficient supercapacitor energy storage.

unstable yields, and a continued feed of electric power supply to the electricity supply grid of a community could lead to grid instability. This brings up economic, political, sociocultural, technological, and environmental challenges [1]. These challenges are therefore best addressed through smart storage using innovative technologies that allow harnessing and storing energy.

Smart energy storage will have a direct economic impact. Applications of smart energy storage technologies will improve energy access and its sustainability. This phenomenon may be overcome by developing technologies that store electrical power. The storage of electrical energy in smart energy storage systems balances the electrical load, thereby advancing efficiency in the smart generation and use of energy. Thus, smart energy storage technologies serve as backup power sources in times of need or high energy demand [2].

Current trends in smart energy storage are characterized by practices developed from scientific and industrial research targeting to employ technologies that will provide cost-effective energy for an cost-effective economy without straining resources of the national, regional, and global environment [3–6]. Charge-discharge innovations include molybdenum disulfide/mesoporous carbon nanocomposites, molybdenum sulfide/tellurium hetero-composite, $NiSe_2$ sheet-like nano-architectures (**Figure 3**), all structured into high-performance supercapacitor electrode material. Similarly, materials such as crude detonated nanodiamonds/titania, two-dimensional metal chalcogenides analogous $NiSe_2$ nanosheets, defect-rich MoS ultrathin nanosheets, and Cu_2ZnSnS_4 ultrathin film have been used to make supercapacitors, solar cells, and counter electrodes in the photoelectrochemical smart energy storage systems [7, 8].

South Africa is pursuing opportunities in an energy system enhanced economically, socially, and environmentally. Thus, technologies are being explored within timeframes that best fit sustainable solutions in electricity supply, energy efficiency, renewable energy sources, and cleaner, green fossil fuels. Thus, the current trends are drawn along hydroelectric power reserves, fossil fuels, coal gasification, biogas, sewerage works, liquefied natural gas, and pebble-bed nuclear energy reactors. Future prospects of smart storage include low-temperature superconductors, fast breeder

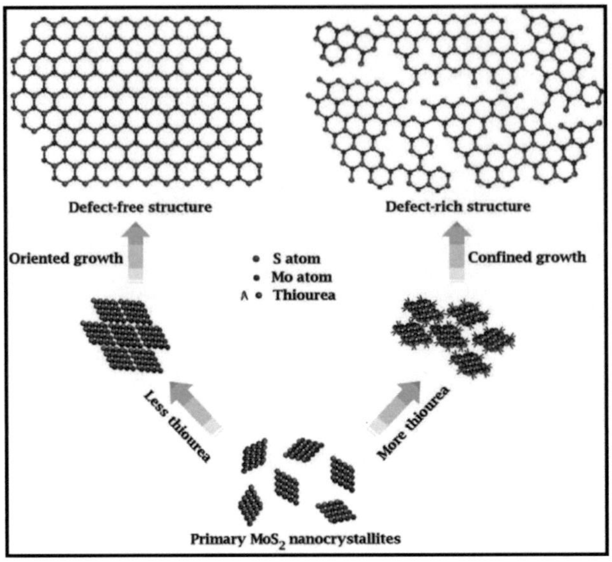

Figure 3.
Structural models of defect-free and defect-rich structures and the designed synthetic pathways for the MoS_2 nanocrystallites.

nuclear reactors, hydrogen fuel, advanced solar technologies, molten salt power towers, and artificial photosynthesis of sunlight into stored energy [1, 5].

Contemporary research in the energy focus area seen the development of innovative technologies aimed at providing alternative sources of energy and energy storage above and below what the current market has to offer. Competing technologies specialize in one or more of the following energy storage technologies, including lithium-ion batteries, compressed air and liquid fuel energy storages (CAES and LFES), vanadium/zinc flow batteries (VFB and ZFB), and solar photovoltaics and supercapacitors. These technologies have engineering designs that have two-step mechanisms of photoelectrochemical energy quanta packaging and efficient energy dissipation through electro-current discharge [9–12].

The use of smart energy storage systems and smart grids, serve converting inefficient energy distribution processes, by incorporating two-way charge-discharge technology is fundamental to revolutionizing energy conversion. This is one of the adopted strategies to mitigate the risks of the development of smart energy storage technologies. The digital technology in a smart energy grid permits communication between the smart energy storage unit and its distribution along with the sensors in the transmission lines. The technologies that power the Internet, satellite stations, space technologies, and the smart electricity grid consist of controls, computers, automation, and new technologies and equipment working together for a sustainable energy system in what is called IoT smart grid technology [13]. **Figure 4** shows a sustainable energy conversion system incorporating photovoltaic solar energy panels efficiently powering the harnessing of energy from the sun, converting it into electrical energy, and transmitting it through the distribution grid to an air-compressing motor. The compressed air is stored as stored potential energy, which, upon sensory digital technology activation, is then pumped through a generator to produce electrical energy. These technologies work with the electrical grid to respond digitally to the rapid dynamical electricity demand. The energy that is conventionally used to get

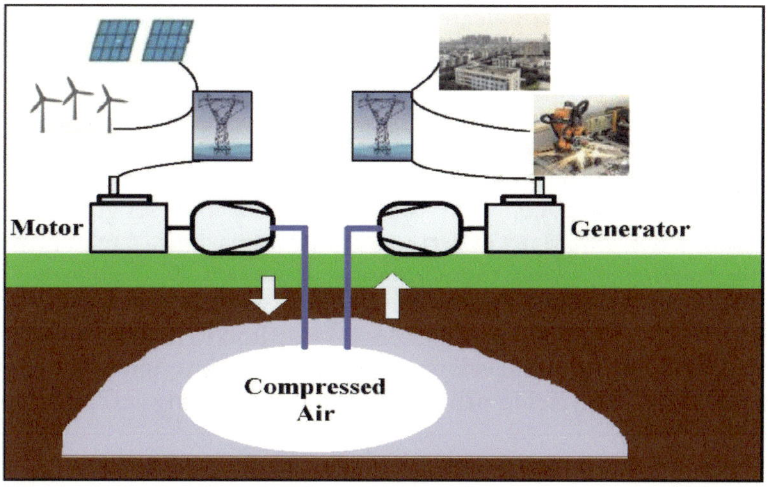

Figure 4.
Overview of compressed air energy storage and technology development.

lost along transmission networks is converted into some form of stored energy and smartly stored for easy conversion and access [14].

Supercapacitor batteries have innovatively been revolutionized to industrial-scale capacity for giant smart energy storage equipment. Their industrial application serves both regulatory and smart storage purposes and is a testament to revolutionizing energy conversion. **Figure 5** shows an industrial-scale supercapacitor for energy storage converted from different energy systems. However, there is a risk of smart grid failure resulting in stored energy losses and limiting utilities to detect problems and redirect power automatically, leading less easily to outages and increasing the period of those affected. A smarter energy grid is envisioned to make it easier to connect

Figure 5.
Smart energy system: Super- and ultracapacitors and hybrid electricity storage.

renewable energy sources to the electricity grid, reducing greenhouse gas emissions. The smart energy grid also allows consumers to monitor and adjust their energy usage away from peak periods, thus further benefiting the consistency of the grid. This should also assist the power utility in managing power demands by customers and monitoring power distribution.

Eventually, the smart energy grid offers grid operators, utilities, and consumers considerably more data about energy usage, and each stakeholder group can use this data to improve efficiency. Many experts agree that a smart grid can improve energy efficiency by nearly 10%, saving the economy as much as $42 billion in energy costs. However, the development and application of these fine technologies pose economic challenges as the use of such smart technologies may lead to increased costs for consumers, especially in underdeveloped communities. Thus, photoelectrochemical technologies have played a fundamental role in revolutionizing energy conversion and its sustainability [6].

Recent developments in smart energy storage technology are making batteries a viable asset that is easy to incorporate as a key part of the power grid. The use of batteries, whether at your home or at an electric utility site, can be used to store energy when the power demand is low, and production is high. Energy generated from solar panels, wind power, biogas, hot springs, ocean waves, hydropower generation, or conventional coal-powered electricity is being used to charge up battery storage packs.

The photoelectrochemical technologies allow increased efficiency of energy packing in smart storage devices through widespread connection of battery storage technology to mitigate the fluctuations of supply and demand. Various smart energy storage technologies are being developed to advance and innovate on the current regimes of storage packs. For example, Leaper Innovative Green Energies (LIGE) is a South African company that recently developed a new type of CAES air battery, as shown in **Figure 6** [14]. Innovative research is being carried out at various levels

Figure 6.
LiGE compressed air energy storage system (CAES) for smart electricity.

so that energy storage will undergo a major transformation in the coming decades through revolutionizing energy conversion from different smart energy storage systems. Sustainable energy sources such as wind and solar will replace coal and gas in the near future. Storing energy is a fundamental solution, but it is not an easy phenomenon. Great leaps in research and innovation have been made in smart energy storage by developing photoelectrochemical materials for the fabrication of high-end, high-efficiency smart energy storage devices or technologies that are economical, sustainable, environmentally friendly, and technologically viable [15]. Thus, it is important to understand the mechanisms of photoelectrochemical materials in energy conversion and smart energy storage technologies that allow the accumulation and storage of energy. This chapter explores the two-step innovative smart energy storage using photoelectrochemical materials for the fabrication of high-end, high-efficiency smart energy storage devices.

2. Photoelectrochemical technologies

Photoelectrochemistry propagates the concept of electrochemical current produced when ultraviolet light is irradiated onto an electrode in solution. Photoelectrochemistry thus involves the absorption of incoming light in the valence band of a semiconductor material [16]. This results in excitation of electrons in that band to the conduction band, which in a photoanode is followed by an electrochemical interfacial reaction that uses electrons from ions in the interface to fill the holes in the valence band created by the departure of the excited electron. This state of charge separation and sustaining the potential difference is the adopted mechanism of packing electricity in the photoelectrochemical semiconducting material [17]. The photoelectrochemical materials that exhibit superiority in the sustenance of the electronic excited state in the conduction band are used to fabricate solar cells and supercapacitors. Detailed studies have been done on different materials, especially metal chalcogenides such as Ni-MoS_2 and Bi_2WO_6 and CZTS. The molybdenum disulfide (MoS_2) in the above matrix exhibits a graphene-like layered structure, as shown in **Figure 7**, in which S–Mo–S covalent bonds exist within the plane and van der Waals forces hold the plane [18].

In the photoelectron-activated system of the chemical material (**Figure 8**), it is the p-doped semiconductors that provide cathodic electrons when irradiated with light of sufficient energy and the n-doped type semiconductors that yield the holes to act as photoanodes when semiconductors are used. In cells involving semiconductors but driven by an outside power source rather than by incident photons, the situation is reversed; the n-type emits electrons and the p-type receives them [12, 17]. However, the production of current in a photo electro chemical cell (PEC-cell) is not limited to UV light. An n- or p-type semiconductor is generally used as a working electrode with a platinum (Pt) counter electrode. Electron-hole pairs are generated on the working electrode by photon absorption with an energy level that is equal to or higher than the bandgap (E_g) of the photoanode semiconductor. In addition, the structure of the PEC cells is not limited to photoanode, and we can have photocathode as well. If an n-type semiconductor is used, electrons are collected in the photoanode and are subsequently transported to the counter electrode through an external circuit. The photogenerated electrons are consumed to reduce H^+ into H_2 at the cathode, while holes take part in the oxidation of water into O_2 and H^+ at the anode. In a photocathodic PEC system, a p-type semiconductor is employed as the working electrode, and photogenerated electrons are used to reduce H^+ into H_2, while in the counter electrode, water is

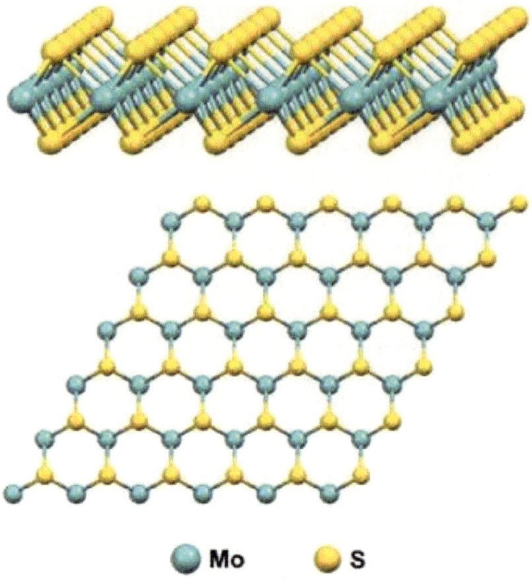

Figure 7.
Electronic structure of a single MoS_2 monolayer.

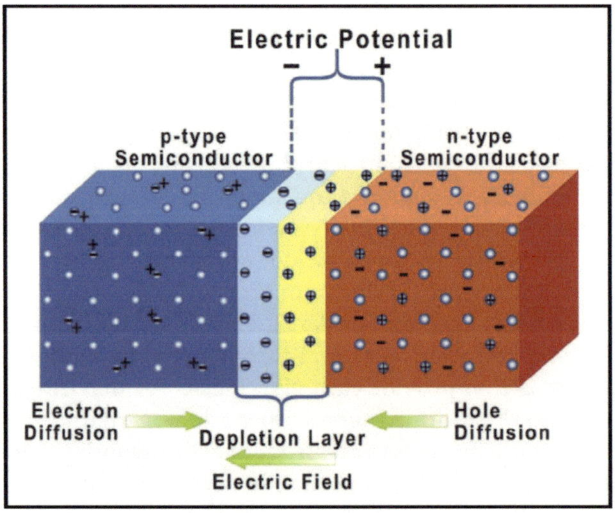

Figure 8.
Semiconductor physics—Electric field inside a p-n junction.

oxidized into O_2 and H^+. In summary, n-type semiconductors produce anodic photocurrent in which holes are transferred toward the electrolyte, while p-type semiconductors generate a cathodic photocurrent by transmitting electrons toward the electrolyte [19].

The choice of photoelectrochemical materials for the fabrication of high-end, high-efficiency smart energy storage devices is based on the material inherently exhibiting excellent metal conductivity, abundant surface functional terminals, exposed terminal active metal sites, good dispersion in water, and large specific surface area [20]. This quality has been predominantly found in materials called MXenes

DOI: http://dx.doi.org/10.5772/intechopen.1005424

and has found wide applications in capacitor preparation, electrocatalysis, energy storage batteries, electrochemical sensors, photothermal conversion, water purification, biomedicine, water splitting, photocatalysis, and other fields [20].

The MXene is a unique two-dimensional (2D) material made up of carbides, nitrides, or carbonitrate of transition metal (**Figure 9**). Presently, MXene attracts the attention of researchers due to its unique properties including its excellent electrical metal conductivity, charger mediator, high thermal conductivity, environmental flexibility, and tunable band gap ranging from 0.92 to 1.75 eV [22]. Previous studies reported hybrid composites engineered in an *in situ* heterojunction construction using partially oxidized $Ti_3C_2T_x$ sheets and photo-active $NiWO_4$ nanoparticles used for partial surface oxidation of $Ti_3C_2T_x$ to form $Ti_3C_2T_x$-TiO_2/$NiWO_4$ hybrid composite with a general matrix of MX-$NiWO_4$. Similar work was done in research designed to unpack the process of synthesis, mechanisms, challenges, and future prospects of applications of Ti_3C_2 MXene and its heterojunctions for photocatalytic dye degradation efficiency (**Figure 10**) [23]. Fabrication of smart energy storage materials allows access to both surface and internal architectures of the photoelectrochemical materials, which, in turn, enhance band gaps that facilitate visible light absorption. Material nucleation during crystallization is manipulated such that the surface is composed of O—functional groups. This makes the material inherently have greater light absorption capacity through enhanced reflectance and absorption in the visible range. Therefore, the O—functionary surface exceedingly outperforms the -OH and -F functionalities [24]. This phenomenon constitutes the first of the two-step innovative smart energy storage.

Figure 10 shows how the electronic transitions are set off by UV light. The mechanism illustrates how the material undergoes electron charge shifts, setting potential differences that instigate charge flow, hence electricity. The pre-eminent photoelectrochemical technology innovation for smart energy storage involves the use of semiconductors formed by taking binary building blocks and cross-substituting them to form ternary and quaternary systems. The access to internal architecture of these structures is used to tune the physiochemical properties while retaining the morphological structure of the material during high-end fabrication of the energy storage device [26].

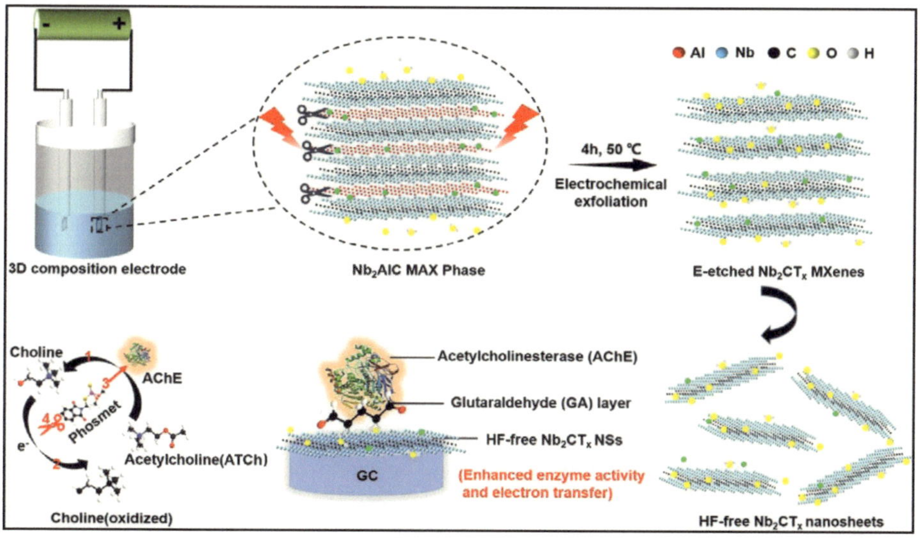

Figure 9.
Schematic diagram of fluorine-free electro-etching, separation, and application of Nb_2C [21].

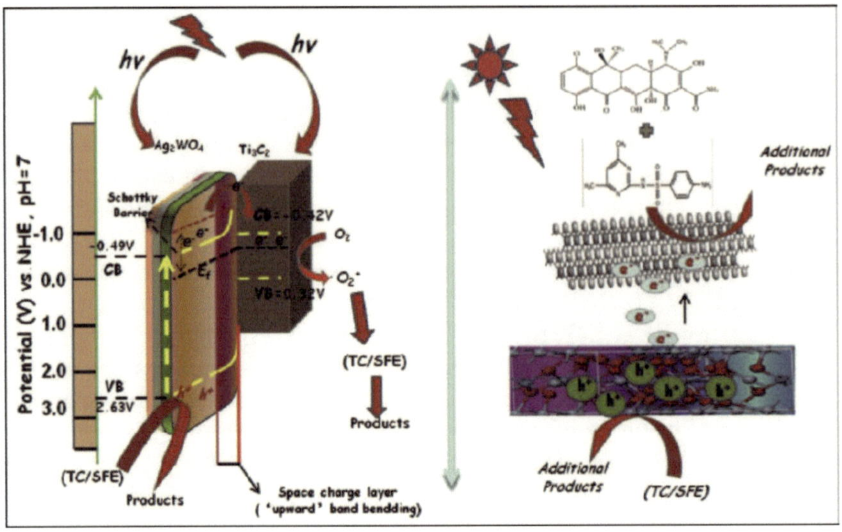

Figure 10.
Schematic of the mechanism of titan tetracycline hydrochloride (TC) and sulfadimidine (SFE) photodegradation over the Ag_2WO_4/Ti_3C_2 Schottky catalyst and its interfacial photocarrier transfer process [25].

3. Smart energy storage technologies: batteries

The second step of the innovative smart energy storage is the packing of the energy into a battery system. An electric battery is a source of electric power consisting of one or more electrochemical cells with external connections for powering electrical devices [28]. When a battery is supplying power, its positive terminal is the cathode, and its negative terminal is the anode. Photovoltaic cells have been widely used in the assembly of photovoltaic system battery storage including dye sensitized solar cells, Si photovoltaics, Li-ion batteries, and photoelectrochemical photovoltaic cells. However, the photoelectrochemical photovoltaic contains two interfaces at which charge transport has to switch from electronic to ionic and vice versa, as in battery cells. In contrast, the conventional electrochemical cells without dyes both have the semiconductor electrode and the counter electrode immersed in the redox electrolyte. **Figure 11** illustrates the mechanism of energy conversion, and it involves the incident light exciting the semiconductor electrode and the photogenerated electrons and holes being separated in the space charge region [9, 12].

The performance of photoelectrochemical is reportedly modified using a wide range of semiconducting materials. The interfacial modification of semiconductors with $Ti_3C_2T_X$ MXene provides internal control of the transfer direction of photogenerated positive holes and realizes an efficiently spatial directed distribution of photogenerated charges [27–29]. The effectiveness of the development of green and renewable energy technologies for the continuous supply of secure and reliable energy is increasingly becoming more dependent on energy storage devices that can fulfill the requirements of short-term and long-term durable energy outputs. The innovative smart energy storage using photoelectrochemical materials is revolutionizing energy conversion toward sustainability and efficiency. However, batteries are highly attractive due to their higher energy density, but they possess limited power density compared to that required for the prolonged use of energy. Capacitors have

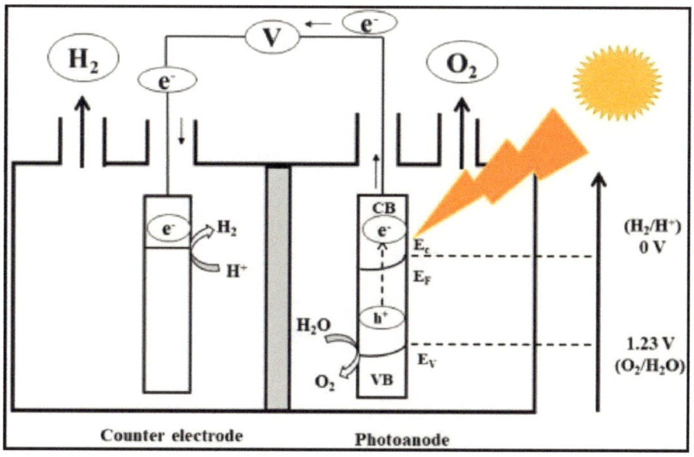

Figure 11.
The basic principles of water splitting for a photoelectrochemical cell. E_F, Fermi level.

countered the batteries' shortcomings by inherently having a high-power density delivery but suffering a low-energy storage capacity [30].

The advent of innovative photoelectrochemical technologies enhanced the role in sustainability of both batteries and conventional capacitors by revolutionizing energy conversion. The performance gaps of batteries can be sustained with the rigorous, innovative development of superior models of two-step innovative photoelectrochemical smart energy storage materials and techniques, in the form of electrochemical capacitors, supercapacitors, and ultracapacitors, fabricated as high-end high-efficiency devices [12, 30]. The choice of photoelectrochemical materials for the development of batteries is significantly drawn from the exhibition of its novel physical and chemical characteristics, including uniform planer structure, strong metallic conductivity, effective functional groups, and consistency in the structural architecture of the material matrix. MXene-based materials have been found to possess excellent photoelectrochemical performance and long-term stability [24].

A tandem PEC is a technology developed to sustainably revolutionize energy conversion by water splitting for hydrogen production in a device architecture based on different photoelectrode absorbers, and there are two main models including photoanode/photocathode (PEC/PEC) and photoelectrode/photovoltaic (PEC/PV) tandem cells. Light absorption and energy band matching are energy conversion mechanisms for enhanced solar-to-hydrogen (STH) efficiency [31]. This two-step, innovative smart energy storage promotes the performance of standalone semiconductor material and finding new materials. **Figure 12** shows a tandem PEC cell composed of photoelectrochemical materials with applications in the fabrication of high-end, high-efficiency absorber materials connected in series and therefore operating at the same current density with an optimized configuration where the two absorbers are placed back-to-back, the photovoltaic electron collector and wired to the photocathode. Its hole collector is then wired to a water oxidation anode. It has been proved that in a dual-absorber PEC–PV tandem, the photoelectrode and the photovoltaic utilize photons from different regions of the solar spectrum to enable broad sunlight harvesting [32].

In an innovative technological development similar to Cu_2O-perovskite-IrO_2 tandem PEC cell fabrication, vanadium oxide compounds – based PEC cells have had several applications in electrochromic devices, including as cathode material

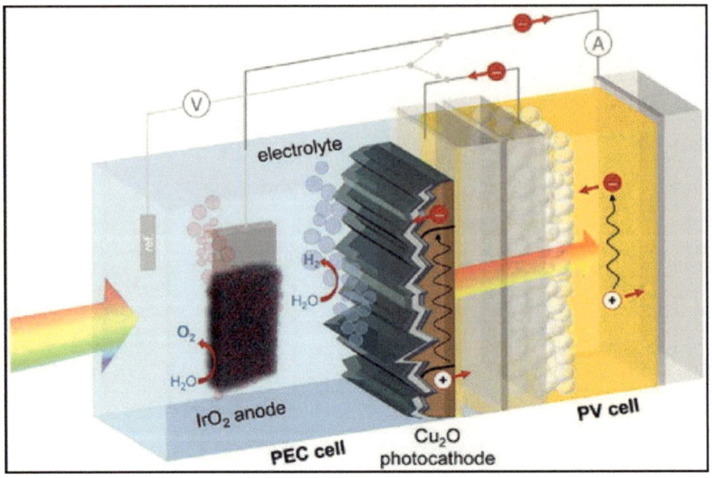

Figure 12.
Schematic illustration of the Cu_2O-perovskite-IrO_2 tandem PEC cell [31].

in batteries. Detailed research revealed that Co-based vanadium oxide materials possess superior characteristics such as magnetic field-induced transition, strong anisotropic character, and quantum criticality nature which are their unique attributes. Researchers have synthesized perovskite composites of Mg, Fe, Te, W, and Ni-based V_2O_6. However, studying the chemical and structural properties of these materials did not comprehensively cover the knowledge gap in understanding the physical properties of these compounds [33].

The structural, electronic, and optical properties of nickel vanadium oxide, NiV_2O_6, have been scrutinized using the plane-wave pseudopotential technique based on density functional theory (DFT) with generalized gradient approximation (GGA). The parameters determined in one comprehensive research study include optimized lattice constants of the material, the electronic band structure, total and partial densities of states, and the optical properties using computational methodology. Theoretical measurements including dielectric function, refractive index, reflectivity, absorption coefficient, loss function, and the photoconductivity of NiV_2O_6 were calculated. It was found that the band structure analysis corresponds to a compound of a direct band gap semiconductor with a band gap of 0.172 eV, and the Ni-3d states are substantively near the Fermi level. The photoelectrochemical analysis of optical performances reveals that NiV_2O_6 is a good dielectric material, and the large reflectivity in the low-energy region makes it a good candidate for the two-step innovative smart energy storage in a solar energy conversion system [12, 29, 33].

4. Innovative photoelectrochemical energy conversion technologies: supercapacitors

The characteristic feature of supercapacitors is their inherent consolidation of the two-step innovative smart energy storage with special manipulation of the matrix materials to be photoelectrochemical in a revolutionary energy conversion system that is both sustainable and efficient in performance. The fabrication of the high-end, high-efficiency sustainable smart energy storage supercapacitor device requires that the material be able to pack high-energy density along with high-power density in its matrix [34]. Previous studies reported on the utilization of the features of 2D-layered structure and high surface area of materials such as MoS_2 as electrodes for supercapacitors to achieve high-energy density and capacitance retention [7, 30, 35, 36].

Supercapacitor electrodes have been fabricated using WS_2 and MoS_2, and their energy storage and energy conversion performances were measured against same commercial materials. The electrochemical analysis revealed the pseudocapacitive character of disulfide-based supercapacitor electrodes. The materials exhibited a strong influence on the scan rate of the specific capacitance found, which was due to the diffusion of ions and the pseudocapacitive nature of charge storage. Thus, the MoS_2 and WS_2 are prepared by thermal decomposition for energy storage applications by means of supercapacitors and energy conversion through water electrolysis and hydrogen generation [35]. Similarly, high-performance MoS_2 thin film supercapacitor electrodes were prepared using a direct magnetron sputtering technique. The innovative and novel 3D porous structure of the MoS_2 film exhibited an excellent capacitance of approx.330 F cm^{-3} along with a high volumetric power and energy density of 40–80 W cm^{-3} and 1.6–2.4 mW h cm^{-3}, respectively. The high-end, high-efficiency smart energy storage device's optimized MoS_2 electrode shows outstanding cyclic stability, yielding capacitance retention of over 97% after 5000 cycles of charging/discharging. **Figure 13** shows an industrial-grade supercapacitor power bank modeled from the specifications articulated above [7, 34, 36].

In an innovative development of an effective electrode material for high-performance supercapacitors, MoS_2/Te nanocomposite was prepared. The two-step innovative smart energy storage device had electrochemical characterization of the as-fabricated nanocomposite electrode material showing a high specific capacitance of 402.53 F g^{-1} from a galvanostatic charge-discharge (GCD) profile conducted at 1 A g^{-1} current density. The electrode material exhibited significant rate performance with high cyclic stability reaching up to 92.30% under 4000 cycles of galvanostatic charge-discharge profile at a current density of 10 A g^{-1} [30]. Lithium-ion capacitors are the latest revolutionary energy conversion devices of two-step innovative smart energy storage, which combine the high-end and high-efficiency capacity of both lithium batteries and supercapacitors. The photoelectrochemical cathode materials

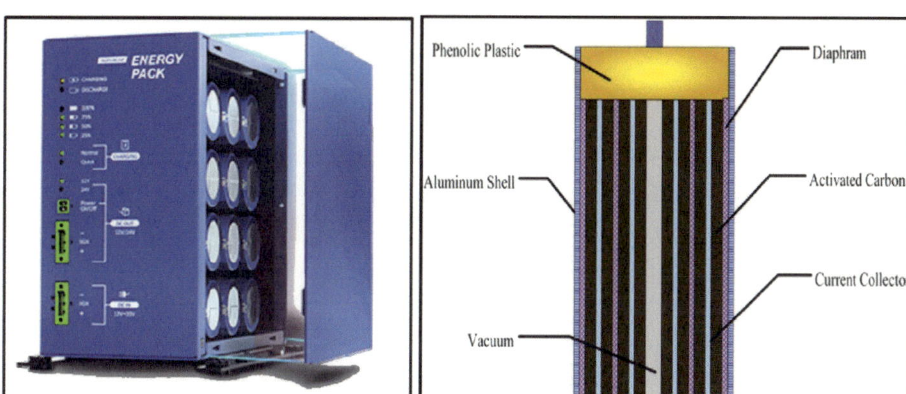

Figure 13.
Industrial-grade Supercapacitor power backup device and schematic for the working principle of the supercapacitor.

of lithium-ion capacitors have a large surface area that enhances adsorb/desorb Li+, while the anode materials have high lithium storage strength [29].

5. Conclusions

The ability of materials to facilitate high efficiency in energy conversion through electrochemical mechanisms of charge separation and subsequent charge transfer with photogenerated holes transfer in the opposite direction to the photogenerated electron in the conduction band. Thus, the use of solar power and other renewable energy sources have unstable yield, and a continued feed of electric power supply to the entire grid could lead to grid instability and hence the need to embark on revolutionizing energy conversion through the development of photoelectrochemical technologies with an advanced role in sustainability. The two-step innovative smart energy storage provides for sustainable storage of solar energy converted into electrical energy and is able to be discharged efficiently in a high-end charge/discharge electrochemical system. Photoelectrochemical materials for the fabrication of high-end, high-efficiency smart energy storage devices have been synthesized in the form of $Ti_3C_2T_X$ MXenes, In_xSey, Mo_xS_y, $In_xTi_ySe_z$, Mo_xSe_y, perovskite-based materials such as MoS_2/Te, Cu_2O-perovskite-IrO_2, and perovskite composites of Mg, Fe, Te, W, and Ni-based V_2O_6. These materials exhibited various degrees of efficiency in the fabrication of high-end, high-efficiency smart energy storage devices such as batteries, electrochemical capacitors, supercapacitors, and ultracapacitors.

Acknowledgements

The authors would like to acknowledge the Durban University of Technology for supporting the publication of this work through its Office of the Research and Postgraduate Support Directorate by funding this project. Gratitude is extended to the Department of Applied Chemistry in the Faculty of Applied Sciences at the Durban University of Technology for the environment and material resources for executing this research work.

Author details

Farai Dziike
Department of Applied Chemistry, Durban University of Technology, Durban, South Africa

*Address all correspondence to: faraid1@dut.ac.za

References

[1] Winkler H. Energy policies for sustainable development in South Africa. Energy for Sustainable Development. 2006;**11**:26-34

[2] Azwar E, Wan Mahari WA, Chuah JH, Vo DVN, Ma NL, Lam WH, et al. Transformation of biomass into carbon nanofiber for supercapacitor application – A review. International Journal of Hydrogen Energy [Internet]. 2018;**43**(45):20811-20821. DOI: 10.1016/j.ijhydene.2018.09.111

[3] Energy Research Centre University of Cape Town. Energy policies for sustainable development in South Africa. Energy for Sustainable Development. 2006;**11**:1-223

[4] Federation R. Introducing BRICS Youth Energy Outlook 2019. The One and Only International Research on Energy Development of the BRICS Countries Prepared by Young Researchers and Scientists. Brics Youth Energy Agency; 2019

[5] Banno N. Low-temperature superconductors: Nb3Sn, Nb3Al, and NbTi. Superconductivity. 2023;**6**(January):1-19

[6] Wenzel P. Public-sector transformation in South Africa. Progress in Development Studies. 2007;7(1):47-64

[7] Choudhary N, Patel M, Ho YH, Dahotre NB, Lee W, Hwang JY, et al. Directly deposited MoS_2 thin film electrodes for high performance supercapacitors. Journal of Materials Chemistry A. 2015;**3**(47):24049-24054

[8] Ahmad R, Brandl M, Distaso M, Herre P, Spiecker E, Hock R, et al. A comprehensive study on the mechanism behind formation and depletion of Cu_2ZnSnS_4 (CZTS) phases. CrystEngComm. 2015;**17**(36):6972-6984

[9] Shi J, Zhao P, Wang X. Piezoelectric-polarization-enhanced photovoltaic performance in depleted-heterojunction quantum-dot solar cells. Advanced Materials. 2013;**25**(6):916-921

[10] Zhou H, Hsu WC, Duan HS, Bob B, Yang W, Bin ST, et al. CZTS nanocrystals: A promising approach for next generation thin film photovoltaics. Energy Environmental Sciences. 2013;**6**(10):2822-2838

[11] Barkhouse DAR, Gunawan O, Gokmen T, Todorov TK, Mitzi DB. Yield predictions for photovoltaic power plants: Empirical validation, recent advances and remaining uncertainties. Progress in Photovoltaics: Research and Applications. 2015;**20**(1):6-11. DOI: 10.1002/pip.1160

[12] Wei D, Amaratunga G. Photoelectrochemical cell and its applications in optoelectronics. International Journal of Electrochemical Science. 2007;**2**(12):897-912

[13] Krishna TG, Chandra Babu Naidu P, Cherukupalli K, Reddy AV, Sravani M. Smart energy grid using Iot. International Journal for Research in Applied Science & Engineering Technology (IJRASET). 2022;**10**(2):629. DOI: 10.22214/ijraset.2022.4029

[14] Renugadevi N, Saravanan S, Naga Sudha CM. IoT based smart energy grid for sustainable cites. Materials Today: Proceedings [Internet]. 2021;**81**(2):98-104. DOI: 10.1016/j.matpr.2021.02.270

[15] Clay C. Biocomposites from Renewable Resources: Preparation and

Complimentary Contributor Copy. Nova Science Publishers, Inc; 2017

[16] Kunzelmann V, Sharp I. Semiconductor photoelectrochemistry advanced lab course. Walter Schottky Institute. 2020;**4**(1):1-25

[17] Bockris JOM, Reddy AKN. Photoelectrochemistry 10.1. In: Photoelectrochemistry [Internet]. Boston: Springer; 2004. pp. 1539-1540. DOI: 10.1007/0-306-48036-0_1

[18] Lin SJ, Ting JM, Hsu KC, Fu YS. A composite photocatalyst based on hydrothermally-synthesized Cu_2ZnSnS_4 powders. Materials (Basel). 2018;**11**(1):1-13

[19] Li J, Wu N. Semiconductor-based photocatalysts and photoelectrochemical cells for solar fuel generation: A review. Catalysis Science & Technology [Internet]. 2014;**4**(12):4440. DOI: 10.1039/C4CY90047B

[20] Liang J, Zhou D, Liu D, Song H. Applications of MXenes in photoelectrochemistry: A review. Materials Research Bulletin [Internet]. 2024;**169**(July 2023):112536. DOI: 10.1016/j.materresbull.2023.112536

[21] Song M, Pang SY, Guo F, Wong MC, Hao J. Fluoride-free 2D niobium carbide MXenes as stable and biocompatible nanoplatforms for electrochemical biosensors with ultrahigh sensitivity. Advancement of Science. 2020;**7**(24):1-8

[22] Mathew S, Ramachandra M, Devi KRS, Pinheiro D, Manickam S, Pang CH, et al. Synthesis, mechanisms, challenges, and future prospects of Ti3C_2 MXene and its heterojunctions for photocatalytic dye degradation efficiency: A comprehensive review. Materials Today Sustainability [Internet]. 2023;**24**(October):100568. DOI: 10.1016/j.mtsust.2023.100568

[23] Soomro RA, Jawaid S, Zhang P, Han X, Hallam KR, Karakuş S, et al. $NiWO_4$-induced partial oxidation of MXene for photo-electrochemical detection of prostate-specific antigen. Sensors and Actuators B: Chemical. 2021;**328**(June 2020):129074

[24] Solangi NH, Karri RR, Mazari SA, Mubarak NM, Jatoi AS, Malafaia G, et al. MXene as emerging material for photocatalytic degradation of environmental pollutants. Coordination Chemistry Reviews [Internet]. 2023;**477**:214965. DOI: 10.1016/j.ccr.2022.214965

[25] Fang Y, Cao Y, Chen Q. Synthesis of an Ag2WO_4/Ti3C_2 Schottky composite by electrostatic traction and its photocatalytic activity. Ceramics International [Internet]. 2019;**45**(17):22298-22307. DOI: 10.1016/j.ceramint.2019.07.256

[26] Chen S, Gong XG, Walsh A, Wei SH. Defect physics of the kesterite thin-film solar cell absorber Cu_2 $ZnSnS_4$. Applied Physics Letters. 2010;**96**(2):8-10

[27] Jiang Q, Wang H, Wei X, Wu Y, Gu W, Hu L, et al. Efficient $BiVO_4$ photoanode decorated with Ti_3C_2TX MXene for enhanced photoelectrochemical sensing of Hg(II) ion. Analytica Chimica Acta. 2020;**1119**:11-17

[28] Fichtner M. Recent research and progress in batteries for electric vehicles. Batteries & Supercaps. 2022;**5**(2):e202100224

[29] David Gaima Kafadi A, Yusuf Hafeez H, Mohammed J, Ndikilar CE, Suleiman AB, Isah AT. A recent prospective and progress

on MXene-based photocatalysts for efficient solar fuel (hydrogen) generation via photocatalytic water-splitting. International Journal of Hydrogen Energy [Internet]. 2024;**53**(November 2023):1242-1258. DOI: 10.1016/j.ijhydene.2023.12.111

[30] Karki HP, Kim H, Jung J, Oh J. Synthesis of molybdenum sulfide/tellurium hetero-composite by a simple one-pot hydrothermal technique for high-performance supercapacitor electrode material. Nanomaterials. 2021;**11**(9):1-13

[31] Chen Q, Fan G, Fu H, Li Z, Zou Z. Tandem photoelectrochemical cells for solar water splitting. Advances in Physics: X [Internet]. 2018;**3**(1):863-884. DOI: 10.1080/23746149.2018.1487267

[32] Dias P, Schreier M, Tilley SD, Luo J, Azevedo J, Andrade L, et al. Transparent cuprous oxide photocathode enabling a stacked tandem cell for unbiased water splitting. Advanced Energy Materials. 2015;**5**(24):1-17

[33] Rahman MA, Rahaman MZ, Khatun MA, Sarker MAR. First principles investigation of structural, electronic and optical properties of NiV_2O_6. Computational Condensed Matter [Internet]. 2018;**15**(August 2017):95-99. DOI: 10.1016/j.cocom.2017.10.004

[34] Gupta H, Chakrabarti S, Mothkuri S, Padya B, Rao TN, Jain PK. High performance supercapacitor based on 2D-MoS_2 nanostructures. Materials Today: Proceedings [Internet]. 2020;**26**:20-24. DOI: 10.1016/j.matpr.2019.04.198

[35] Ondrejka P, Kemény M, Dubina M, Novák P, Hotový I, Mikolášek M. Supercapacitors and energy conversion structures based on WS_2 and MoS_2 disulfides. Journal of Electrical Engineering. 2021;**72**(4):256-261

[36] Shyyko L, Kotsyubynsky V, Budzulyak I, Rachiy B. Molybdenum disulfide/mesoporous carbon nanocomposite as electrode material for supercapasitors. Energetika. 2016;**61**(3-4):91-99

Section 2

Microbial Fuel Cells (MFCs)

Chapter 9

3D Printing in Microbial Fuel Cell

Ryan Yow Zhong Yeo, Krishan Balachandran, Irwan Ibrahim, Mimi Hani Abu Bakar, Manal Ismail, Wei Lun Ang, Eileen Hao Yu and Swee Su Lim

Abstract

The rise of additive manufacturing (AM), commonly known as 3D printing (3DP), is attributed to its ability to fabricate complex 3D structures swiftly and accurately from computer-aided design (CAD) models with minimal labor involvement. Given the heightened popularity in 3DP, researchers have explored its potential in microbial fuel cell (MFC) technology, utilizing it for the production of various MFC elements such as reactor bodies, separators, and membranes. Over the last decade, innovative electrode designs and cell arrangements have emerged, contributing to the enhanced performance of MFCs. This is largely owing to the capability of 3DP, allowing for individual optimization of each MFC component by facilitating independent design for reactors and components. Moreover, a significant attribute of 3DP technology lies in its consistent production capabilities, enabling the scalability of MFC systems by creating multiple stacks of MFC units while ensuring minimal material wastage and eliminating human errors. The forthcoming book chapter discusses the application of 3DP in MFCs.

Keywords: microbial fuel cell (MFC), 3D printing (3DP), additive manufacturing (AM), reactor design, electrode design

1. Introduction

Microbial fuel cell (MFC) is the most researched technology among other microbial electrochemical technologies (MET). Owing to the extensive efforts accomplished by researchers, the primary purpose of generating electricity using MFC has now expanded to various applications, including wastewater treatment, bioremediation, desalination, energy conversion for green fuel synthesis, carbon sequestration, and biosensing [1, 2]. MFCs are bioelectrochemical systems (BESs) that utilize exoelectrogens as biocatalysts, which are electron-producing microorganisms that oxidize organic and inorganic components to produce electrons to the anode [3]. These electrons are then transferred to the cathode *via* the load to facilitate the oxygen reduction reaction (ORR). Till date, the MFC technology has reached a fascinating point, with enormous potential for practical applications, especially in pursuing the United Nations' sixth (clean water and sanitation) and seventh (affordable and clean energy) sustainable development goals (SDGs). In order to realize such implementations on a larger scale, it is critical to improve the technology's efficiency relating to the MFC's performance, manufacturing and operating costs, stability, and lifespan [4]. This can be achieved by optimizing the design and fabrication process of each MFC component.

Additive manufacturing (AM), often known as 3D printing, is a fabrication technique that generates complex and sophisticated three-dimensional objects from computer-aided design (CAD) files [5]. Compared with conventional MFC modification techniques, AM enables a higher degree of design freedom and opens up various possibilities for the type of materials used [6], which is significant for the performance of electrodes and the durability of reactor bodies. Since exoelectrogens grow and form a biofilm on the anode, the anode material and its design have been extensively studied for the past two decades. In terms of electrode design, three-dimensional electrode structures possess higher performance compared to those of two-dimensional structures due to their higher surface area, lower internal resistances, lower activation energy, and lower mass transfer overpotential [7]; while stainless steel is the most feasible electrode material for MFC as other metals tend to corrode easily [8]. Therefore, a combination of carbon-based catalysts with stainless steel electrodes also enhances the biocompatibility of the MFC anodes. By controlling the structure design and printing material, novel MFC components, including electrodes, reactor bodies, and membranes, can be fabricated.

2. 3D printed MFC electrodes

Microbial fuel cell (MFC) electrodes with optimal biocompatibility for biofilm development and excellent electrochemical characteristics for high current density production have always been vastly pursued by MET researchers. Notably, anodes with 3D surface offer high surface area for efficient biofilm formation thanks to the easier access of substrate to the exoelectrogens, which remarkably reduces mass transfer restrictions [9]. Owing to AM's flexibility, 3D electrodes with macroporous structures can be fabricated with ease. Based on **Figure 1a**, different sizes and designs of electrodes were fabricated using the fused deposit modeling (FDM) printing technique. These structures were printed layer-by-layer using conductive polylactic acid (PLA) filaments as printing materials. The researchers found that the electrodes

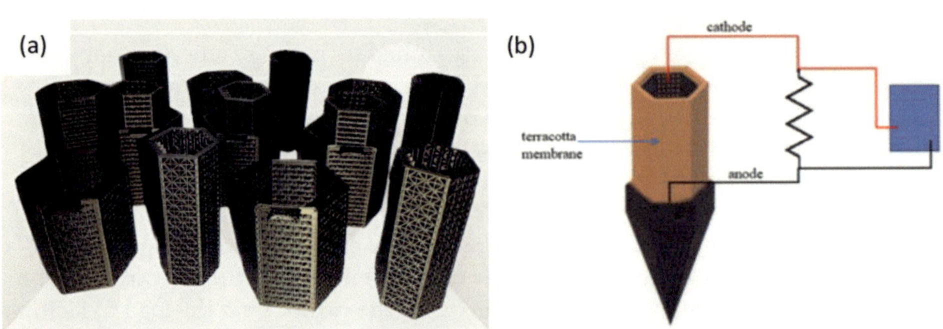

Figure 1.
(a) Different sizes and designs of 3D printed conductive PLA electrodes. Reprinted with permission from Ref. [10] under an open-access creative commons CC BY-SA license. Copyright 2023, CBIORE. (b) Assembled PMFC with 3D printed anode and cathode. Reprinted with permission from Ref. [11] under an open-access creative commons CC BY-SA license. Copyright 2023, CBIORE.

had minimal degradation and fouling after conducting the experiment for 58 days in a plant MFC (PMFC) system [10], which proved that 3D printed electrodes do possess excellent physical and cycling stability [12]. Stackable PMFC electrodes were also fabricated and tested for their electricity generation efficiency. It was found that stacking the 3D printed PMFCs in series caused voltage loss while stacking the cells in parallel preserved the voltage and current [11], which means parallel stacks of PMFCs are more feasible for scaling up. Also, different unique designs and configurations of MFC can be achieved through 3D printing (**Figure 1b**).

Despite the superior mechanical properties of the 3D printed electrodes, conductive PLA as the sole electrode material still suffers from low electrical conductivity [10, 11]. Generally, modification of the electrode surface with suitable catalysts can enhance the electrochemical properties of the electrode. This was proven when nonconductive and conductive PLA anodes synthesized through FDM were both modified with carbon coating and tested for 25 days. The peak power generated by the carbon-coated nonconductive and conductive PLA anodes was relatively close, indicating that the performance was mainly attributed to the catalyst coating regardless of the electrode base material [13]. Following the success of surface modification on 3D printed electrodes, filament manufacturers were driven to develop composite filaments that consist of conductive PLA and other catalysts. Indeed, conductive PLA electrode infused with 8 wt% graphene powder were successfully printed *via* FDM and used in a *P. aeruginosa* MFC system. It was found that the graphene-containing conductive PLA electrode produced a power output of 110.74 ± 14.63 μW m^{-2}, which was comparable to the highly conductive carbon cloth electrode (93.49 ± 5.17 μW m^{-2}) [14]. Apart from commercial FDM filaments, researchers have also tried to synthesize their own filaments with desired compositions. Carbon black powders were mixed with acrylonitrile butadiene styrene (ABS), which is a common 3D printing material and extruded into filaments using a commercial filament extruder. The carbon black-ABS anodes performed significantly better than the plain PLA anodes and exhibited comparable power outputs with the carbon cloth anodes [15]. This is due to the addition of carbon black powders, which greatly enhanced the biocompatibility of the electrodes.

Inspired by the customized filaments, researchers have discovered a more streamlined approach to 3D print composite materials without synthesizing the filaments. Through direct ink writing (DIW) method, composite inks in the form of viscous liquid can be extruded directly layer-by-layer using a dispensing machine with various adjustable nozzle sizes. A PTFE-free alginate-activated carbon electrode was synthesized by manual extrusion of the thick carbon-alginate paste (**Figure 2**) and left to solidify for 24 hours. The 3D printed PTFE-free alginate-activated carbon electrode marked the highest power output up to 268 μW as compared to the commercial carbon block (85 μW) and the non-3D printed PTFE-activated carbon (98 μW) when the MFCs were tested for 28 days [16]. Indeed, biofilm development on the electrode surface remains one of the most time-consuming processes in MFC. Recently, the direct printing of a living MFC anode was reported. Living bacteria (*Shewanella Oneidensis* MR-1) ink was prepared by mixing its growth medium with cellulose, acetylene carbon black, and sodium alginate. The as-prepared ink was then extruded through a nozzle to form the desired electrode design and cured using a calcium chloride solution to form a solidified calcium chloride matrix containing the bacteria. A peak current density of 9.233 μA cm^{-2} was reached within 4 days upon stabilization of the MFC [17]. It was observed that there were only a small number of bacterial cells embedded in the electrode structure prior to any tests; after the stability test

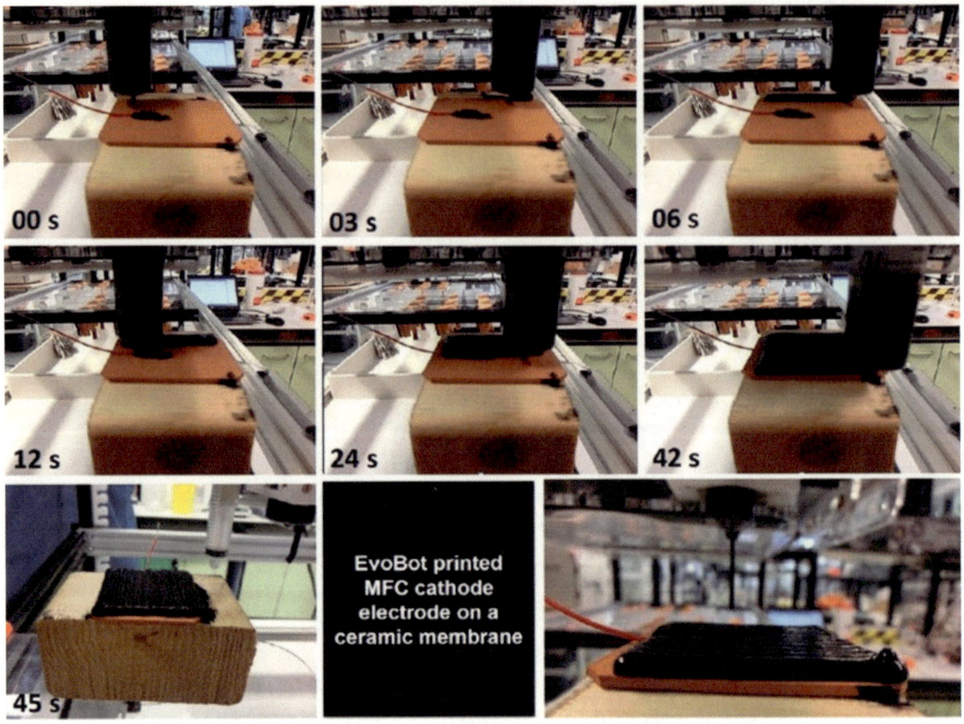

Figure 2.
Direct ink writing of PTFE-free alginate-based carbon electrode under 1 minute of fabrication time. Reprinted with permission from Ref. [16] under an open-access creative commons CC BY license. Copyright 2020, MDPI.

was carried out, the number of bacteria increased significantly while retaining their physical characteristics. This indicates that 3D printed live microbial electrodes are feasible for MFC technologies and the growth of the bacteria can be restored by providing sufficient nutrients.

Among other 3D printing techniques, digital light processing (DLP) has been drawing immense interest thanks to its rapid printing speed, flexibility in structure design, and high-resolution printing capabilities. DLP works by projecting the shape of an entire layer of a sliced-3D model on the printing platform and curing that entire layer at the same time before moving to the next layer [18]. In short, DLP prints 3D structures based on a projection layer-by-layer moving only on one axis (the z axis, up and down), while FDM and DIW fabricates by moving in all three axes (the x and y axis of a plane, followed by the z axis) to form the 3D structure. By using UV curable resins, 3D printed porous structures were fabricated and served as the base material for copper electroless plating. As a result, 3D printed porous copper electrode was successfully produced (**Figure 3**), which recorded a maximum voltage and power density of 62.9 mV and 6.45 mW m^{-2}, respectively, which were 8.3 and 12.3 times higher than the commercial copper mesh electrodes. However, the authors have identified the toxicity of copper ions to the *Shewanella Oneidensis MR-1* culture after 15 days of operation, which hindered the performance of the MFC [19]. Despite the drawbacks of this research, the potential of DLP in MFC was shown and could be further explored with more biocompatible materials. This phenomenon was later realized when a 3D porous carbon structure with high biocompatibility was prepared by carbonization of a DLP printed 3D porous structure. In this work, 3D porous

Figure 3.
FESEM image of DLP printed anode structure, insert shows the actual high-resolution 3D printed carbon electrode. Reprinted with permission from Ref. [19] under an open-access creative commons CC BY license. Copyright 2018, Frontiers.

structures with various pore sizes were printed, mainly 100, 200, 300, 400, and 500 µm. The substitution of copper with carbon led to a maximum voltage and power density obtained was 453.4 mV and 233.5 mW m^{-2} [20], respectively, which were 7.2 and 36.2 times higher than their DLP-printed-copper electrode counterpart.

Another 3D printing technology similar to that of DLP that uses light projection is masked stereolithography (MSLA). MSLA creates the whole layer of the 3D structure as well, but instead of using a projector that projects light in a certain shape, MSLA utilizes an array of ultraviolet light-emitting diodes (LEDs) as its light source that remains unchanged regardless of the shape of the layer. The shape of the printing layer is formed by an LCD panel acting as a photomask that blocks light surrounding the shape where nothing is to be printed. One of the greatest benefits of MSLA is that it allows for high-resolution printing of complex structures rapidly, making it more suitable for large-scale applications [21]. By using MSLA, a 3D printed electrode decorated with graphene oxide (GO) and polypyrrole (PPy) was synthesized. Upon carbonization, average pore sizes of 0.4 mm were achieved for the electrode, which provided large specific surface area for biofilm development and ensuring efficient mass transfer of the electrode. When used as the MFC bioanode, the PPy/GO electrode recorded a high peak power density of 22.4 W m^{-2}, which was approximately 19 times higher than that of commercial carbon felt anodes [22].

In essence, 3D printing methods have demonstrated effectiveness in creating electrodes with enhanced structures and improved properties. Among the various 3D printing techniques, DIW and FDM methods stand out due to their flexibility and simplicity of operation. As 3D printing continues to evolve, it is anticipated that more sophisticated printing techniques will be discovered in electrodes fabrication. Lastly, 3D printing also represents a significant revolution in electrode fabrication. In contrast to the traditional 2D planar electrode structure, 3D printing offers a novel approach to address challenges in microbial electrochemical technologies by employing three-dimensional electrode architecture designs that greatly boost the performance of the MFC, as well as biocompatibility with the electron producing microbes.

3. 3D printed MFC reactor bodies

One of the greatest challenges of commercializing MFCs in practical applications is the scale-up of the reactor components such as reactor bodies, membranes, and electrodes as the inconsistent efficiency of MFCs is still unable to compensate for the huge capital costs. Furthermore, the scaling-up process for the practical implementation of MFCs requires substantial expenses mainly due to the materials used, namely platinum electrodes and Nafion membranes. Hence, researchers have strived to obtain a more stable energy production from MFCs and reduce their costs. For these reasons, scientists have explored the use of additive manufacturing to fabricate MFC parts, including the reactor bodies.

In the development of METs, biocompatible materials are always preferred as they provide better platforms for biofilm attachment and growth. By using FDM, nontoxic biocompatible PLA was used to construct the MFC reactors with a high level of accuracy up to 5 μm. It can be observed that the 3D printed single-chamber MFC reactor does not differ from the conventional acrylic fiber material in terms of built quality. Next, complementary reactor shapes can be further printed to construct double-chamber MFCs, indicating that the 3D printing has indeed opened up various potentials for reactor up-scaling or stacking MFC configurations. Conductive PLA with low resistance values (30 $\Omega\ cm^{-1}$) was also fabricated in this study. It was found that both configurations were greatly affected by the high internal resistance, which hindered the power generation. However, the double-chamber MFC produced a better result compared to the single-chamber MFC, which was due to the use of cation exchange membrane (CEM) as well as potassium permanganate at the cathode side, which balanced the redox reactions [23].

Plant membrane fuel cells (PMFCs) show promising prospects for generating bioelectricity as they represent renewable and environmentally friendly energy sources. However, their capacity for power generation remains a challenge because their power output has not yet reached a level that makes them a cost-effective option for renewable energy. These lower power densities are a result of various internal resistances within the system, including ohmic, activation, bacterial metabolic, and concentration losses. In order to improve the stacking efficiency, unique stake PMFC design was 3D printed and studied, where conductive PLA was used to print the electrodes and normal PLA was used to print the separator. A significant increase of average power from 1.06×10^{-9} W to 5.47×10^{-7} W was documented when the individual PMFC was arranged in three-stacked PMFCs connected in series, hinting that the performance of PMFCs was more efficient when they were stacked together. Additionally, PMFCs with a nine-stack parallel connection produced the highest power at 1.62×10^{-4} W with a power density of 9.18 mW m^{-2} and a voltage of 4.54 V. Apart from the superior electrochemical performance, the design of PMFC as watering stakes enhanced the transport of water and oxygen to the plant roots, which were crucial for strong root development. A healthy root environment provides a good supply of rhizodeposits that serve as valuable nutrients for microbes, which allows the biofilm to consume and metabolize in order to produce electrons [24].

Lately, soil microbial fuel cells (SMFCs) have garnered significant interest as an eco-friendly innovation with extensive possibilities in generating renewable energy, treating environmental pollution, and enabling sensing capabilities. In the last 10 years, scientists primarily concentrated on creating SMFCs for low-power sensor surveillance, cleaning contaminated soil, and exploring novel sensing applications

[25, 26]. Nonetheless, the main setback of SMFCs originates from the low power output and high operating costs that limit their real field application. In another study, a SMFC reactor was 3D printed using PLA filaments to generate electricity from rice washing wastewater (RWW). This is because RWW represents a huge portion of domestic wastewater, especially in Asian countries and it also contains starch, minerals, and vitamins that are useful for microbial oxidation. 3D printed SMFC with carbon felt electrodes were plugged into the soil with constant addition of 15 mL RWW every 48 hours. Two types of soil (muddy and sandy) were used in this experiment. It was reported that SMFC in muddy soil with RWW had generated a maximum power density of 485.2 mW m^{-2}, while SMFC in sandy soil only generated 112 mW m^{-2} [27]. These results indicated that 3D printed MFCs are able to operate in different types of soil.

3D printing stands out as one of the most optimal choices in creating micro-devices with exact and accurate dimensions [28]. A miniaturized 3D printed MFC reactor enclosure was created to allow oxygen flow at the air-cathode and fuel supply at the anode side. As a result, this miniaturized MFC achieved a maximum power density of 192 μW cm^{-2} and a maximum current density of 1.3 mA cm^{-2}. Furthermore, consistent power generation was observed where repeatable voltage production of 410 mV was recorded, proving that the 3D printed enclosure possessed excellent stability [29]. This means that the enclosure could be reused by just replacing the electrodes on both terminals, thereby providing a cost-effective solution. Similar to the aforementioned work, a small-scale MFC tank with a mixer design on the anode chamber was fabricated *via* 3D printing as well to promote uniform mixing of the microbial culture and the nutrient substrate. As a result, this high-efficiency mixing area had contributed toward a maximum power density of 199.24 mW m^{-2} using graphene-coated carbon cloth as the anode. Moreover, the internal resistance of the system was lowered to only 0.66 kΩ, which indicated a low anode impedance due to the mixing anode chamber design [30].

4. 3D printed MFC separators

Every MFC consists of a separator or membrane that provides a physical separation between the electrodes and in some cases, their electrolytes. This is due to the different reactions of interest at each terminal, which require different electrolytes to suit their conditions. The performance of the separator is influenced by the type of material, thickness, pore size, and surface conditions. Furthermore, membranes help to reduce oxygen and substrate diffusion during microbial oxidation [31], which ultimately helps in improving the MFC's efficiency. Despite the merits of using a separator, the most commonly used PEM in MFCs such as Nafion 117 membrane suffers from poor conductivity at low humidity and structural changes at high humidity [32], which disrupt the ion transfer process. Hence, the optimization of separators' parameters can be addressed by using additive manufacturing.

In a novel proton exchange membrane (PEM) development work, TangoPlus acrylate photopolymer resin and natural rubber latex membranes were successfully printed and compared with the commercially available CMI-7000 cation exchange membrane (CEM). Thinner membranes were fabricated by using a 3D printer; the TangoPlus polymer (**Figure 4**) recorded a membrane thickness of 116 μm and the latex recorded a membrane thickness of 100 μm, which were significantly lower

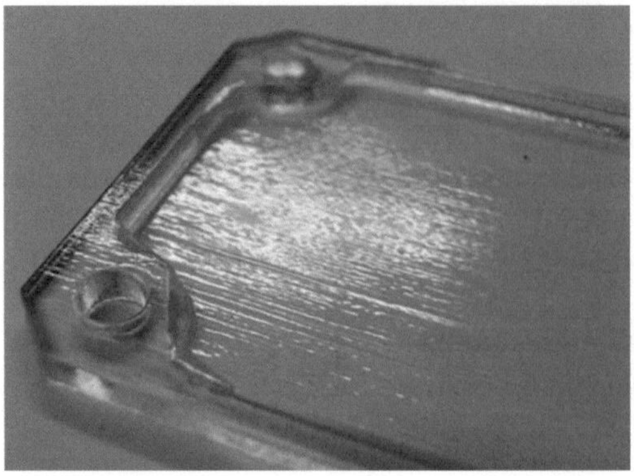

Figure 4.
3D printed TangoPlus MFC separator. Reprinted with permission from Ref. [33] under an open-access creative commons CC BY license. Copyright 2021, MDPI.

than that of conventional CEM with a membrane thickness of 450 μm. Generally, a thinner membrane is desired as it reduces internal resistance, which enhances proton conductivity [33]. However, this study reported that the peak power production for TangoPlus, latex, and CEM was 0.92, 10.51, and 11.39 μW, respectively. Therefore, it is highly necessary to explore materials with ionically conductive properties so that the ion exchange process can be carried out more efficiently [34].

In search of an alternative to the costly Nafion 117 membrane and CMI-7000 CEM, multiple 3D printed separators were synthesized using low-cost extrudable clay as printing materials. Fimo™ air-dry, terracotta air-dry, and terracotta modeling clays were printed using a robotic liquid dispensing machine. The MFCs were then constructed using the 3D printed separators in a membrane electrode assembly (MEA) configuration. The electrochemical studies demonstrated that the Fimo™ air-dry membrane recorded the lowest average surface resistance of 135 Ω, which was even lower than the commercial CEM with a resistance of 143 Ω. Similarly, the terracotta air-dry and terracotta modeling clay membranes possessed slightly higher resistance values of 158 and 169 Ω, respectively. Despite the similar average surface resistance values of all four membranes (150 ± 15 Ω), all three 3D printed ceramic-based membranes produced significantly better results compared to the CEM. Membrane Terracotta air-dry membrane obtained the highest power output at 130 μW followed by Fimo™ air-dry membrane with 111 μW, indicating the superiority of air-dry clay as MFC membranes. Terracotta modeling clay membrane underperformed at 73 μW, and lastly, the commercial CEM only generated 66 μW [35]. In short, separator materials remain a crucial element within the MFC system and require more exploration to identify novel and efficient membrane materials.

5. Conclusion

In essence, the incorporation of 3D printing technology in MFCs has opened avenues for innovation, allowing for meticulous design and customization of every crucial MFC components. While significant progress has been made, continuous exploration of novel materials and optimization of 3D printing parameters in this field ensures further advancements in MFC technology.

Acknowledgements

The authors would like to express their gratitude to the Universiti Kebangsaan Malaysia and Indah Water Research Center (IWRC) for the financial support of this work via Geran Translasi (UKM-TR-010), Dana Pecutan (PP-SELFUEL-2022), and Indah Water Konsortium Sdn. Bhd. (KK-2022-015). Ryan Yow Zhong Yeo would also like to thank Universiti Kebangsaan Malaysia for partially supporting his study through Peruntukan Khas Institut SELFUEL 2022 (BG-H-RA0100-00-0101071).

Conflict of interest

The authors declare no conflict of interest.

Nomenclature

3DP	3D printing
ABS	acrylonitrile butadiene styrene
AM	additive manufacturing
BES	bioelectrochemical system
CAD	computer-aided design
CEM	cation exchange membrane
DIW	direct ink writing
DLP	digital light processing
FDM	fused deposit modeling
GO	graphene oxide
LED	light-emitting diode
MEA	membrane electrode assembly
MET	microbial electrochemical technologies
MSLA	masked stereolithography
MFC	microbial fuel cell
ORR	oxygen reduction reaction
PEM	proton exchange membrane
PLA	polylactic acid
PMFC	plant microbial fuel cell
PPy	polypyrrole
PTFE	polytetrafluoroethylene
RWW	rice washing wastewater
SDG	sustainable development goal
SMFC	soil microbial fuel cell

Author details

Ryan Yow Zhong Yeo[1,2], Krishan Balachandran[1], Irwan Ibrahim[1], Mimi Hani Abu Bakar[1], Manal Ismail[1], Wei Lun Ang[2], Eileen Hao Yu[3] and Swee Su Lim[1*]

1 Fuel Cell Institute, Universiti Kebangsaan Malaysia, Bangi, Selangor, Malaysia

2 Faculty of Engineering and Built Environment, Department of Chemical and Process Engineering, Universiti Kebangsaan Malaysia, Bangi, Selangor, Malaysia

3 Department of Chemical Engineering, Loughborough University, Loughborough, United Kingdom

*Address all correspondence to: limss@ukm.edu

References

[1] Boas JV, Oliveira VB, Simões M, Pinto AMFR. Review on microbial fuel cells applications, developments and costs. Journal of Environmental Management. 2022;**307**:114525. DOI: 10.1016/j.jenvman.2022.114525

[2] Santoro C, Arbizzani C, Erable B, Ieropoulos I. Microbial fuel cells: From fundamentals to applications. A review. Journal of Power Sources. 2017;**356**:225-244. DOI: 10.1016/j.jpowsour.2017.03.109

[3] Kumar R, Singh L, Wahid ZA, Din MFM. Exoelectrogens in microbial fuel cells toward bioelectricity generation: A review. International Journal of Energy Research. 2015;**39**(8):1048-1067. DOI: 10.1002/er.3305

[4] You J, Preen RJ, Bull L, Greenman J, Ieropoulos I. 3D printed components of microbial fuel cells: Towards monolithic microbial fuel cell fabrication using additive layer manufacturing. Sustainable Energy Technologies and Assessments. 2017;**19**:94-101. DOI: 10.1016/j.seta.2016.11.006

[5] Jadhav A, Jadhav VS. A review on 3D printing: An additive manufacturing technology. Materials Today: Proceedings. 2022;**62**:2094-2099. DOI: 10.1016/j.matpr.2022.02.558

[6] Calignano F, Tommasi T, Manfredi D, Chiolerio A. Additive manufacturing of a microbial fuel cell—A detailed study. Scientific Reports. 2015;**5**(1):17373. DOI: 10.1038/srep17373

[7] Sonawane JM, Yadav A, Ghosh PC, Adeloju SB. Recent advances in the development and utilization of modern anode materials for high performance microbial fuel cells. Biosensors & Bioelectronics. 2017;**90**:558-576. DOI: 10.1016/j.bios.2016.10.014

[8] Pocaznoi D, Calmet A, Etcheverry L, Erable B, Bergel A. Stainless steel is a promising electrode material for anodes of microbial fuel cells. Energy & Environmental Science. 2012;**5**(11):9645-9652. DOI: 10.1039/C2EE22429A

[9] Liu Y, Harnisch F, Fricke K, Schröder U, Climent V, Feliu JM. The study of electrochemically active microbial biofilms on different carbon-based anode materials in microbial fuel cells. Biosensors & Bioelectronics. 2010;**25**(9):2167-2171. DOI: 10.1016/j.bios.2010.01.016

[10] Palmero DPE, Pamintuan KRS. Characterization of plant growth promoting potential of 3D-printed plant microbial fuel cells. International Journal of Renewable Energy Development. 2023;**12**(5):11. DOI: 10.14710/ijred.2023.52291

[11] Constantino GPP, Dolot JMC, Pamintuan KRS. Design and testing of 3D-printed stackable plant-microbial fuel cells for field applications. 2023;**12**(2):10. DOI: 10.14710/ijred.2023.44872

[12] Chu T, Park S, Fu K. 3D printing-enabled advanced electrode architecture design. Carbon Energy. 2021;**3**(3):424-439. DOI: 10.1002/cey2.114

[13] You J, Fan H, Winfield J, Ieropoulos IA. Complete microbial fuel cell fabrication using additive layer manufacturing. Molecules. 2020;**25**(13):3051. DOI: 10.3390/molecules25133051

[14] Slate AJ, Hickey NA, Butler JA, Wilson D, Liauw CM, Banks CE, et al. Additive manufactured graphene-based electrodes exhibit beneficial performances in Pseudomonas

aeruginosa microbial fuel cells. Journal of Power Sources. 2021;**499**:229938. DOI: 10.1016/j.jpowsour.2021.229938

[15] Callanan CJ, Hsu L, McGee A, editors. Formulation and evaluation of carbon black 3D printing materials. In: OCEANS 2018 MTS/IEEE, Charleston, SC, USA. IEEE; 22-25 October 2018. 2018. DOI: 10.1109/OCEANS.2018.8604767

[16] Theodosiou P, Greenman J, Ieropoulos IA. Developing 3D-printable cathode electrode for monolithically printed microbial fuel cells (MFCs). Molecules. 2020;**25**(16):3635. DOI: 10.3390/molecules25163635

[17] Freyman MC, Kou T, Wang S, Li Y. 3D printing of living bacteria electrode. Nano Research. 2020;**13**(5):1318-1323. DOI: 10.1007/s12274-019-2534-1

[18] Zhao Z, Tian X, Song X. Engineering materials with light: Recent progress in digital light processing based 3D printing. Journal of Materials Chemistry C. 2020;**8**(40):13896-13917. DOI: 10.1039/D0TC03548C

[19] Bian B, Wang C, Hu M, Yang Z, Cai X, Shi D, et al. Application of 3D printed porous copper anode in microbial fuel cells. Frontiers in Energy Research. 2018;**6**:50. DOI: 10.3389/fenrg.2018.00050

[20] Bian B, Shi D, Cai X, Hu M, Guo Q, Zhang C, et al. 3D printed porous carbon anode for enhanced power generation in microbial fuel cell. Nano Energy. 2018;**44**:174-180. DOI: 10.1016/j.nanoen.2017.11.070

[21] Valizadeh I, Al Aboud A, Dörsam E, Weeger O. Tailoring of functionally graded hyperelastic materials via grayscale mask stereolithography 3D printing. Additive Manufacturing. 2021;**47**:102108. DOI: 10.1016/j.addma.2021.102108

[22] He Y, Yang J, Fu Q, Li J, Zhang L, Zhu X, et al. Structure design of 3D hierarchical porous anode for high performance microbial fuel cells: From macro-to micro-scale. Journal of Power Sources. 2021;**516**:230687. DOI: 10.1016/j.jpowsour.2021.230687

[23] Jannelli E, Di Trolio P, Flagiello F, Minutillo M. Development and performance analysis of biowaste based microbial fuel cells fabricated employing additive manufacturing technologies. Energy Procedia. 2018;**148**:1135-1142. DOI: 10.1016/j.egypro.2018.08.029

[24] Chua MAM, De Los Santos AJT, Pamintuan KRS, editors. On-site stacking efficiency performance of a novel full-3D-printed plant microbial fuel cell electrode assembly. In: Trends in Environmental Sustainability and Green Energy. Cham: Springer International Publishing; 2023

[25] Sherafatmand M, Ng HY. Using sediment microbial fuel cells (SMFCs) for bioremediation of polycyclic aromatic hydrocarbons (PAHs). Bioresource Technology. 2015;**195**:122-130. DOI: 10.1016/j.biortech.2015.06.002

[26] Yu B, Tian J, Feng L. Remediation of PAH polluted soils using a soil microbial fuel cell: Influence of electrode interval and role of microbial community. Journal of Hazardous Materials. 2017;**336**:110-118. DOI: 10.1016/j.jhazmat.2017.04.066

[27] Nguyen H-U-D, Nguyen D-T, Taguchi K. A compact, membrane-less, easy-to-use soil microbial fuel cell: Generating electricity from household rice washing wastewater. Biochemical Engineering Journal. 2022;**179**:108338. DOI: 10.1016/j.bej.2022.108338

[28] Ambrosi A, Pumera M. 3D-printing technologies for electrochemical applications. Chemical Society Reviews. 2016;**45**(10):2740-2755. DOI: 10.1039/C5CS00714C

[29] Jayapiriya US, Inoue S, Goel S. A facile technique to develop conductive paper based bioelectrodes for microbial fuel cell applications. Biosensors & Bioelectronics. 2022;**214**:114479. DOI: 10.1016/j.bios.2022.114479

[30] Huang S-J, Ubando AT, Wang C-Y, Su Y-X, Culaba AB, Lin Y-A, et al. Modification of carbon based cathode electrode in a batch-type microbial fuel cells. Biomass and Bioenergy. 2021;**145**:105972. DOI: 10.1016/j.biombioe.2021.105972

[31] Jayapiriya US, Goel S. Influence of cellulose separators in coin-sized 3D printed paper-based microbial fuel cells. Sustainable Energy Technologies and Assessments. 2021;**47**:101535. DOI: 10.1016/j.seta.2021.101535

[32] Reinholdt M, Ilie A, Rouald̀es S, Frugier J, Schieda M, Coutanceau C, et al. Plasma membranes modified by plasma treatment or deposition as solid electrolytes for potential application in solid alkaline fuel cells. Membranes [Internet]. 2012;**2**(3):529-552

[33] Ramirez-Nava J, Martínez-Castrejón M, García-Mesino RL, López-Díaz JA, Talavera-Mendoza O, Sarmiento-Villagrana A, et al. The implications of membranes used as separators in microbial fuel cells. Membranes. 2021;**11**(10):738. DOI: 10.3390/membranes11100738

[34] Philamore H, Rossiter J, Walters P, Winfield J, Ieropoulos I. Cast and 3D printed ion exchange membranes for monolithic microbial fuel cell fabrication. Journal of Power Sources. 2015;**289**:91-99. DOI: 10.1016/j.jpowsour.2015.04.113

[35] Theodosiou P, Greenman J, Ieropoulos I. Toward monolithically printed MFCs: Development of a 3d-printable membrane electrode assembly (mea). International Journal of Hydrogen Energy. 2019;**44**(9):4450-4462. DOI: 10.1016/j.ijhydene.2018.12.163

Chapter 10

Microbial Community

Hajar Rajaei Litkohi and Hosein Yazdi Dehnavi

Abstract

The microbial community employed as biocatalyst in microbial fuel cells (MFC) play a crucial role in degradation of organic substances and bioelectricity generation. While degradation of organic matters and electrical current generation in MFC technology is predominantly depend on metabolic activities of electroactive bacteria such as *Geobacter* and *Proteobacteria*, these bacteria engage in mutual interactions with non-electroactive counterparts within the microbial community. These mutual interactions can modify system performance, which is widely depended on operational conditions, the source of the initial microbial inoculum, substrate diversity and system's components. Consequently, it is essential to gain a comprehensive understanding of the ecological behavior of microbial communities under diverse conditions to optimize system efficiency. Numerous research studies have delved into the microbial communities under varying circumstances, and the objective of this research is to elucidate the distinctions among microbial communities and investigate the factors that impact their composition.

Keywords: microbial fuel cells, microbial consortium, biocatalyst, bioelectricity generatione, electroactive bacteria

1. Introduction

The growing needs of our modern world, such as energy, water, and food, have caused significant harm to our environment. As our demand for electricity continues to rise sharply, there's a pressing need to explore renewable energy sources [1]. One of the emerging technologies garnering substantial interest as a renewable energy source is microbial Fuel Cells (MFC) that use biological catalysts, primarily bacteria, to generate electric energy from organic matter, whether in the environment or waste [2]. MFC work with microorganisms as bio-catalysts that can extract energy from various waste materials and organic compounds [3]. The microorganisms oxidize organic and inorganic substrates at the anode, leading to the generation of electrons. Efficient transfer of these electrons from within the cells to the anode, in anoxic conditions, is vital for electric current production and involves mechanisms such as electron shuttles, mediators, and microbial nanowires [4, 5].

Diverse microorganisms, primarily bacteria from different phylogenetic groups, exhibit the capacity to generate electricity in MFCs without necessitating a mediator. Known as exoelectrogens or electrogenic microorganisms, several bacterial species like *Geobacter* spp., *Shewanella* spp., *Rhodoferax ferrireducens*, among others, have been identified for their electricity-producing abilities within MFCs. Additionally, microalgae, yeast, and fungi have also been reported in these systems [5].

Various factors significantly impact the operational efficiency of MFCs, including reactor configuration, anodic and cathodic materials, membrane type, substrate characteristics, internal resistance, loading rate, pH, buffer solutions, anodophilic populations, temperature, and salinity [5–10]. Moreover, biological parameters such as inoculation source, pure culture versus consortium, Extracellular Electron Transfer (EET) rate, biofilm growth rate, and composition/activity of exoelectrogen populations play crucial roles since biocatalytic activity drives the MFC process [11–17].

Mixed microbial communities, particularly in bacterial MFCs, offer several advantages of adaptability and robustness to changing operating conditions [18, 19]. These communities demonstrate resilience against alterations in operating conditions, enabling detailed investigations of individual factors and fostering synergistic interactions among diverse microbes to enhance MFC stability and performance.

The utilization of mixed microbial communities within MFCs presents numerous advantages due to their ability to adapt and endure fluctuations in operating and environmental conditions. Conversely, the initiation phase of MFCs heavily relies on the establishment of biofilms and an enrichment process. The composition and structure of the resultant biofilm and microbial community formation significantly hinge upon various influencing factors. Nevertheless, comprehending the precise impact of these factors on the microbial community's composition during optimal MFC performance poses a challenging endeavor. This study seeks to investigate and unravel the intricate influence of these factors on the behavior and dynamics of the microbial community.

2. Source of inoculum

Exoelectrogenic communities can be sourced from a variety of origins, including forest soil, anaerobic and aerobic sludge derived from wastewater treatment facilities, sediments, and compost [20–24]. In addition, isolated strains of model exoelectrogenes like *Geobacter spp.* and *Shewanella spp.* can also be utilized [25]. The impact of these diverse bacterial sources on the performance of such systems has been the subject of investigation in numerous studies. The presence of higher concentrations of exoelectrogenes and bacteria that promote biofilm formation can accelerate the development of biofilms on the anode [26]. However, the presence of electron acceptors, such as metal oxides and methanogenic microorganisms within the initial sludge can significantly influence the composition of bacterial communities on the anode and subsequently affect the performance of microbial fuel cells [27, 28]. Therefore, the careful selection of an inoculum source with a high concentration of active exoelectrogenes is of utmost importance.

The diversity of the initial microbial population and the concentrations of exoelectrogenic and non-exoelectrogenic bacteria are significantly influenced by the choice of inoculum source. Nonetheless, it is feasible to improve the inoculum to mitigate potential negative impacts. In a study conducted by Ishii et al., the inoculum was obtained from three distinct sources: (1) Anaerobic sludge from a wastewater treatment plant, (2) Paddy field soil, and (3) Coastal wetland sediments. The bioelectrochemical behavior of these three inocula was examined in a single-chamber fuel cell, and microbial population analysis was carried out using the 16S rRNA method. Paddy field soil displayed the most diverse initial microbial population in comparison to coastal wetland sediments and anaerobic sludge. In this study, an initial external resistance of 510 Ohms was applied, followed by a stepwise reduction to 100 and 25 Ohms to promote the enrichment of exoelectrogenic microorganisms. After 30 days of operation, the highest electric current, measuring 5.4 mA, was produced by the inoculum from coastal lagoon sediments. The analysis of the microbial population revealed that the stepwise enrichment process resulted in a reduction in microbial diversity within these three reactors. As a result, seven dominant bacterial groups emerged, comprising 90% of the microbial population in these cells. Consequently, by enriching the microbial population, it becomes possible to reduce the dependency of microbial fuel cell performance on the specific source of inoculum preparation [29].

While the source of inoculum can initially influence microbial community, over time and with consistent operational conditions, dominant microbial populations tend to converge. In a study conducted by Gao et al., two distinct inocula were examined. One was derived from the aerobic tank of a wastewater treatment plant, while the other originated from the anaerobic treatment reactor handling high sulfur wastewater. Both inocula underwent anaerobic treatment before being inoculated into the system. Bioelectrochemical analyses revealed that the inoculum from aerobic sludge reached maximum voltage more rapidly and exhibited higher voltage and current generation compared to the anaerobic sludge. Examination of the biofilm morphology on the anode biofilm indicated a higher presence of spherical bacteria in the aerobic sludge. Microbial population analysis in the initial stages showed that the aerobic sludge had greater diversity than the anaerobic sludge. However, as time passed and the system operated continuously, the microbial populations in both cells underwent progressive changes, ultimately leading to a convergence in dominant bacterial groups. In both inocula, dominant microbial populations became similar. The microbial population of the aerobic sludge inoculum was characterized by greater complexity and diversity, whereas the microbial population of the anaerobic sludge had a simpler composition. Notably, the anaerobic inoculum contained bromate-reducing bacteria, which are associated with the degradation of complex organic matter and sulfur reduction, reflecting the high sulfur content in the wastewater source. The similarity in dominant microbial populations in both inocula was attributed to their shared substrate. Specifically, *Geobacter* exoelectrogen species, biofilm-forming *Zoogloea*, and *Acinetobacter* species were prominent in both inocula [30].

Nutrient availability in the primary inoculum source significantly influences the diversity and microbial composition of the initial inoculum. In sedimentary Microbial Fuel Cells (SMFC), the sediments serve a dual role as both the substrate and the source of the inoculum, making sediment properties a crucial factor in SMFC performance. In a research study, seven lakes (denoted as S1 to S7) were assessed using SMFC technology. Of these, S1, S2, and S3 were large lakes, while S4 to S7 were smaller local lakes. The research aimed to investigate the geochemical characteristics and the initial microbial communities of each lake and their correlation with the biochemical performance, microbial composition, and microbial population in the SMFC systems. The findings revealed that the highest power output, 25.2 and 23.9 mW/m^2, was observed in S4 and S5, respectively, and these two systems exhibited the quickest startup times, at 7 and 9 days, respectively. The analysis of sediment characteristics indicated that the high relative yield in S4 and S5 could be attributed to the abundance of carbon and nitrogen sources, signifying higher nutrient levels in these sediments. Furthermore, the removal and degradation of organic matter in S4 and S5 were approximately 1.5 times higher than in the other systems. The initial microbial population in S4 and S5 displayed greater diversity compared to the other systems, and the types of bacteria comprising the populations in S4 and S5 differed from those in the other lakes. All seven SMFC systems exhibited similar voltage levels, ranging from 500 to 600 mV. After 60 days, an examination of the microbial populations on the anode indicated that *Proteobacteria* were predominant in all samples. However, in S4, only 43% of the population consisted of *Proteobacteria*, with *Pseudomonas* bacteria playing a significant role in the microbial population of S4 and S5 [31].

In the selection of the inoculum source, it is essential to consider the specific objectives of employing a fuel cell. For instance, when the aim is the remediation or elimination of a particular pollutant, it becomes possible to isolate microorganisms from the contaminated environment containing the target substance and employ them within a fuel cell system. In a study conducted by Sharma et al., samples were obtained from soil contaminated with hydrocarbons for the purpose of addressing phenanthrene pollution through the use of fuel cells. Following the preparation of the soil sample, it was dissolved in heptamethylnonane, and after an incubation period of 300 days, the mixture underwent filtration and subsequent centrifugation. The centrifuged material was utilized as the primary source of inoculum and introduced into a two-chamber fuel cell. Within this cell, phenanthrene was immobilized on the anode electrode, serving as a substrate for the formation of a biofilm by bacteria capable of decomposing organic matter. The investigation revealed that an increase in the loading concentration of this carbon material on the anode led to a reduction in current production. The maximum power output achieved was 37 mW/m^2 at a concentration of 2 mg/cm^2. Upon the analysis of the microbial population, it was observed that electroactive bacteria, notably *Pseudomonas*, were prevalent within this system, and these bacteria exhibited the capability to degrade phenanthrene, contributing to the pollutant's removal process [32].

3. Anolyte as substrate

The electrical energy generated by MFC significantly relies on the composition of the organic matter within the anolyte. Microorganisms exhibit the ability to metabolize an array of substrates, spanning from inorganic compounds to intricate organic compounds, utilizing them as sources of energy or carbon. MFC studies have explored various substrates, including simple organic compounds like glucose, ethanol, and acetate, which are easily metabolized, as well as complex organic compounds like starch, chitin, and cellulose. However, while these organic materials are effective, they might not be economically efficient. Alternatively, wastewaters present a viable and cost-effective option. Nevertheless, wastewaters encompass a diverse range of complex organic materials, influencing the microbial population based on the specific organic constituents present within them [33–37].

In research conducted by sotres et al., a comparison was made between synthetic wastewater and pig slurry as an anolyte. The synthetic wastewater exhibited a power output of 2138 mW/m^3, while pig slurry showed 5623 mW/m^3. Analysis of microbial diversity revealed that the type of substrate significantly influenced the microbial population. In the synthetic wastewater where acetate was utilized, bacterial groups like *Alcaligenaceae*, *Pseudomonadaceae* and *Clostridiaceae* were among the dominant families. Conversely, in the pig slurry, the *Flavobacteriaceae*, *Chitinophagaceae*, *Comamonadaceae* and *Nitrosomonadaceae* families stood out as dominant among the microbial groups [38]. Indeed, the composition of the anolyte exerts a significant

Inoculum	Anolyte	Power output	Microbial community	Ref.
Wastewater sludge	The organic fraction of municipal solid waste (OFMSW)	116 mW/m^2		[16]
moist manure		123 mW/m^2		
sediment sample	brewery wastewater (inlet)	8.001 μW/cm^2	*Geobacter* *Shewanella* *Clostridium*	[39]
	brewery wastewater (outlet)	1.843 μW/cm^2		
sludge	glucose,	0.11 W/m^2	*Clostridium* *Azospirillum*	[40]
	sodium acetate	0.098 W/m^2	*Moheibacter* *Azospirillum*	
	food waste hydrolysate	0.173 W/m^2	*Rummeliibacillus* *Burkholderia, Enterococcus* *Clostridium*	
anaerobic consortia	orange peel waste	358.8 mW/	*firmicutes*	[41]
sludge	Food waste	422 mW/m^2	*Firmicutes* *Bacteroidetes* *Proteobacteria*	[42]
sludge	sea food processing wastewater	05 mW/m^2	*Stenotrophomonas*	[43]
Swine wastewater	swine wastewater	1.92 kWh/m^3	*Arcobacter* *Pseudomonas* *Acinetobacter*	[44]

Table 1.
Anolytes and their effect on microbial community.

influence on the microbial community, regardless of the source of inoculum. The specific constituents and composition of the anolyte play a crucial role in shaping and determining the microbial population within the system. **Table 1** provides an overview of various anolytes and their influence on the microbial community.

4. Additives

Modifying the anolyte by introducing a range of additives can alter its environmental conditions, consequently enhancing cell performance. This manipulation not only impacts bacterial metabolism but also contributes to improving overall cell efficiency. However, these alterations also exert an influence on the microbial population. Additives such as Cu^{2+}, CD^{2+}, S^{2-}, and surfactants (TWEEN80) have been employed to manipulate the environmental conditions [45–47].

In a study conducted by Chen et al., they incorporated sodium citrate as an additive, serving dual roles in pH adjustment and electron transfer. This combined addition led to a notable enhancement of 58% compared to the control group, with the output power peaking at a maximum of 785 mW/m^2. Investigation of the microbial population revealed an increase in microbial diversity following the introduction of citrate, attributed to two primary reasons. Firstly, maintaining a neutral pH balance promotes the growth of a wider array of microorganisms. Secondly, the citrate additive itself serves as a carbon source for microorganisms, stimulating their growth. In the control condition without citrate, *Proteobacteria* and *Bacteroidetes* were the dominant species. However, with the inclusion of citrate, the relative abundance of *Proteobacteria* decreased, while *Bacteroidetes* and *Firmicutes* exhibited higher prevalence. Among the *Proteobacteria*, *gammaproteobacteria* were predominant in the control condition, yet the addition of citrate diminished their dominance and elevated the relative abundance of alphaproteobacterial [48].

Exoelectrogens transfer the produced electrons to the anode using two methods: direct transfer and transfer facilitated by electron mediators. Additionally, the introduction of artificial electron mediators can enhance electron transfer within exoelectrogenes [49]. These mediators affect the bioelectrochemical properties of MFC and thus affect the microbial population as well. In Lin's study, they compared two electron mediators, neutral red (NR) and potassium ferricyanide (PFC), in research involving a two-chamber fuel cell fueled by toluene. The introduction of these electron mediators resulted in an increase in voltage, elevating it from 5.53 mV to 109 mV and 2.88 mV for NR and PFC, respectively. Analysis of microbial diversity indicated that these electron mediators play a role in shaping the microbial community's development. Seven bacterial groups were consistently found in all microbial fuel cells (MFCs). However, the reduced variability in the presence of these bacterial groups in MFCs utilizing NR and PFC respectively could be attributed to potential toxicity linked with these mediators [50].

5. Components

In general, the prevailing belief is that the current generation in an MFC is contingent on both the quantity of biofilm formation on the anode electrode and the level of bioelectrical activity within this biofilm [51]. Furthermore, the anode electrode stands as a pivotal component in determining the overall cost of fuel cell technology, typically constituting about 10% of the total expenses for constructing an MFC [52]. Consequently, the selection of the anode material plays a critical role in influencing both the performance and cost considerations of the battery. Carbon-based materials, such as carbon paper, carbon cloth, graphite, carbon brush, and similar options, are commonly employed in research due to their advantageous electrochemical properties and cost-effectiveness [53]. Variations among these materials pertain to their surface area, porosity, and conductivity, factors that directly impact the absorption rate and growth of electroactive bacteria.

In this investigation, four distinct carbon materials—granular activated carbon (GAC), granular semicoke (GS), carbon felt cube (CFC), and granular graphite (GG)—each possessing respective surface areas of 686 m2/g, 236 m2/g, 0.845 m2/g, and 0.623 m2/g, were assessed within a dual-chambered MFC. The power output was determined as 31.5, 27.9, 26.1, and 23.8 W/m^2 for GAC, CFC, GG, and GS, respectively. The researchers scrutinized the microbial community present on each electrode using the 16S rRNA method. Notably, GAC demonstrated the highest power output and displayed a dominant presence of 95% *Geobacter* in its microbial composition. Similarly, the CFC anode exhibited a dominance of *Geobacter* (43%) and *Ruminobacillus* (16%). On the GG electrode, the most prevalent phylum was *Azospira*, constituting 65% of the community. Conversely, the GS anode displayed an abundance of *Bacteroidetes* (19%), *Sinorhizobium* (15%), and *Synergistes* (13%). The outcomes underscored that the greater porosity and improved bioelectrochemical behavior of GAC resulted in a higher power output and an enrichment of electroactive bacteria [54].

The physical and chemical properties of the anode material significantly impact the composition of the electroactive biofilm formed by microorganisms. Altering these properties directly influences the microbial community, its diversity, and enhances the enrichment of microbial populations. Numerous studies have demonstrated that stabilizing metals or metal oxides on carbon materials leads to a notable increase in conductivity, often by two to threefold. This heightened conductivity plays a pivotal role in improving the enrichment of microbial populations [55, 56].

In a study by Xu et al., nanoparticles of MNO^2, Fe^3O^4, and Pd were loaded on carbon cloth treated with 5% Nafion. These modified carbon cloth materials served as anodes in a two-chamber cell. They were then compared against a control anode consisting of bare carbon cloth. The initial inoculum for the MFC was obtained from activated sludge sourced from a wastewater treatment plant. Synthetic wastewater was used as the substrate for the MFC. Electrochemical indicated that Pd, MNO2, and Fe3O4 exhibited power output of 824, 782, and 728 mW/m^2, respectively. These values surpassed the power output of the control, which utilized bare carbon cloth with a power output of 680 mW/m^2. Anodes treated with MNO2 and Fe3O4 particularly facilitated the enrichment of *Geobacter*, constituting for 73% of the microbial coverage. On the Pd-treated anode, *Geobacter* and *sphaerochaeta* bacteria were the predominant species, comprising 38% and 16% of the microbial composition, respectively [57].

In a different study, bio-synthesized gold nanoparticles (bio-Au) were utilized independently as well as in combination with multi-walled carbon nanotubes (MWCNTs) to modify carbon cloth electrodes. Overall, the incorporation of gold nanoparticles significantly enhanced the bioelectric behavior of the MFC. Specifically, BioAu/MWCNT showed the best performance compared to unmodified carbon cloth. The BioAu/MWCNT modification resulted in a 140% improvement in system startup time and a 56% increase in system output power, with the maximum power output reaching 178 mw. In this anode modified with gold nanoparticles and MWCNT, the bacterial groups *Gammaproteobacteria* and *Negativicutes* were among the dominant species [58].

In conclusion, the modification of the anode improves its physiochemical characteristics, creating a significant preference for EAB over fermentative bacteria. Despite not notably impacting bacterial diversity, the increased presence of electroactive bacteria on the modified electrodes resulted in better MFC system performance. **Table 2** provides a summary of the modifications made to the anodes.

Certain environmental conditions, such as elevated COD levels, high salinity, and the toxicity of wastewater, can hinder the performance of microbial biofilms. To enhance performance under such challenging circumstances, bacteria can be securely immobilized on the anode surface to mitigate the adverse effects of these harsh conditions and prevent a decline in MFC performance. Various methods for immobilizing bacteria on the anode have been reported, including adsorption, covalent binding, and entrapping. Through these methods, a substantial concentration of microorganisms can be firmly anchored to the anode's surface. Among these different stabilization techniques, the entrapping method stands out due to its stability and enhanced strength [64].

In a study conducted by Leo et al., they employed the entrapping method to immobilize inoculum obtained from a municipal wastewater treatment plant onto carbon cloth using agarose gel. They examined the impact of varying concentrations of sodium acetate (ranging from 1 to 20 g/l) and compared it to non-immobilized

Anode modification	Inoculum	Power output	Microbial community	Ref.
Graphene oxide (GO)	active sludge	124.58 mW/ m^2	increase in the relative ratio of *Proteobacteria*, but a decrease of *Firmicutes*	[59]
tungsten carbide	anaerobic sludge	3.26 W/m^2	*Geobacter* *Geothrix* *Pseudomonas*	[60]
CF/N-CNT/PANI		3.8 W/m ^{3}	*Desulfuromonas* *Geobacter*	[61]
poly(3,4-ethylenedioxythiophene)	sludge		*Geobacter*	[62]
Graphene oxide (GO) and carbon nanotubes (CNTs)	—	—	*Pseudomonas Thauera* *Diaphorobacter* *Tumebacillus Lysobacter*	[63]

Table 2.
Anode modifications and dominant microorganism.

carbon cloth as control. The stabilization of bacteria led to a 60% and 70% increase in output power for concentrations of 5 and 10 g/liter, respectively. Through an analysis of the microbial community, they demonstrated that stabilizing the inoculum on carbon cloth resulted in a 60% prevalence of *Geobacteria*, whereas the control anode exhibited a *Geobacteria* prevalence of 40% [65].

The membrane constitutes a pivotal factor that influences both the bioelectrochemical behavior and the microbial community within the developed biofilm on the anode electrode of a MFC [66, 67]. Serving as a barrier between the electrodes and compartments of the MFC, it plays a crucial role in upholding the anaerobic condition on the anode and preventing the undesired leakage of organic substances from the anolyte to the catholyte [68, 69]. Any leakage of anolyte substrate to the catholyte adversely impacts the kinetics of oxygen reduction. Furthermore, oxygen leakage into the anolyte promotes the proliferation of aerobic or facultative anaerobic bacteria, leading to the wastage of substrate and influencing the kinetics of electron production [70]. Additionally, the adhesion of microorganisms to the membrane's surface results in biofouling affecting overall performance of the system [71]. Consequently, the type of membrane emerges as a key determinant affecting the overall performance of the system, consequently influencing the microbial population dynamics.

They modified a polymeric membrane (SPEEK) with nano silver, ranging from weight percentages of 2.5 to 10 mg/cm-2, and assessed its performance in a tubular Microbial Fuel Cell (MFC). The treatment with nano silver showed no substantial difference in water absorption levels, yet it notably enhanced proton ion transfer rates and reduced internal resistance and oxygen transfer rates. Examining various weight percentages for the treated membranes, it was found that the membrane treated with 7.5% by weight of silver exhibited the highest output power compared to other treatments. Subsequently, a comparison was made between the output power and microbial population of SPEEK7.5 with Nafion 117 and the non-treated SPEEK membrane, where the maximum output power recorded was 156 mw/m2. The presence of silver nanoparticles exhibited an antibacterial property, resulting in reduced biofouling on the membrane, with only silver-resistant bacteria thriving in the environment. The analysis identified 33 bacterial groups, with over 6000 bacterial species falling under *Proteobacteria* and *Firmicutes* [72].

6. Operational conditions

The oxygen present in the anolyte can significantly impact the functioning of MFC. As the performance of MFCs relies on exoelectrogenic bacteria, the presence of oxygen and aeration can alter their performance by influencing these microorganisms. Exoelectrogens respond differently to the presence of oxygen. For instance, *Geobacter*, being an obligate anaerobe, is highly susceptible to oxygen exposure. Even minimal contact with oxygen can inhibit its respiration and decrease power generation [73]. Conversely, facultative anaerobes like *Shewanella* can generate electricity under both aerobic and anaerobic conditions, although the quantity produced varies [74]. While several studies have demonstrated that even minimal oxygen ingress to the anode can diminish MFC performance, higher concentrations of oxygen can actually enhance performance [75].

In an investigation by Quan et al., a MFC was employed, and a combination of aerobic and anaerobic sludge served as the inoculum. The researchers noted that the output voltage was greater under anaerobic conditions compared to aerobic conditions. Additionally, they observed prolonged cycles in the anaerobic setup. These findings suggest that the presence of aerobic conditions encouraged the proliferation of bacteria detrimental to the system's performance. Analysis of the microbial population revealed that aeration led to a decline in diversity, with species such as *Bacteroidetes* and others becoming dominant [76].

External resistance is one of the factors influencing the performance of MFC. In some studies, it has been shown that the use of external resistance has increased the removal of COD [77]. Given that the anode functions as the final electron acceptor, adjusting the external resistance can impact system performance by directing electron flow towards the cathode and facilitating the regeneration of the electron acceptor. This electron flow alteration can influence the diversity and composition of the microbial community [78]. Understanding the effect of external resistance on the microbial community provides a holistic understanding of exoelectrogen performance. External resistance prompts bacterial growth, with higher resistance leading to increased biomass. However, as resistance intensifies, there's a reduction in current generation [79]. Consequently, it's crucial to explore the appropriate amount of external resistance to optimize system performance.

Katuri et al., conducted a study exploring the influence of external resistance on the composition of the anode biofilm and microbial population. They employed a two-chamber MFC, utilizing glucose as a substrate, and varied the resistances from 100 to 50 kΩ. Their observations revealed that as resistance increased, the current production in the system decreased, and the system reached peak performance at a slower rate. Interestingly, they found that the highest production of biomass was achieved with 25 and 50 kΩ resistors, surpassing even the conditions of open circuit voltage. The investigation of microbial diversity showed that with the increase in resistance from 100 ohms to 50 kΩ, bacterial groups increased from 15 to 27. 100 Ohm resistance showed the simplest microbial diversity [80].

Continuous operation stands as a vital requirement for effective and economical wastewater treatment processes. In this context, the Organic Loading Rate (OLR) holds importance, and its control is influenced by the Hydraulic Retention Time (HRT). HRT, a critical operational parameter, directly affects OLR, subsequently impacting the production of electricity in MFCs [81, 82].

The study conducted by Haavisto et al. utilized a continuous two-chamber system supplied with xylose. Their observations revealed a significant impact of

HRT variations on electricity generation. Decreasing the HRT from 3.5 days to 1 day resulted in an increase in system voltage from 344 mV to 480 mV. However, further reduction in HRT to 0.75 days and 0.17 days led to decreased voltages of 218 mV and 156 mV, respectively. Moreover, when adjusting the external resistance, the highest current production of 2460 mA/m^2 was achieved with an HRT of 1 day. Additionally, HRTs exceeding 1 day displayed Chemical Oxygen Demand (COD) levels below 0.4 g. L^{-1}/day, indicating insufficient substrate availability for the metabolism of microorganisms in the system. The decrease in Hydraulic Retention Time (HRT) led to the elimination of bacteria that were not attached to the biofilm. Consequently, there were notable shifts in microbial population diversity throughout the experiment. When the HRT was reduced to 0.5 days, there was an increase in the xylose content within the anolyte. This increase correlated with a significant rise in *Christensenella*, a type of bacteria adept at fermenting xylose. During the fermentation process of xylose, acetate, propionate, and butyrate are generated as byproducts, serving as carbon sources for exoelectrogenic bacteria. However, the heightened concentration of the substrate led to excessive proliferation of xylose-fermenting bacteria. This overgrowth, in turn, resulted in decreased current production within the system [83].

Studies have demonstrated that fluctuations in temperature significantly affect the initiation and continuous operation of MFCs. For instance, MFCs operating below 15°C experienced prolonged startup times and limited electricity production within a month of operation. However, recent research has highlighted successful startup and sustained functioning of both MFCs and Microbial Electrolysis Cells (MECs) even at extremely low temperatures, such as 4°C. These studies uncovered the existence of distinct psychrophilic populations within the anode biofilms, supporting extracellular electron transfer in MECs at lower temperatures [84, 85].

In the study conducted by Mei et al., MFC reactors were initially operated at 25°C to establish a biofilm on the anode surface. Following biofilm formation, the performance of the system was evaluated across temperatures of 10, 20, and 30°C over six cycles. Their findings demonstrated a correlation between decreasing temperature and increased cycle duration, as well as a decrease in power output. The highest output power, reaching 894 mW/m^2, was achieved at 30°C. Moreover, the maximum COD removal of 93% was observed at 30°C, with reduced COD removal at lower temperatures.

Throughout all tested temperatures, *Proteobacteria*, *Bacteroidetes*, and *Firmicutes* were the dominant bacterial groups. Specifically, *Proteobacteria* accounted for 61%, 65%, and 45% at temperatures of 10, 20, and 30°C, respectively. *Bacteroidetes* prevalence was 17%, 20%, and 25.5% at 10, 20, and 30°C, while *Firmicutes* constituted 6% at 10 and 20°C, rising to 11% at 30°C [86]. The study indicated that biofilm formation, metabolic activity, and extracellular electron transfer in MFCs are influenced by temperature.

7. Biodegradation

The potential of MFC technology extends beyond bioenergy production, delving into bioremediation, marking a promising avenue for efficient and sustainable environmental cleanup. Biodegradation serves as a sustainable resolution to combat persistent pollutant contamination. While physical-chemical methods exhibit rapid treatment efficacy, they often entail high costs and the potential generation of harmful by-products, diverging from the principles of Green Remediation. Bioremediation, on the other hand, harnesses the inherent capability of natural

Pollutant	Efficiency (%)	Power Output	Dominant Bacteria	Ref.
Food wastes	95.9	56 mW/m2	*Geobacter* *Bacteroides*	[90]
vanadium	76.8	19 ± 11 mW/m	*Deltaproteobacteria* *Bacteroidetes Spirochaete*	[91]
Copper	98.3%	10.2 W m-3	*Proteobacteria* *Bacteroidetes* *Actinobacteria* *Acidobacteria*	[92]
sulfamethoxazole	87.52	1186.2 mW/m2	*Proteobacteria* *Bacteroidetes*	[93]
nitrogen	100	6.3 W/m3	*Nitrosomonas* *Nitratireductor* *Acidovorax*	[94]
p-nitrophenol	81		*Corynebacterium* *Comamonas* *Chryseobacterium* *Rhodococcus*	[95]
bisphenol and ibuprofen	78.7–63.2, respectively	28.63 mW m2	*Chlorobi* *Proteobacteria* *Firmicutes* *Bacteroidetes*	[96]
Congo red decolorization	97%	41.1 mW/m2	*Proteobacteria* *Bacteroidetes* *Actinomycetes and Firmicutes*	[97]
polychlorinated biphenyls	(37.5	108.89 mW/m2	*Geobacter and Pseudomonas* *Longilinea Desulfofustis)*	[98]

Table 3.
Pollutant removal and dominant bacterial community.

microbial communities to break down contaminants. This approach aligns with environmental sustainability, offering economic and social advantages while restoring ecosystems [86].

The composition of the microbial community is impacted by the substances that need to be biologically degraded. The recognition of resistance genes and clarification of degradation mechanisms, such as efflux pumps and modifying factors, emphasize the importance of biodegradation within MFCs. It underscores the complex interplay between the primary microorganisms present on the anode and the particular substrates utilized in MFCs. Grasping these associations is crucial for enhancing the efficiency of MFCs in breaking down pollutants, highlighting the essential role of biodegradation in the sustainable remediation of polluted water [87–89]. **Table 3** summarizes various research studies conducted in the field of biodegradation, focusing on the dominant microbial populations responsible for reducing pollution related to specific pollutants.

8. Conclusion

Microbial fuel cells have a dual function of electrical power generation and treating a variety of wastewater that contain both organic and inorganic pollutants.

As a form of biological technology, these cells utilize microbial catalysts to drive bioelectrochemical reactions. Mixed microbial populations are predominantly employed due to their exceptional adaptability to diverse environmental conditions. While certain bacterial groups may enriched and dominated based on system function and the type of wastewater utilized, this is contingent upon various factors. The primary source of the initial microbial community significantly influences microbial community development and can be selected based on the intended application of the MFCs. Additionally, organic and inorganic substances within the anolyte create conditions favoring specific species that metabolize these substances, influencing bacterial metabolism and their dominance. Alterations in MFC components, such as electrodes and membrane modifications can influence electrochemical properties, which impact biological properties and the enrichment of the microbial community. Environmental parameters like temperature, light, aeration and external resistance also exert considerable influence over the microbial community. Thus, it is essential to perceive and explore MFCs as a form of biological technology in order to enhance their performance and target specific applications.

Conflict of interest

The authors declare that they have no known competing financial interests or personal relationships that could have appeared to influence the work reported in this paper.

Author details

Hajar Rajaei Litkohi[1*] and Hosein Yazdi Dehnavi[2]

1 Department of Nano Biotechnology, College of Biotechnology, Amol University of Special Modern Technologies, Amol, Iran

2 Department of Microbial Biotechnology, College of Biotechnology, Amol University of Special Modern Technologies, Amol, Iran

*Address all correspondence to: h.rajaei@ausmt.ac.ir; rajaei1361@gmail.com

References

[1] Himri Y, Malik AS, Boudghene Stambouli A, Himri S, Draoui B. Review and use of the Algerian renewable energy for sustainable development. Renewable and Sustainable Energy Reviews. 2009;**13**:1584-1591. DOI: 10.1016/j.rser.2008.09.007

[2] Mendes FB, Bordignon SE. Renewable Energy and the Role of Biofuels in the Current World. In: Recent Developments in Bioenergy Research. Amsterdam, The Netherlands: Elsevier; 2020. pp. 65-84. DOI: 10.1016/B978-0-12-819597-0.00003-9

[3] Cai T, Meng L, Chen G, Xi Y, Jiang N, Song J, Zheng S, et al. Application of advanced anodes in microbial fuel cells for power generation: A review. Chemosphere. 2020;**248**:5. DOI: 10.1016/j.chemosphere.2020.125985

[4] Kouzuma A, Kasai T, Hirose A, Watanabe K. Catabolic and regulatory systems in Shewanella oneidensis MR-1 involved in electricity generation in microbial fuel cells. Frontiers in Microbiology. 2015;**6**:1-11. DOI: 10.3389/fmicb.2015.00609

[5] Kumar R, Singh L, Zularisam AW. Exoelectrogens: Recent advances in molecular drivers involved in extracellular electron transfer and strategies used to improve it for microbial fuel cell applications. Renewable and Sustainable Energy Reviews. 2016;**56**:1322-1336. DOI: 10.1016/j.rser.2015.12.029

[6] Lefebvre O, Tan Z, Kharkwal S, Ng HY. Effect of increasing anodic NaCl concentration on microbial fuel cell performance. Bioresource Technology. 2012;**112**:336-340. DOI: 10.1016/j.biortech.2012.02.048

[7] Borole AP, Reguera G, Ringeisen B, Wang Z-W, Feng Y, Kim BH. Electroactive biofilms: Current status and future research needs. Energy & Environmental Science. 2011;**4**:4813. DOI: 10.1039/c1ee02511b

[8] Pant D, Van Bogaert G, Diels L, Vanbroekhoven K. A review of the substrates used in microbial fuel cells (MFCs) for sustainable energy production. Bioresource Technology. 2010;**101**:1533-1543. DOI: 10.1016/j.biortech.2009.10.017

[9] Logan BE, Rabaey K. Conversion of wastes into bioelectricity and chemicals by using microbial electrochemical technologies. Science. 1979;**337**(2012):686-690. DOI: 10.1126/science.1217412

[10] Jung S, Regan JM. Influence of external resistance on Electrogenesis, Methanogenesis, and anode prokaryotic communities in microbial fuel cells. Applied and Environmental Microbiology. 2011;**77**:564-571. DOI: 10.1128/AEM.01392-10

[11] Kiely PD, Regan JM, Logan BE. The electric picnic: Synergistic requirements for exoelectrogenic microbial communities. Current Opinion in Biotechnology. 2011;**22**:378-385. DOI: 10.1016/j.copbio.2011.03.003

[12] White HK, Reimers CE, Cordes EE, Dilly GF, Girguis PR. Quantitative population dynamics of microbial communities in plankton-fed microbial fuel cells. The ISME Journal. 2009;**3**:635-646. DOI: 10.1038/ismej.2009.12

[13] Freguia S, Teh EH, Boon N, Leung KM, Keller J, Rabaey K. Microbial fuel cells operating on mixed fatty acids. Bioresource Technology. 2010;**101**:1233-1238. DOI: 10.1016/j.biortech.2009.09.054

[14] Ishii S, Suzuki S, Norden-Krichmar TM, Phan T, Wanger G, Nealson KH, et al. Microbial population and functional dynamics associated with surface potential and carbon metabolism. The ISME Journal. 2014;**8**:963-978. DOI: 10.1038/ismej.2013.217

[15] Miceli JF, Parameswaran P, Kang D-W, Krajmalnik-Brown R, Torres CI. Enrichment and analysis of anode-respiring bacteria from diverse anaerobic Inocula. Environmental Science & Technology. 2012;**46**:10349-10355. DOI: 10.1021/es301902h

[16] El-Chakhtoura J, El-Fadel M, Rao HA, Li D, Ghanimeh S, Saikaly PE. Electricity generation and microbial community structure of air-cathode microbial fuel cells powered with the organic fraction of municipal solid waste and inoculated with different seeds. Biomass and Bioenergy. 2014;**67**:24-31. DOI: 10.1016/j.biombioe.2014.04.020

[17] Yates MD, Kiely PD, Call DF, Rismani-Yazdi H, Bibby K, Peccia J, et al. Convergent development of anodic bacterial communities in microbial fuel cells. The ISME Journal. 2012;**6**:2002-2013. DOI: 10.1038/ismej.2012.42

[18] Yan X, Lee H-S, Li N, Wang X. The micro-niche of exoelectrogens influences bioelectricity generation in bioelectrochemical systems. Renewable and Sustainable Energy Reviews. 2020;**134**:110184. DOI: 10.1016/j.rser.2020.110184

[19] Nevin KP, Richter H, Covalla SF, Johnson JP, Woodard TL, Orloff AL, et al. Power output and columbic efficiencies from biofilms of Geobacter sulfurreducens comparable to mixed community microbial fuel cells. Environmental Microbiology. 2008;**10**:2505-2514. DOI: 10.1111/j.1462-2920.2008.01675.x

[20] Ishii S, Shimoyama T, Hotta Y, Watanabe K. Characterization of a filamentous biofilm community established in a cellulose-fed microbial fuel cell. BMC Microbiology. 2008;**8**:6. DOI: 10.1186/1471-2180-8-6

[21] Ishii S, Suzuki S, Norden-Krichmar TM, Nealson KH, Sekiguchi Y, Gorby YA, et al. Functionally stable and phylogenetically diverse microbial enrichments from microbial fuel cells during wastewater treatment. PLoS One. 2012;**7**:e30495. DOI: 10.1371/journal.pone.0030495

[22] Kouzuma A, Kasai T, Nakagawa G, Yamamuro A, Abe T, Watanabe K. Comparative metagenomics of anode-associated microbiomes developed in Rice Paddy-field microbial fuel cells. PLoS One. 2013;**8**:e77443. DOI: 10.1371/journal.pone.0077443

[23] Dunaj SJ, Vallino JJ, Hines ME, Gay M, Kobyljanec C, Rooney-Varga JN. Relationships between soil organic matter, nutrients, bacterial community structure, and the performance of microbial fuel cells. Environmental Science & Technology. 2012;**46**:1914-1922. DOI: 10.1021/es2032532

[24] Cercado B, Byrne N, Bertrand M, Pocaznoi D, Rimboud M, Achouak W, et al. Garden compost inoculum leads to microbial bioanodes with potential-independent characteristics. Bioresource Technology. 2013;**134**:276-284. DOI: 10.1016/j.biortech.2013.01.123

[25] Madjarov J, Prokhorova A, Messinger T, Gescher J, Kerzenmacher S.

The performance of microbial anodes in municipal wastewater: Pre-grown multispecies biofilm vs. natural inocula. Bioresource Technology. 2016;**221**:165-171. DOI: 10.1016/j.biortech.2016.09.004

[26] Logan BE. Exoelectrogenic bacteria that power microbial fuel cells. Nature Reviews. Microbiology. 2009;**7**:375-381. DOI: 10.1038/nrmicro2113

[27] An J, Kim B, Nam J, Ng HY, Chang IS. Comparison in performance of sediment microbial fuel cells according to depth of embedded anode. Bioresource Technology. 2013;**127**:138-142. DOI: 10.1016/j.biortech.2012.09.095

[28] Song T-S, Cai H-Y, Yan Z-S, Zhao Z-W, Jiang H-L. Various voltage productions by microbial fuel cells with sedimentary inocula taken from different sites in one freshwater lake. Bioresource Technology. 2012;**108**:68-75. DOI: 10.1016/j.biortech.2011.11.136

[29] Ishii S, Suzuki S, Yamanaka Y, Wu A, Nealson KH, Bretschger O. Population dynamics of electrogenic microbial communities in microbial fuel cells started with three different inoculum sources. Bioelectrochemistry. 2017;**117**:74-82. DOI: 10.1016/j.bioelechem.2017.06.003

[30] Gao C, Wang A, Wu W-M, Yin Y, Zhao Y-G. Enrichment of anodic biofilm inoculated with anaerobic or aerobic sludge in single chambered air-cathode microbial fuel cells. Bioresource Technology. 2014;**167**:124-132. DOI: 10.1016/j.biortech.2014.05.120

[31] Song N, Jiang H-L. Effects of initial sediment properties on start-up times for sediment microbial fuel cells. International Journal of Hydrogen Energy. 2018;**43**:10082-10093. DOI: 10.1016/j.ijhydene.2018.04.082

[32] Sharma M, Nandy A, Taylor N, Venkatesan SV, Ozhukil Kollath V, Karan K, et al. Bioelectrochemical remediation of phenanthrene in a microbial fuel cell using an anaerobic consortium enriched from a hydrocarbon-contaminated site. Journal of Hazardous Materials. 2020;**389**:121845. DOI: 10.1016/j.jhazmat.2019.121845

[33] Ren Z, Ward TE, Regan JM. Electricity production from cellulose in a microbial fuel cell using a defined binary culture. Environmental Science & Technology. 2007;**41**:4781-4786. DOI: 10.1021/es070577h

[34] Rezaei F, Richard TL, Logan BE. Analysis of chitin particle size on maximum power generation, power longevity, and Coulombic efficiency in solid–substrate microbial fuel cells. Journal of Power Sources. 2009;**192**:304-309. DOI: 10.1016/j.jpowsour.2009.03.023

[35] Logan B, Cheng S, Watson V, Estadt G. Graphite Fiber brush anodes for increased power production in air-cathode microbial fuel cells. Environmental Science & Technology. 2007;**41**:3341-3346. DOI: 10.1021/es062644y

[36] Cheng S, Liu H, Logan BE. Increased performance of single-chamber microbial fuel cells using an improved cathode structure. Electrochemistry Communications. 2006;**8**:489-494. DOI: 10.1016/j.elecom.2006.01.010

[37] Liu H, Cheng S, Logan BE. Production of electricity from acetate or butyrate using a single-chamber microbial fuel cell. Environmental Science & Technology. 2005;**39**:658-662. DOI: 10.1021/es048927c

[38] Sotres A, Tey L, Bonmatí A, Viñas M. Microbial community dynamics in

continuous microbial fuel cells fed with synthetic wastewater and pig slurry. Bioelectrochemistry. 2016;**111**:70-82. DOI: 10.1016/j.bioelechem.2016.04.007

[39] Çetinkaya AY, Köroğlu EO, Demir NM, Baysoy DY, Özkaya B, Çakmakçı M. Electricity production by a microbial fuel cell fueled by brewery wastewater and the factors in its membrane deterioration. Chinese Journal of Catalysis. 2015;**36**:1068-1076. DOI: 10.1016/S1872-2067(15)60833-6

[40] Xin X, Hong J, Liu Y. Insights into microbial community profiles associated with electric energy production in microbial fuel cells fed with food waste hydrolysate. Science of The Total Environment. 2019;**670**:50-58. DOI: 10.1016/j.scitotenv.2019.03.213

[41] Miran W, Nawaz M, Jang J, Lee DS. Conversion of orange peel waste biomass to bioelectricity using a mediator-less microbial fuel cell. Science of The Total Environment. 2016;**547**:197-205. DOI: 10.1016/j.scitotenv.2016.01.004

[42] Asefi B, Li S-L, Moreno HA, Sanchez-Torres V, Hu A, Li J, et al. Characterization of electricity production and microbial community of food waste-fed microbial fuel cells. Process Safety and Environmental Protection. 2019;**125**:83-91. DOI: 10.1016/j.psep.2019.03.016

[43] Jayashree C, Tamilarasan K, Rajkumar M, Arulazhagan P, Yogalakshmi KN, Srikanth M, et al. Treatment of seafood processing wastewater using upflow microbial fuel cell for power generation and identification of bacterial community in anodic biofilm. Journal of Environmental Management. 2016;**180**:351-358. DOI: 10.1016/j.jenvman.2016.05.050

[44] Li M, Zhou M, Tian X, Tan C, Gu T. Enhanced bioenergy recovery and nutrient removal from swine wastewater using an airlift-type photosynthetic microbial fuel cell. Energy. 2021;**226**:120422. DOI: 10.1016/j.energy.2021.120422

[45] Xu Y-S, Zheng T, Yong X-Y, Zhai D-D, Si R-W, Li B, et al. Trace heavy metal ions promoted extracellular electron transfer and power generation by Shewanella in microbial fuel cells. Bioresource Technology. 2016;**211**:542-547. DOI: 10.1016/j.biortech.2016.03.144

[46] Ieropoulos I, Gálvez A, Greenman J. Effects of sulphate addition and sulphide inhibition on microbial fuel cells. Enzyme and Microbial Technology. 2013;**52**:32-37. DOI: 10.1016/j.enzmictec.2012.10.002

[47] Wen Q, Kong F, Ma F, Ren Y, Pan Z. Improved performance of air-cathode microbial fuel cell through additional tween 80. Journal of Power Sources. 2011;**196**:899-904. DOI: 10.1016/j.jpowsour.2010.09.009

[48] Chen W, Liu Z, Li Y, Xing X, Liao Q, Zhu X. Improved electricity generation, coulombic efficiency and microbial community structure of microbial fuel cells using sodium citrate as an effective additive. Journal of Power Sources. 2021;**482**:228947. DOI: 10.1016/j.jpowsour.2020.228947

[49] Matsena MT, Nkhalambayausi Chirwa EM. Advances in microbial fuel cell technology for zero carbon emission energy generation from waste. In: Biofuels and Bioenergy, Elsevier: Biofuels and Bioenergy. Amsterdam, The Netherlands: Elsevier; 2022. pp. 321-358. DOI: 10.1016/B978-0-323-85269-2.00013-7

[50] Lin C-W, Wu C-H, Chiu Y-H, Tsai S-L. Effects of different mediators on

electricity generation and microbial structure of a toluene powered microbial fuel cell. Fuel. 2014;**125**:30-35. DOI: 10.1016/j.fuel.2014.02.018

[51] Rabaey K, Rozendal RA. Microbial electrosynthesis — Revisiting the electrical route for microbial production. Nature Reviews. Microbiology. 2010;**8**:706-716. DOI: 10.1038/nrmicro2422

[52] Rozendal RA, Hamelers HVM, Rabaey K, Keller J, Buisman CJN. Towards practical implementation of bioelectrochemical wastewater treatment. Trends in Biotechnology. 2008;**26**:450-459. DOI: 10.1016/j.tibtech.2008.04.008

[53] Fan X, Zhou Y, Jin X, Song R, Li Z, Zhang Q. Carbon material-based anodes in the microbial fuel cells, carbon. Energy. 2021;**3**:449-472. DOI: 10.1002/cey2.113

[54] Sun Y, Wei J, Liang P, Huang X. Electricity generation and microbial community changes in microbial fuel cells packed with different anodic materials. Bioresource Technology. 2011;**102**:10886-10891. DOI: 10.1016/j.biortech.2011.09.038

[55] Jung K-W, Cho S-K, Yun Y-M, Shin H-S, Kim D-H. Rapid formation of hydrogen-producing granules in an up-flow anaerobic sludge blanket reactor coupled with high-rate recirculation. International Journal of Hydrogen Energy. 2013;**38**:9097-9103. DOI: 10.1016/j.ijhydene.2013.05.059

[56] Khilari S, Pandit S, Ghangrekar MM, Das D, Pradhan D. Graphene supported α-MnO2 nanotubes as a cathode catalyst for improved power generation and wastewater treatment in single-chambered microbial fuel cells. RSC Advances. 2013;**3**:7902. DOI: 10.1039/c3ra22569k

[57] Xu H, Quan X, Xiao Z, Chen L. Effect of anodes decoration with metal and metal oxides nanoparticles on pharmaceutically active compounds removal and power generation in microbial fuel cells. Chemical Engineering Journal. 2018;**335**:539-547. DOI: 10.1016/j.cej.2017.10.159

[58] Wu X, Xiong X, Owens G, Brunetti G, Zhou J, Yong X, et al. Anode modification by biogenic gold nanoparticles for the improved performance of microbial fuel cells and microbial community shift. Bioresource Technology. 2018;**270**:11-19. DOI: 10.1016/j.biortech.2018.08.092

[59] Chen J, Hu Y, Zhang L, Huang W, Sun J. Bacterial community shift and improved performance induced by in situ preparing dual graphene modified bioelectrode in microbial fuel cell. Bioresource Technology. 2017;**238**:273-280. DOI: 10.1016/j.biortech.2017.04.044

[60] Liu D, Chang Q, Gao Y, Huang W, Sun Z, Yan M, et al. High performance of microbial fuel cell afforded by metallic tungsten carbide decorated carbon cloth anode. Electrochimica Acta. 2020;**330**:135243. DOI: 10.1016/j.electacta.2019.135243

[61] Wang Y, Chen Y, Wen Q, Zheng H, Xu H, Qi L. Electricity generation, energy storage, and microbial-community analysis in microbial fuel cells with multilayer capacitive anodes. Energy. 2019;**189**:116342. DOI: 10.1016/j.energy.2019.116342

[62] Kang YL, Pichiah S, Ibrahim S. Facile reconstruction of microbial fuel cell (MFC) anode with enhanced exoelectrogens selection for intensified electricity generation. International

Journal of Hydrogen Energy. 2017;**42**:1661-1671. DOI: 10.1016/j.ijhydene.2016.09.059

[63] Liang Y, Zhai H, Liu B, Ji M, Li J. Carbon nanomaterial-modified graphite felt as an anode enhanced the power production and polycyclic aromatic hydrocarbon removal in sediment microbial fuel cells. Science of The Total Environment. 2020;**713**:136483. DOI: 10.1016/j.scitotenv.2019.136483

[64] Yang X-Y, Tian G, Jiang N, Su B-L. Immobilization technology: A sustainable solution for biofuel cell design. Energy & Environmental Science. 2012;**5**:5540-5563. DOI: 10.1039/C1EE02391H

[65] Luo H, Yu S, Liu G, Zhang R, Teng W. Effect of in-situ immobilized anode on performance of the microbial fuel cell with high concentration of sodium acetate. Fuel. 2016;**182**:732-739. DOI: 10.1016/j.fuel.2016.06.032

[66] Jang JK, Pham TH, Chang IS, Kang KH, Moon H, Cho KS, et al. Construction and operation of a novel mediator- and membrane-less microbial fuel cell. Process Biochemistry. 2004;**39**:1007-1012. DOI: 10.1016/S0032-9592(03)00203-6

[67] Ghangrekar MM, Shinde VB. Performance of membrane-less microbial fuel cell treating wastewater and effect of electrode distance and area on electricity production. Bioresource Technology. 2007;**98**:2879-2885. DOI: 10.1016/j.biortech.2006.09.050

[68] Modestra JA, Reddy CN, Krishna KV, Min B, Mohan SV. Regulated surface potential impacts bioelectrogenic activity, interfacial electron transfer and microbial dynamics in microbial fuel cell, renew. Energy. 2020;**149**:424-434. DOI: 10.1016/j.renene.2019.12.018

[69] Behera M, Jana PS, More TT, Ghangrekar MM. Rice mill wastewater treatment in microbial fuel cells fabricated using proton exchange membrane and earthen pot at different pH. Bioelectrochemistry. 2010;**79**:228-233. DOI: 10.1016/j.bioelechem.2010.06.002

[70] Erable B, Duţeanu NM, Ghangrekar MM, Dumas C, Scott K. Application of electro-active biofilms. Biofouling. 2010;**26**:57-71. DOI: 10.1080/08927010903161281

[71] Noori MT, Ghangrekar MM, Mukherjee CK, Min B. Biofouling effects on the performance of microbial fuel cells and recent advances in biotechnological and chemical strategies for mitigation. Biotechnology Advances. 2019;**37**:107420. DOI: 10.1016/j.biotechadv.2019.107420

[72] Kugarajah V, Dharmalingam S. Effect of silver incorporated sulphonated poly ether ether ketone membranes on microbial fuel cell performance and microbial community analysis. Chemical Engineering Journal. 2021;**415**:128961. DOI: 10.1016/j.cej.2021.128961

[73] Lovley DR, Ueki T, Zhang T, Malvankar NS, Shrestha PM, Flanagan KA, et al. Geobacter: The microbe electric's physiology, Ecology, and practical applications. In: Advances in Microbial Physiology. Cambridge, Massachusetts: Academic Press; 2011. pp. 1-100. DOI: 10.1016/B978-0-12-387661-4.00004-5

[74] Biffinger JC, Byrd JN, Dudley BL, Ringeisen BR. Oxygen exposure promotes fuel diversity for Shewanella oneidensis microbial fuel cells. Biosensors & Bioelectronics. 2008;**23**:820-826. DOI: 10.1016/j.bios.2007.08.021

[75] Biffinger JC, Ringeisen BR. Air-exposed microbial fuel cells and screening techniques. Journal of Biotechnology. 2010;**150**:24-25. DOI: 10.1016/j.jbiotec.2010.08.074

[76] Quan X, Quan Y, Tao K. Effect of anode aeration on the performance and microbial community of an air–cathode microbial fuel cell. Chemical Engineering Journal. 2012;**210**:150-156. DOI: 10.1016/j.cej.2012.09.009

[77] Aelterman P, Versichele M, Marzorati M, Boon N, Verstraete W. Loading rate and external resistance control the electricity generation of microbial fuel cells with different three-dimensional anodes. Bioresource Technology. 2008;**99**:8895-8902. DOI: 10.1016/j.biortech.2008.04.061

[78] Lyon DY, Buret F, Vogel TM, Monier J-M. Is resistance futile? Changing external resistance does not improve microbial fuel cell performance. Bioelectrochemistry. 2010;**78**:2-7. DOI: 10.1016/j.bioelechem.2009.09.001

[79] Picioreanu C, Head IM, Katuri KP, van Loosdrecht MCM, Scott K. A computational model for biofilm-based microbial fuel cells. Water Research. 2007;**41**:2921-2940. DOI: 10.1016/j.watres.2007.04.009

[80] Katuri KP, Scott K, Head IM, Picioreanu C, Curtis TP. Microbial fuel cells meet with external resistance. Bioresource Technology. 2011;**102**:2758-2766. DOI: 10.1016/j.biortech.2010.10.147

[81] Hashemi J, Samimi A. Steady state electric power generation in up-flow microbial fuel cell using the estimated time span method for bacteria growth domestic wastewater. Biomass and Bioenergy. 2012;**45**:65-76. DOI: 10.1016/j.biombioe.2012.05.011

[82] Lay C-H, Kokko ME, Puhakka JA. Power generation in fed-batch and continuous up-flow microbial fuel cell from synthetic wastewater. Energy. 2015;**91**:235-241. DOI: 10.1016/j.energy.2015.08.029

[83] Haavisto JM, Kokko ME, Lay C-H, Puhakka JA. Effect of hydraulic retention time on continuous electricity production from xylose in up-flow microbial fuel cell. International Journal of Hydrogen Energy. 2017;**42**:27494-27501. DOI: 10.1016/j.ijhydene.2017.05.068

[84] Lu L, Ren N, Zhao X, Wang H, Wu D, Xing D. Hydrogen production, methanogen inhibition and microbial community structures in psychrophilic single-chamber microbial electrolysis cells. Energy & Environmental Science. 2011;**4**:1329. DOI: 10.1039/c0ee00588f

[85] Cheng S, Xing D, Logan BE. Electricity generation of single-chamber microbial fuel cells at low temperatures. Biosensors & Bioelectronics. 2011;**26**:1913-1917. DOI: 10.1016/j.bios.2010.05.016

[86] Boas JV, Oliveira VB, Simões M, Pinto AMFR. Review on microbial fuel cells applications, developments and costs. Journal of Environmental Management. 2022;**307**:114525. DOI: 10.1016/j.jenvman.2022.114525

[87] Xue W, Li F, Zhou Q. Degradation mechanisms of sulfamethoxazole and its induction of bacterial community changes and antibiotic resistance genes in a microbial fuel cell. Bioresource Technology. 2019;**289**:121632. DOI: 10.1016/j.biortech.2019.121632

[88] Chen J, Yang Y, Liu Y, Tang M, Wang R, Tian Y, et al. Bacterial community shift and antibiotics resistant genes analysis in response to biodegradation of oxytetracycline in

dual graphene modified bioelectrode microbial fuel cell. Bioresource Technology. 2019;**276**:236-243. DOI: 10.1016/j.biortech.2019.01.006

[89] Liu R, Gao C, Zhao Y-G, Wang A, Lu S, Wang M, et al. Biological treatment of steroidal drug industrial effluent and electricity generation in the microbial fuel cells. Bioresource Technology. 2012;**123**:86-91. DOI: 10.1016/j.biortech.2012.07.094

[90] Jia J, Tang Y, Liu B, Wu D, Ren N, Xing D. Electricity generation from food wastes and microbial community structure in microbial fuel cells. Bioresource Technology. 2013;**144**:94-99. DOI: 10.1016/j.biortech.2013.06.072

[91] Zhang B, Tian C, Liu Y, Hao L, Liu Y, Feng C, et al. Simultaneous microbial and electrochemical reductions of vanadium (V) with bioelectricity generation in microbial fuel cells. Bioresource Technology. 2015;**179**:91-97. DOI: 10.1016/j.biortech.2014.12.010

[92] Wu Y, Zhao X, Jin M, Li Y, Li S, Kong F, et al. Copper removal and microbial community analysis in single-chamber microbial fuel cell. Bioresource Technology. 2018;**253**:372-377. DOI: 10.1016/j.biortech.2018.01.046

[93] Jiang J, Wang H, Zhang S, Li S, Zeng W, Li F. The influence of external resistance on the performance of microbial fuel cell and the removal of sulfamethoxazole wastewater. Bioresource Technology. 2021;**336**:125308. DOI: 10.1016/j.biortech.2021.125308

[94] Park Y, Park S, Nguyen VK, Yu J, Torres CI, Rittmann BE, et al. Complete nitrogen removal by simultaneous nitrification and denitrification in flat-panel air-cathode microbial fuel cells treating domestic wastewater. Chemical Engineering Journal. 2017;**316**:673-679. DOI: 10.1016/j.cej.2017.02.005

[95] Zhao H, Kong C-H. Enhanced removal of p-nitrophenol in a microbial fuel cell after long-term operation and the catabolic versatility of its microbial community. Chemical Engineering Journal. 2018;**339**:424-431. DOI: 10.1016/j.cej.2018.01.158

[96] Li H, Zhang S, Yang X-L, Yang Y-L, Xu H, Li X-N, et al. Enhanced degradation of bisphenol and ibuprofen by an up-flow microbial fuel cell-coupled constructed wetland and analysis of bacterial community structure. Chemosphere. 2019;**217**:599-608. DOI: 10.1016/j.chemosphere.2018.11.022

[97] Li C, Luo M, Zhou S, He H, Cao J, Luo J, et al. Study on synergistic mechanism of PANDAN modification, current and electroactive biofilms on Congo red decolorization in microbial fuel cells. International Journal of Hydrogen Energy. 2020;**45**:29417-29429. DOI: 10.1016/j.ijhydene.2020.07.246

[98] Wu M, Xu X, Lu K, Li X. Effects of the presence of nanoscale zero-valent iron on the degradation of polychlorinated biphenyls and total organic carbon by sediment microbial fuel cell. Science of The Total Environment. 2019;**656**:39-44. DOI: 10.1016/j.scitotenv.2018.11.326

Chapter 11

Operational Principles of MFCs

Ali Shirpay

Abstract

In this chapter, after describing how the microbial fuel cell operates and how the chemical energy resulting from the oxidation of a substrate through oxidation/reduction reactions is converted into electricity with chemical reactions, the factors affecting the performance of the fuel cell including the effect of temperature, the effect pH, external resistance, type of electrode, size, and distance of electrodes, type, and composition of microorganisms, as well as the shape, structure, and size of the chamber have been investigated. Since the purpose of fuel cell design is to produce electric current from microorganisms, therefore, the current density criterion and how to calculate it are briefly explained.

Keywords: microbial fuel cell, various fuel cell structures, size and distance of electrodes, microorganism, power density

1. Introduction

The microbial fuel cell is an electrochemical device that converts the chemical energy contained in organic and inorganic materials in the environment or waste into electrical energy by using microorganisms as a biocatalyst. The driving force in these cells is the oxidation-reduction reactions of a raw material in which a microorganism is used as a catalyst [1].

The microbial fuel cell has a structure more or less similar to polymer fuel cell and includes an anode and anode chamber, cathode and cathode chamber, proton transfer membrane, precursor, and microorganism. In microbial fuel cells, a microorganism is used as a biocatalyst, which produces electrons and protons by decomposing and oxidizing organic materials. Only in its anode, the chemical transformations are done in a different way, as a result, its anode structure is different from the polymer cell anode structure. The working principles of these cells are the same as those governing other fuel cells. The main difference is that the catalysts in these cells are microorganisms or enzymes, and as a result, there is no need for precious metals. The reaction conditions are very simple and to carry out the reaction, ambient temperature, air pressure, and neutral pH are needed. The fuel cell system can be operated for a long time [2]. The advantages of this type of cell are:

- Does not need a catalyst.
- It has good stability during long working time.

- Reactions are automatically adjusted and do not require recharging.
- There is no need for a heat exchanger or a cooler to carry out the reaction.

Biocatalysts are used in microbial fuel cells and these biocatalysts have the following prominent features:

- Can be used in moderate and low temperatures.
- Biocatalysts are cheaper than metal catalysts.
- They are renewable and can be used for different types of cells with different substrates [3].

Microorganisms have the ability to produce active substances that can be produced in a specific environment, then transferred to the cell as fuel; the microbial fermentation process can take place directly in the anode part and the same chamber [4].

The microbial fuel cell must be designed to provide anaerobic conditions in the anode chamber because the presence of oxygen in the anode chamber prevents the production of electrons or by absorbing electrons and protons, prevents the formation of electricity in the external circuit. Also, the air injected in the cathode should provide dissolved oxygen for the reaction [5]. Based on their performance, microbial fuel cells are designed as single-chamber and double-chamber. Designing two-chamber systems that can operate in two continuous and discontinuous modes is difficult due to its complex design. Designing single chamber systems is easier and cost-saving. Usually, these types of systems only include an anode chamber, and the cathode electrode is directly exposed to air. The outline of a microbial fuel cell device for both cases is shown in **Figure 1**.

The substrate is oxidized as fuel by microorganisms in the anaerobic chamber where the anode electrode is located and releases electrons and protons. Protons are transferred to the cathode compartment through the proton exchange membrane and are converted into water vapor in the reaction with oxygen. The presence of oxygen in the anodic chamber prevents the production of electrons or cuts off the flow of electricity by consuming electrons and protons. Therefore, the device must be designed in such a way as to provide anaerobic conditions in the anodic chamber. Two electrodes are connected by a wire that contains a load (a device that consumes energy), but in the laboratory, they use a resistor as a load. Basically, the membrane is permeable for the protons produced in the anode, and they can move to the cathode, where they combine with oxygen and electrons transferred through the wire to form water. The electric current produced by a microbial fuel cell can be measured in the laboratory by observing the voltage drop across a resistor using a voltmeter connected to a computer [6].

A wide range of organic substances from simple molecules such as carbohydrates and proteins to complex mixtures of organic substances found in food, human and animal wastes, chocolate and beer wastes as precursors as materials available to different microorganisms for use in the system Microbial fuel cells are used [7]. Microbial fuel cells are divided into two groups based on the mechanism of electron transfer, with and without a mediator. Mediated microbial fuel cells are electrochemically inactive and electron transfer is carried out by mediators such as *thionine*, *methylviologen*, *humicacid*, neutral red, etc. Mediator molecules receive electrons from

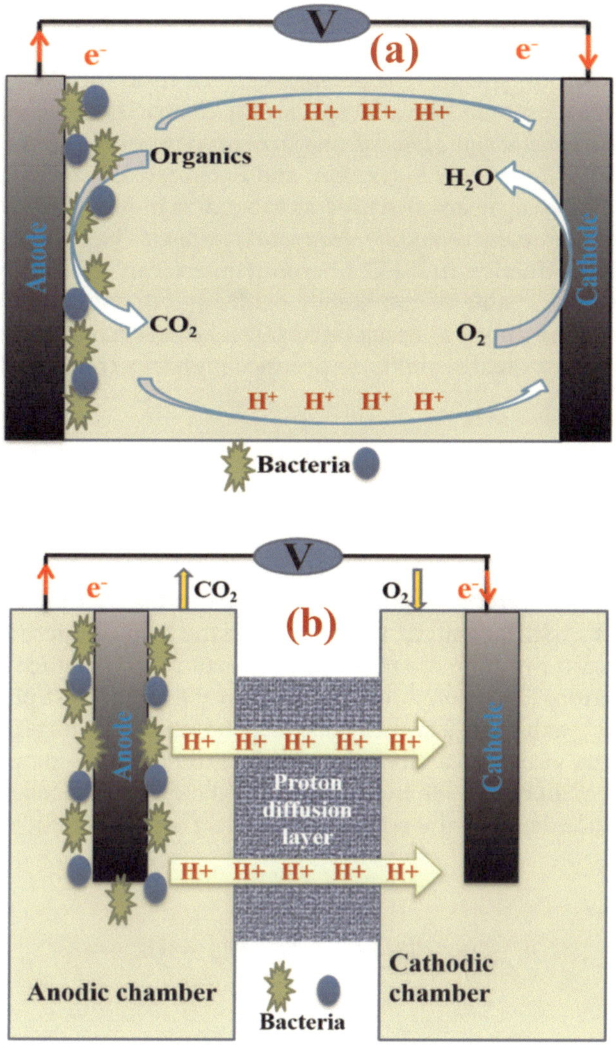

Figure 1.
A view of the performance of the microbial fuel cell (a) single chamber and (b) double chamber.

microorganisms and discharge them on the surface of the anode electrode. Using these intermediaries is costly. In some cases, the toxicity of these substances is also a problem. In unmediated microbial fuel cells, electrochemically active bacteria such as *Shewanella Putrefaciens*, *Aeromonas hydrophila*, and *Geobacter* are used for electron transfer. Using *pili* and membrane *cytochromes*, these bacteria deliver electrons directly to the anode electrode and do not need an intermediary [8].

Some microalgae, yeasts, and fungi have also been reported to produce electric current in MFCs, which are used as a substrate or as an aid to the anode or cathode. The nomenclature of the microorganisms used has not yet been standardized. However, some terms have been proposed for microorganisms that can exogenously transfer electrons to the anode without using any artificial mediators. Some of these terms include exoelectrogens, electrogenic microorganisms, electrochemically active bacteria, anodeophiles, anode-respiring bacteria, and electrophiles. In addition, microorganisms can be named according to their performance in MFC. Microorganisms that donate electrons to the electrode (anode) in MFCs are called electrode reducers, while microorganisms that accept electrons from the electrodes are called electrode oxidizers. Common bacterial species for electricity generation in MFCs are *Geobacter spp.*, *Shewanella spp.*, *Rhodoferax ferrireducens*, *Aeromonas hydrophila*, *Pseudomonas aeruginosa*, *Clostridium butyricum*, and *Enterococcus gallinarum*. Apart from microorganisms, microalgae are also used as substrates in microbial fuel cells. In yeast, in addition to *Saccharomyces cerevisiae*, *Hansenula anomala* has also successfully demonstrated current production in MFC. Microorganisms can effectively help in the production of energy capable of completely oxidizing organic compounds and transferring electrons to the anode at accelerated rates. Studies have shown that mixed-culture biofilms have a greater ability to produce higher current density than pure-culture biofilms [9].

2. Factors affecting the performance of microbial fuel cells

To design a fuel cell with industrial applications, in order to produce electricity from municipal waste water and simultaneous purification, various factors are involved. Fuel cells with a catalytic reaction, with the help of microorganisms, oxidize organic matter or biomass in the anode section and produce carbon dioxide, protons, and electrons. Microorganisms formed in the anode part play the role of catalyst in analogy with chemical fuel cells. Depending on the different feeds in the fuel cell electrolyte, various reactions can occur in the fuel cell components. For example, if the feed in the anode is an affluent and oxygen acts as an electron acceptor in the cathode, then the reactions in the cathode and anode will be as follows [10]:

In the anode:

$$C_{12}H_{22}O_{11} + 13H_2O \rightarrow 12C_2O + 48H^+ + 48e^{-\cdot}$$

In the cathode:

$$48H^+ + 48e^- + 12O_2 \rightarrow 24H_2O$$

General reaction:

$$C_{12}H_{22}O_{11} + 12O_2 \rightarrow 12C_2O + 11H_2O\&\triangle G^0 = -5792.2kJ/mole$$

Electrons reach the cathode through an external circuit and in the presence of proton (H^+) which comes to this part through the membrane, they combine with the oxygen blown in the cathode and produce water; In this way, this process continues many times with the completion of the circuit. Much researches has been done to optimize the parameters and increase the efficiency of microbial fuel cells in order to overcome the limitations of proton and electron transfer, increasing the surface area and reactivity of the electrodes. Several factors including the effect of temperature, effect of pH, external resistance, type of electrode, size and distance of electrodes, type and composition of microorganisms, shape, structure, and size of the chamber as well as the effect of various chemical additives have been investigated. In this study, we discuss some effective factors as follows:

2.1 Effect of temperature

Temperature is an important parameter in the decomposition process and has a major effect on the initial biolayer formation process and on its bioelectrocatalytic performance. Also, biofilms grown at higher temperatures tend to have higher electrochemical activity than those grown at lower temperatures [11]. Therefore, operating temperature provides an effective strategy to reduce MFC start-up time and improve energy efficiency. The optimal temperature of microorganisms is known to be in the range of 35 to 40 °C. When the temperature drops below this value, the activity of microorganisms gradually decreases. The performance of microbial fuel cells based on the production of electricity from wastewater treatment is generally improved in a higher temperature range, although it has been proven that microbial fuel cells are not as sensitive as conventional anaerobic analyzers in a relatively low-temperature range (i.e. between 4 and 15 °C). The advantage that microbial fuel cells have compared to other conventional decomposers is the ability to decompose at low flow rates and low temperatures, to the extent that this characteristic causes that in cold regions where the rate of urban sewage is relatively low, it can be Microbial fuel cell was used as a suitable device for wastewater treatment.

In biochemistry, a measure in the form of Q10 is expressed as the rate of change of a biological system as a result of an increase in temperature by 10 °C in the temperature coefficient or Eq. (1):

$$Q_{10=(R_1/R_2)^{10/(T_2-T_1)}} \tag{1}$$

where R is the rate and T is the temperature in degrees Celsius or Kelvin. Generally, the Q10 value is between 1 and 3.

Focusing on MFCs, temperature changes in the system in terms of kinetics (activation energy, mass transfer coefficients, solution conductivity...), thermodynamics (Gibbs free energy, electrode potentials...), system outputs (i.e. output power and amount removal of COD) and affects the nature and type of microbes. These changes are given in Eqs. (2)–(4):

$$\Delta G = -nFE_{rev} \tag{2}$$

$$V_{MFC} = E^0_{MFC} - \eta_{anode} - \eta_{cathode} - I\,R_{int} \tag{3}$$

$$C_xH_yO_z + pH_2O \rightarrow xC_2O + qH + qe^- \tag{4}$$

Which is the relationship Eq. (2) energy potentially available, relationship Eq. (3) Electromotive force, and relationship Eq. (4) Organic matter oxidation. Although the research on MFC has increased greatly in recent years, systematic information on the effect of temperature on microbial fuel cells is very scarce and has only been conducted in a range of different temperatures from 4 to 35°C. In addition, different conditions and systems have been used in different studies, leading to large variations in the reported parameters for the target temperature. This feature makes it difficult to determine the effect of temperature on performance from the various available data. When the temperature in an MFC system changes, the composition of the anodic microbial consortium changes correspondingly. This change along with the modification of other parameters such as diffusion coefficients, activation energies, etc. will affect the system outputs including I, P, and COD_{rem} rate [12].

2.2 Effect of pH

The pH value in the anode chamber can affect the metabolic activity of microorganisms and also the electron and proton (H^+) production mechanism and thus the efficiency of microbial fuel cells. In general, microorganisms respond to environmental pH changes by carrying out processes such as proton exchange and amino acid decomposition. Changes in pH lead to changes in parameters such as ion concentration and membrane capacity. For example, in a tested microbial fuel cell, the pH of the anodic solution reached its minimum value after 30 to 40 hours and after feeding. During this experiment, under two temperature ranges, i.e. (8 to 22 °C) and (20 to 35 °C), when the pH of the input feed was 7.4, the pH of the anodic solution reached its lowest value, i.e. 6.2, after the mentioned time. This expression shows that the proton exchange through the membrane was slower than the production rate in the anodic chamber. Therefore, protons are only limited to the reaction in the cathode chamber where oxygen is present as an oxidant, and as a result, there is an accumulation of protons in the anode chamber, which leads to a drop in pH. Therefore, the performance of a fuel cell is at its best when the pH of the anode chamber is kept constant using a buffer solution, and it causes the same amount of proton produced in the anode to actually pass through the membrane and reach the cathode [13].

2.3 External resistance

External resistance can affect the power performance of the MFC by regulating the electron flow from the anode to the cathode. According to Jacobi's law, the optimal external resistance to produce maximum energy in MFC should be close to its internal resistance. External resistance can also affect the structure of the anode biofilm. The external resistance controls the flow of electrons from the anode to the cathode and affects the output potential and current of MFCs according to Ohm's law. As a result, the output power is affected by the external resistance. Changes in external resistance in addition to electricity production also affect the relative diversity of the microbial community [14]. One study showed that anodic biofilms developed under 100 Ω external resistive pressure showed better biofilms compared to samples subjected to controlled pressure of 1 MΩ [15]. This is not surprising, since an external charge changes the rate and speed of electrons exiting the anode. This change in bacterial metabolism always has an effect and as a result, causes a change in the microbes present in the anode biolayer.

2.4 Electrode type

The anode material is very important in the overall performance of MFC. Anode characteristics affect biofilm formation, electron transfer, and substrate oxidation. To achieve higher performance of MFCs, the selected electrode materials should have good biocompatibility, large surface area, low electrical resistance, good corrosion resistance, and chemical stability [16]. Among the properties of anodic materials, cross-sectional area, pore structure, and surface hydrophobicity have a significant effect on the formation of anodic biofilm. Du et al. [17] reported that polydopamine (superhydrophobic semiconductor) accelerated the formation of anodic biofilm and improved the performance of MFCs. This is in agreement with a previous report showing that anode hydrophobicity can facilitate the attachment of microbes, which in turn increases biofilm formation and power generation [18]. In this study, it has also been shown that such properties increase the proportion of electron-generating elements (eg, bacterial protein) in the biofilm. The most important features to improve the surface interaction of materials used in the anode electrode are: i) Electrical conductivity ii) Corrosion resistance iii) High mechanical strength iv) Developed surface v) Environmentally friendly and vi) Low cost. Metal materials and all kinds of carbon materials, including carbon cloth, carbon brush, carbon rod, carbon mesh, carbon veil, carbon paper, carbon felt, granular activated carbon, granular graphite, carbonized cardboard, graphite plate, and reticulated vitreous carbon, all of these have the characteristics. Among the metal materials, stainless steel plate, stainless steel mesh, stainless steel scrubber, silver plate, nickel plate, copper plate, gold plate, and titanium plate were used as anode electrode materials. Carbon cloth is a carbon material that is mostly used as an anode material due to its high surface area and relatively high porosity. Also, carbon fabric shows high electrical conductivity, flexibility, and mechanical strength in forming more complex three-dimensional structures. The negative aspect is the cost, which is generally very high. Carbon brush is a very interesting material in which carbon fibers are twisted together with a very high surface area and an optimal area-to-volume ratio. Carbon mesh is also commercially available at a relatively low cost and relatively low electrical conductivity. The main problem of carbon mesh is low mechanical strength, which can lead to low durability in high flow conditions. It can also be folded to make a 3D electrode, but the porosity is low. Carbon veil is a very cheap carbon material with relatively high electrical conductivity and high porosity, but its single layer is quite fragile. Because this material is versatile, it can be folded to form a solid, porous three-dimensional electrode. Carbon paper is a flat, relatively porous but expensive, and fragile carbonaceous material that is mainly used on a laboratory scale. Carbon felt performance is usually due to its high porosity and high electrical conductivity characteristics, similar to a carbon veil, large pores allow bacteria to penetrate through the structure. The cost is relatively low and the mechanical strength is high depending on the thickness of the material. Granular activated carbon is also biocompatible, low cost, and very porous but has very low electrical conductivity. In this case, the surface area is very large, but the surface area available for the interaction of bacteria is small because most of the available surface area is on the nanometer scale. The properties of granular graphite are similar to Granular activated carbon, except for a much lower level due to lack of activation. As a result, its electrical conductivity is much higher like Granular activated carbon. Therefore, it is used as a packing material instead of an independent anode. Carbonized cardboard is also a three-dimensional material that is made by single-walled corrugated cardboard from recycled paper subjected to heat treatment (1000 °C) for 1 hour in an inert atmosphere, which has very low cost, high electrical conductivity, and high porosity. A graphite sheet or sheet is a very simple electrode that ensures high electrical conductivity and has a low surface area and surface-to-volume ratio, resulting in lower output levels than porous materials. Reticulated glassy carbon is highly conductive and has high porosity, which allows the biofilm to penetrate the entire structure and colonize the entire electrode. But it is too fragile and too expensive to use in MFCs. Other carbon materials, such as carbonized plant stems, activated carbon nanofibers, and electrospun carbon fibers, have also been used as carbon-based anode electrodes.

Several metallic materials have been used as anode electrodes in MFCs. The most important of which is stainless steel with the main feature of being highly conductive, resistant, and cheap. Other metals such as copper, nickel, silver, gold, and titanium were also investigated as anode electrode materials. Surface chemistry and surface morphology play a key role in the interaction between the biofilm and the anode electrode. Three factors can affect the interaction between the single-cell surface, at the biofilm level and, the integrated system level: 1) Bacterial surface adhesion can be improved by changing the surface chemistry. 2) Surface hydrophilicity/hydrophobicity during bacterial attachment, 3) Oxygen or nitrogen functional groups that facilitate bacteria/surface interactions, and 4) Immobilized mediators. Theoretically, an increase in level should lead to an increase in flow. It was shown that the current produced by the carbon-felt anode was higher than that produced by the flat carbon graphite anode, but there was no correlation between the actual surface and the current. This indicates that, among other reasons, not all the available area has been successfully colonized by the biofilm. The transition from a smooth surface to a more complex three-dimensional surface requires the possibility of encountering the limitations caused by the phenomena of diffusion transfer of products and reactants, as well as the relative pH gradient. These parameters seem to be of great importance for long-term operation. Finally, at the system level, the anode electrodes must be designed to avoid clogging or "dead zones" [19].

2.5 Size and distance of electrodes

In order to achieve better performance in microbial fuel cells, the materials used in the electrode should be more active surfaces with relatively lower activation losses.

The amount of produced current is proportional to the active surface of the electrode. It has been shown in a study that graphite felt with an area of 200 cm^2 produces three times more current than graphite rods with an area of 65 cm^2 [20]. Due to these characteristics, Granular activated carbon is mainly used as a packing material instead of an independent anode. Granular activated carbon must be packed to increase the conductivity, and this may lead to possible blockages in the flow-through configuration of the MFC.

Reducing the distance between the anode and cathode electrodes, by reducing the internal resistance, facilitates the transfer of protons from the cation exchange membrane. In such a situation, microbial fuel cells have shown good performance when the two electrodes are as close as possible so that short circuit does not occur. For example, if the distance between the electrodes is reduced from 4 to 2 cm, the production power increases by 168%, and if this distance is reduced from 2 to 1 cm, the production power in the fuel cell decreases from 811 mW/m^2 to 423 mW/m^2. This paradoxical behavior is probably due to the damage caused by oxygen penetration from the cathode to the microorganism in the anode. It is reminded that if the high internal resistance limits the production power, increasing the anode level does not have a noticeable effect on the output power [21].

Different studies of electrode size in different designs of microbial fuel cells (different scales of electrodes with different volumes) have shown that the size of the electrode has a direct effect on the power density and electric current produced by the fuel cell. By making the electrode and examining the size of the electrode on both small and large scales, the ability to produce electricity has been studied. The results have shown that by changing the size of the electrode in the microbial fuel cell and increasing its cross-section, the intensity of the electric current changes. Also, by reducing the size of the electrode in the small-scale microbial fuel cell, the losses in the activation area are reduced [22]. Therefore, by reducing the size of the electrode in a small-scale microbial fuel cell, the intensity of the electric current and the power density is increased, and the losses in the activation area of the polarization diagram are also reduced to a large-scale. Also, the effect of electrode size by considering two different scales in electricity generation has been studied by Yan et al. [23]. They showed that the power density of the larger anode is higher than that of the smaller one. In a similar study, Seong-Won Im et al. [24] showed the effect of electrode size on the maximum voltage and power density generated, indicating a decrease in these quantities by reducing the scale of the electrode. In another study by Zuraidah Rasep et al. [25] with three electrodes with sizes of 6×6, 7×7, and 8×8 cm, the maximum current produced was 0.72 mA for the 8x8cm electrode and 0.56 mA for 6×6 cm, but in a study, Eduardo et al. [26] examined the electrode-surface area/anode-compartment volume ratio instead of the electrode size and showed that by decreasing the anode-compartment volume, the amount of the electricity generation increases and Also, with the increase of ESAVR from 0.15 to 0.75 $cm^2\ cm^{-3}$, the intensity of currents increases from 583 to 2416 mA m^{-2}. These studies show that the size of electrodes in MFCs plays an important role in electricity generation.

2.6 Microorganisms in the microbial fuel cell

As mentioned, the microbial fuel cell is a device that uses microorganisms to produce electric current through the oxidation of organic materials. According to **Figure 1**, microorganisms play a role by taking electrons from reactants and delivering these electrons from the biofilm to the surface of the anode electrode. It is true to say that microorganisms act as catalysts because they mediate the transfer of electrons from the reactant to the system. However since they absorb energy from the surrounding environment and reactants for their growth and survival, it is not similar to chemical catalysts [27]. The biofilm that includes microorganisms can be in contact with the electrode in two ways, mediated and unmediated. In the intermediate state, substances such as *Brevibacillus*, *SP pseudomonas*, or *supernatant* are used. Microorganisms that can transfer electrons to the electrode replace methanogens. In sediment systems, from 50 to 90% of the microorganisms in the anode are *G-Proteobacteria*, to a lesser extent, cytophages (up to 33%), Firmicutes (11.6%) and J-Proteobacteria (9–10%). which shows the dependence of open systems on the type of sediment community. The composition of microorganisms in open systems can also be influenced by various intermediates that can transfer electrons from bacteria to electrodes. A combination or mixture of biowaste can result in a higher production of recycled stream in any single component MFC. In MFCs operating on complex substrates and wastewaters, a biofilm is formed on the anode, which contains a complex community of microorganisms in addition to known electrogens. A wide range of microorganisms can transfer electrons to the electrode. If a cathode with a ferricyanide applied to its surface was used in a similar system, the predominance of Gram-positive bacteria *Brevibacillus agri*, representatives of Fermicutes, was recorded in the anode compartment.

Few articles described the cases of gram-positive bacteria (MFC - *Clostridium butyricum EG3, Thermincola ferriacetica, Saccharomyces sp., Sarcina lutea, Clostridium beijerinckii, Bacillus cereus, Corynebacterium sp., Paenibacillus*) in microbial fuel cells. While Gram-negative microorganisms used in MFC are: *Pseudomonas fluorescens, Proteus vulgaris, Bacillus violaceus, Escherichia coli, Desulfuromonas acetoxidans, Pseudomonas methanica, Methylovorus dichloromethanicum, Geobacter sulfurreducens, Shewanella putrefaciens, Methylovorus mays, Rhodoferax ferrireducens, Geobacter metallireducens, Shewanella affinis, Shewanella oneidensis, Gluconobacter oxydans, Ochrobactrum sp.,*

Pseudomonas mendocina, Klebsiella oxytoca, Enterobacter sp., Stenotrophomonas acidaminiphila, Pseudomonas pseudoalcaligenes, Klebsiella pneumonae.

Microorganisms-electrogens have the ability to use a wide range of substrates depending on the species. Therefore, the production of electric current by bacteria *Desulfuromonas acetoxidans, Thermincola ferriacetica, Escherichia coli, Bacillus subtilis, Geobacter sulfurreducens* in MFC is used as acetate substrate. *Klebsiella pneumonae* is used as lactose and *Corynebacterium sp.* It is used in a wide range of substrates, from organic acids to sugars and alcohols.

Many species *Saccharomyces sp., Proteus vulgaris, Escherichia coli, Clostridium beijerinckii, Rhodoferax ferrireducens, Clostridium butyricum, Saccharomyces cerevisiae, Pseudomonas mendocina, Gluconobacter oxydans, Pseudomonas pseudoalcaligenes, Paenibacillus lautus, Stenotrophomonas acidaminiphila, Bacillus* subtilis have the ability to produce current in the presence of glucose as a substrate. Based on the absence of oxygen in the anode compartment, anaerobic or facultative anaerobic microorganisms can work there. Anaerobes include *Geobacter sulfurreducens, Rhodoferax ferrireducens, Geobacter metallireducens, Clostridium beijerinckii Enterobacter sp, Clostridium butyricum,* and some as electrogens *Pseudomonas fluorescens, Methylovorus dichloromethanicum, Pseudomonas methanica, Methylovorus mays, Pseudomonas mendocina, Gluconobacter oxydans, Pseudomonas pseudoalcaligenes, Paenibacillus lautus, Bacillus subtilis, Stenotrophomonas acidaminiphila,* were used. Facultative anaerobes also include *Bacillus violaceus, Escherichia coli, Saccharomyces sp., Saccharomyces cerevisiae, Sarcina lutea, Proteus vulgaris, Desulfuromonas acetoxidans, Shewanella putrefaciens, Shewanella oneidensis, Shewanella affinis, Bacillus cereus, Ochrobactrum sp., Klebsiella oxytoca, Klebsiella pneumonae.* The maximum value of current density by a mixed culture in batch mode is 4.3 W/m^2.

Not all bacteria in the population participate in the transfer of electrons to the electrode, because the composition of bacteria in the anode compartment depends on the type of substrate used. Microorganisms used in MFC as biocatalysts are not only species such *as S. oneidensis*, *S. putrefaciens*, *G. sulfurreducens*, but also species of the *genus Pseudomonas, Paracoccus denitrificans*, *Rhodopseudomonas palustris*, *Lautus Bacillus*.

Microorganisms carry out electron transfer in MFCs in two ways, direct and mediated. Certain types of bacteria, metal reducers being the most important, can directly transfer electrons to the anode. Electron transfer mediators are mostly electrochemically inactive and cannot directly transfer electrons to the electrode. In this case, it is possible to use soluble substances known as redox mediators that facilitate electronic transfer. Electrochemically active microorganisms can create intermediate junctions under certain conditions that can play a role in extracellular electron transfer processes. This can happen in two ways by producing secondary and primary metabolites. Secondary metabolites (endogenous intermediates) are redox-active substances and act as reversible terminal electron acceptors that transfer electrons from a bacterial cell to a solid oxidant (MFC anode) or transferred to the aerobic layers of the biofilm, where they are oxidized and can participate again in redox processes. Primary metabolites, for example, sulfide or hydrogen, also act as redox mediators [28].

2.7 Microbial fuel cell structure

One of the most important parameters in the design of microbial fuel cells is the design of the cell chamber structure. Types of microbial fuel cells based on the structure of the cell compartment, we can mention things like H-shaped microbial fuel cells, double-reservoir microbial fuel cells, single-compartment microbial fuel cells, tubular microbial fuel cells, and microbial fuel cells with upward flow [29]. Single-chamber microbial fuel cells have a higher power density due to their lower internal resistance compared to the double-chamber type. In a single-compartment fuel cell, the distance between the anode and the cathode is significantly reduced, and no energy costs are required for cathode active oxygen because the cathode is individually aerated [30]. In addition, microbial fuel cells can be exposed to atmospheric oxygen in both continuous flow and discontinuous flow modes and generate higher output power because there is no problem with limiting the oxygen mass transfer rate to the cathode [29]. Also, the area of anode and cathode can be significantly increased compared to dual systems. Therefore, microbial fuel cells with a single chamber air cathode typically record higher performance and are more stable than their dual counterparts. The weakness of this type of design is its unscalability, when the distance between the electrodes increases, power generation becomes impossible. Due to the fact that the cathode must always be exposed to air, the design of this type of fuel cell on a large scale faces serious complications [29]. In various research, efforts have been made to increase the output power of the fuel cell. For this purpose, various structures were presented for single-compartment microbial fuel cells in order to increase their capabilities in addition to increasing their power and efficiency.

Liu and Logan first developed cubic air-cathode microbial fuel cells in 2004. In addition to its simple design, this type of microbial fuel cell has a flexible structure that can be used to investigate the characteristics that affect the performance of microbial fuel cells. Its simple design has made it possible to test many features affecting the microbial fuel cell and to better understand them. In many researches, this structure is used to investigate the effect of the effective characteristics on the efficiency of the microbial fuel cell. **Figure 2** shows the image of a cubic cathode air microbial fuel cell [31]:

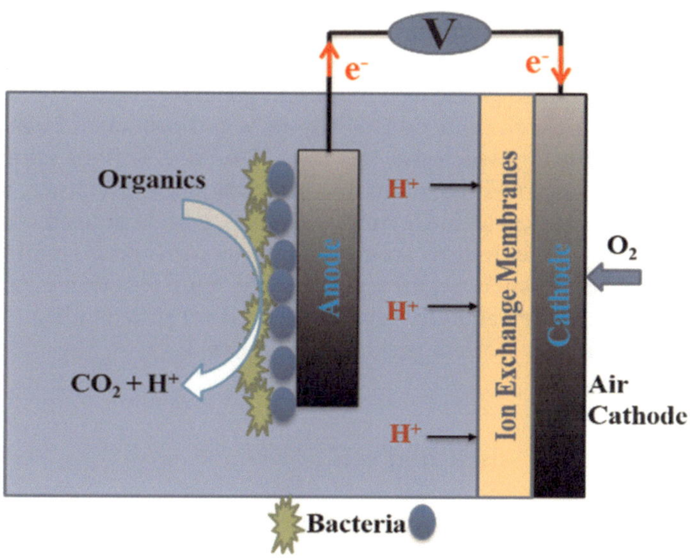

Figure 2.
Air-cathode microbial fuel cell.

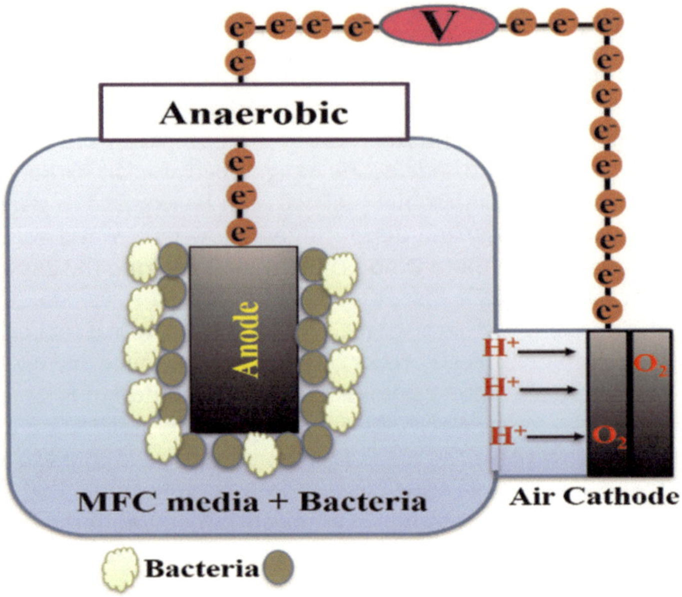

Figure 3.
An example of a single-chamber microbial fuel cell.

In this microbial fuel cell, fabric carbon is usually used to make cathode and anode electrodes. Its ion exchange membrane is Nafion 117 membrane. Its output power using urban sewage as a substrate has been reported as 3.7 W/m^3 [31].

In 2004, Liu and his colleagues invented a structure for single-compartment microbial fuel cells that could be used continuously. This structure consisted of two concentric cylinders made of acrylic. There were holes in the central cylinder for contact with air. The cathode electrode, which was made of carbon fabric, was wrapped around the central cylinder. Eight graphite rods were used as the anode electrode, which surrounded the cathode electrode. They used this structure for the first time to investigate wastewater treatment and generate electricity in continuous mode. This microbial fuel cell reduced COD by 80% and produced 26 MW/m^3 of electricity. Its coulombic efficiency was less than 12% [32]. Various schematics are used to design two-chamber MFCs, cylindrical, rectangular, miniature, high flow configuration with cylindrical shape, cylindrical shape with U-shaped chamber are the most famous of them. These designs are dependent on MFC performance and the use of defined chemical media such as glucose or acetate solutions, wastewater or factory distillates. These designs are generally used for laboratory scale studies. An example of a single-chamber microbial fuel cell is shown in **Figure 3**.

3. Power density in microbial fuel cell

A criterion for measuring the efficiency of a microbial fuel cell is CE (coulombic efficiency), which is a measure of the number of coulombs recovered as a comparison of the electric current with the maximum theoretical number of coulombs, which is defined as follows:

$$E = (Cp/C_T) \times 100 \tag{5}$$

where Cp is the total coulombs calculated by summing the flow over time for lactate consumption, and CT is the theoretical amount of coulombs that can be produced from lactate metabolism, which is calculated as follows:

$$C_T = bF(mol_{\text{lactate}}) \tag{6}$$

where F is Faraday's constant (96,485 C/mol-electron), b represents the number of moles of electrons produced by the conversion of lactate to acetate (b = 4) or CO_2 and H_2O (b = 12) and mol_{lactate} is the moles added to sodium lactate [33].

In order to obtain the highest theoretical amount of energy and removal rate from a fuel (input material), that reactant must be completely oxidized by the relevant microorganisms and converted into carbon dioxide, and its electrons transferred to the electrode. Without oxidation of the fuel, energy leaves the system through unoxidized fuel. For example, the material of Shewanella oneidenss does not completely decompose the lactate reactant and the electron leaving efficiency of this system is low and its CE is around 56.2% [34]. According to studies, the more capable the microorganisms are in fuel oxidation and complete conversion to 2CO, the higher CE is obtained [35]. Bacteria in which high percentages of CE have been reported are:

CE = 94%, Foracetate Grothrix fermentans, [36], Geobacter species has a CE of about 100% for acetate and 83% for glucose, [37] and the highest CE reported for wastewater was between 65% and 89% [38].

The power density produced by a particular microorganism cannot be compared to that of other microorganisms unless the MFC structure, operating conditions, nutrients, and chemical solutions used for the study are considered identical.

The overall performance of an MFC is evaluated in various ways, but mainly through power output and coulombic efficiency. Power is calculated as follows:

$$P = IE_{cell} \tag{7}$$

where the voltage is usually measured across a fixed external resistor, while the current is calculated from Ohm's law ($I = E_{cell}/R_{ext}$)). Therefore, power is usually calculated as follows:

$$P = \frac{E_{cell}^2}{R_{ext}} \tag{8}$$

It is a direct measurement of electrical power. The maximum power is calculated from the polarization curve. The power density is often normalized to some reactor characteristic so that the power output can be compared in different systems. The choice of parameter to use for normalization depends on the application, as many systems are not optimized for power generation. The output power is usually normalized to the predicted anode surface because the anode is where the biological reaction occurs. Therefore, the power density is calculated based on the anode area:

$$P_{An} = \frac{E_{cell}^2}{A_{An}R_{ext}} \tag{9}$$

where A_{An} is the anode area and P_{An} is the power density in (W/m^2). In cases where the cathode reaction is thought to limit the overall power generation or the anode is composed of a material that is difficult to express in terms of surface area (such as granular material), the cathode area (A_{Cat}) can be used to obtain the power density.

The projected surface area of all components as well as the specific surface area and the method of its determination should always be clearly stated.

To make engineering calculations for the size and cost of reactors, and as a useful comparison to chemical fuel cells, power is normalized to reactor volume or

$$P_v = \frac{E_{cell}^2}{vR_{ext}} \tag{10}$$

where P_v is volumetric power (W/m^3) and v is the total volume of the reactor. Using the total bed reactor volume is consistent with the engineering tradition of using the total reactor size as the basis for the calculation. However, the comparison based on total reactor volume is not always the same when comparing double and single-chamber reactors because there is no "second chamber" for the open-air cathode. In such cases, it is useful to compare reactors based on the total volume of the anode compartment. If several reactors operate in unison, for example, a series of stack reactors, the volume used for air space for the cathode is included in the total reactor volume. Therefore, the volume used in the calculation should be clearly stated and the volume of individual rooms should always be clearly stated [24].

The power density diagram that describes the power based on the current is calculated from the polarization curve and the relationship ($P = V^2/R$). We know that there is no current in the open circuit condition, therefore, the power density will not be produced. From this point onwards, the power increases with current and reaches the maximum power point. Beyond this point, the power reaches a limit with increasing ohmic losses and excessive electrode potentials that no more power can be generated (short circuit conditions). In many microbial fuel cells, the ohmic resistance plays an important role in determining the point of maximum achievable power, partly due to the low ionic conductivity of the solution.

4. Conclusion

In this chapter, factors affecting fuel cell performance including temperature, pH effect, external resistance, type of electrode, size and distance of electrodes, type and composition of microorganisms, shape, structure, and size of the chamber as well as the importance of power density in microbial fuel cell are investigated. Because determining the importance of these factors cannot be evaluated, therefore, in the design of the microbial fuel cell, the optimal amount should be extracted from the studies and considered in the design of the fuel cell.

Conflict of interest

The authors declare no conflict of interest.

Author details

Ali Shirpay
Department of Physics, Faculty of Basic Sciences, Khatam Al-Anbia (PBU) University, Tehran, Iran

*Address all correspondence to: a_shirpay@khadu.ac.ir

References

[1] Shirpay A. Bioelectricity generation from human urine in microbial fuel cells: Investigation the effect of additives to urea electrolyte. Biomass Conversion and Biorefinery. 2023:1-2

[2] Chandra R, Castillo-Zacarias C, Mancera-Andrade EI, Mohan SV, Parra-Saldívar R. Fundamentals of biophotovoltaics for conversion of solar energy to bioelectricity. In: Microbial Electrochemical Technology. Elsevier; 2019. pp. 503-523

[3] Shukla A, Suresh P, Berchmans S, Rajendran A. Biological fuel cells and their applications. Current Science. 2004;**87**(4):455-468

[4] He Z, Angenent LT. Application of bacterial biocathodes in microbial fuel cells. Electroanalysis: An International Journal Devoted to Fundamental and Practical Aspects of Electroanalysis. 2006;**18**(19–20):2009-2015

[5] Jafari M, Sedighi KS. Microbial fuel cell: A strategy for bioremediation and energy production. Journal of Environmental Science and Technology. 2018;**20**(3):37-43

[6] Bazina N, Ahmed TG, Almdaaf M, Jibia S, Sarker M. Power generation from wastewater using microbial fuel cells: A review. Journal of Biotechnology. 2023;**374**:17-30

[7] Mishra B, Awasthi SK, Rajak RK. A review on electrical behavior of different substrates, electrodes and membranes in microbial fuel cell. International Journal of Energy Power Engineering. 2017;**11**(9):983-988

[8] Lotfi M, Younesi H, Bahramifar N. Wastewater treatment using dual-chamber microbial fuel cell with saccharomyces cerevisiae. Journal of Water and Wastewater. 2018;**29**(4):101-108

[9] Kumar R, Singh L, Wahid ZA. Role of microorganisms in microbial fuel cells for bioelectricity production. Microbial Factories: Biofuels, Waste treatment. 2015;**1**:135-154

[10] Song HL, Zhu Y, Li J. Electron transfer mechanisms, characteristics and applications of biological cathode microbial fuel cells–a mini review. Arabian Journal of Chemistry. 2019;**12**(8):2236-2243

[11] Liu Y, Climent V, Berna A, Feliu JM. Effect of temperature on the catalytic ability of electrochemically active biofilm as anode catalyst in microbial fuel cells. Electroanalysis. 2011;**23**(2):387-394

[12] Larrosa-Guerrero A, Scott K, Head IM, Mateo F, Ginesta A, Godinez C. Effect of temperature on the performance of microbial fuel cells. Fuel. 2010;**89**(12):3985-3994

[13] Cheng S, Liu H, Logan BE. Increased power generation in a continuous flow MFC with advective flow through the porous anode and reduced electrode spacing. Environmental Science & Technology. 2006;**40**(7):2426-2432

[14] Rismani-Yazdi H, Christy AD, Carver SM, Yu Z, Dehority BA, Tuovinen OH. Effect of external resistance on bacterial diversity and metabolism in cellulose-fed microbial fuel cells. Bioresource Technology. 2011;**102**(1):278-283

[15] Katuri KP, Enright A-M, O'Flaherty V, Leech D. Microbial

analysis of anodic biofilm in a microbial fuel cell using slaughterhouse wastewater. Bioelectrochemistry. 2012; **87**:164-171

[16] Sonawane JM, Yadav A, Ghosh PC, Adeloju SB. Recent advances in the development and utilization of modern anode materials for high performance microbial fuel cells. Biosensors and Bioelectronics. 2017;**90**:558-576

[17] Du Q, An J, Li J, Zhou L, Li N, Wang X. Polydopamine as a new modification material to accelerate startup and promote anode performance in microbial fuel cells. Journal of Power Sources. 2017;**343**:477-482

[18] Liang Y, Feng H, Shen D, Li N, Guo K, Zhou Y, et al. Enhancement of anodic biofilm formation and current output in microbial fuel cells by composite modification of stainless steel electrodes. Journal of Power Sources. 2017;**342**:98-104

[19] Santoro C, Arbizzani C, Erable B, Ieropoulos I. Microbial fuel cells: From fundamentals to applications. A review. Journal of Power Sources. 2017;**356**: 225-244

[20] Gerber M. The Effect of Anode Geometry on Power Output in Microbial Fuel Cells [Dissertation]. Columbus: The Ohio State University; 2014

[21] Fernando E, Keshavarz T, Kyazze G. Complete degradation of the azo dye acid Orange-7 and bioelectricity generation in an integrated microbial fuel cell, aerobic two-stage bioreactor system in continuous flow mode at ambient temperature. Bioresource Technology. 2014;**156**: 155-162

[22] Logan BE. Microbial fuel cells. Hoboken: John Wiley & Sons; 2008

[23] Chaudhuri SK, Lovley DR. Electricity generation by direct oxidation of glucose in mediatorless microbial fuel cells. Nature Biotechnology. 2003; **21**(10):1229

[24] Logan BE, Hamelers B, Rozendal R, Schröder U, Keller J, Freguia S, et al. Microbial fuel cells: Methodology and technology. Environmental Science & Technology. 2006;**40**(17):5181-5192

[25] Shirpay A. Effects of electrode size on the power generation of the microbial fuel cell by Saccharomyces cerevisiae. Ionics. 2021;**27**(9):3967-3973

[26] Wang Y, Jiang Y, Guo D-p, Luo S, Wang X-b, Li Y-F. The influence of the electrode anode sizes of the microbial fuel cell (MFC) on the electrical property and COD removal with the electroplating wastewater to be the cathode. Advanced Materials Research. 2010;**113-116**:2189-2192. DOI: 10.4028/www.scientific.net/AMR.113-116.2189

[27] Im S-W, Lee H-J, Chung J-W, Ahn Y-T. The effect of electrode spacing and size on the performance of soil microbial fuel cells (SMFC). Journal of Korean Society of Environmental Engineers. 2014;**36**(11):758-763. DOI: 10.4491/KSEE.2014.36.11.758

[28] Zuraidah R, Nur SMA, Syazwan M, Ghazali M, Norilhamiah Y, Aida SA, et al. Microbial fuel cell for conversion of chemical energy to electrical energy from food industry wastewater. Journal of Environmental Science and Technology. 2016;**9**(6):481-485

[29] Penteado ED, Fernandez-Marchante CM, Zaiat M, Gonzalez ER, Rodrigo MA. Optimization of the performance of a microbial fuel cell using the ratio electrode-surface area/anode-compartment volume. Brazilian Journal of Chemical Engineering. 2018;**35**:

141-146. DOI: 10.1590/ 0104-6632.20180351s20160411

[30] Konovalova EY, Stom DI, Zhdanova GO, Yuriev DA, Li Y, Barbora L, et al. The microorganisms used for working in microbial fuel cells. In: AIP Conference Proceedings. Vol. 1952, No. 1. AIP Publishing LLC; 2018. p. 020017

[31] Liu H, Logan B. Electricity generation using an air-cathode single chamber microbial fuel cell (MFC) in the absence of a proton exchange membrane. ACS National Meeting Book of Abstracts. 2004;**228**(1):4040-4046

[32] Liu H, Cheng S, Huang L, Logan BE. Scale-up of membrane-free single-chamber microbial fuel cells. Journal of Power Sources. 2008;**179**(1):274-279

[33] Ringeisen BR, Henderson E, Wu PK, Pietron J, Ray R, Little B, et al. High power density from a miniature microbial fuel cell using Shewanella oneidensis DSP10. Environmental Science & Technology. 2006;**40**(8): 2629-2634

[34] Lanthier M, Gregory KB, Lovley DR. Growth with high planktonic biomass in Shewanella oneidensis fuel cells. FEMS Microbiology Letters. 2008;**278**:29-35

[35] Franks AE, Nevin KP. Microbial fuel cells, a current review. Energies. 2010; **3**(5):899-919

[36] Bond DR, Lovley DR. Evidence for involvement of an electron shuttle in electricity generation by Geothrix fermentans. Applied and Environmental Microbiology. 2005;**71**:2186-2189

[37] Nevin KP, Richter H, Covalla SF, Johnson JP, Woodard TL, Orloff AL, et al. Power output and columbic efficiencies from biofilms of Geobacter sulfurreducens comparable to mixed community microbial fuel cells. Environmental Microbiology. 2008;**10**: 2505-2514

[38] Rabaey K, Lissens G, Siciliano SD, Verstraete W. A microbial fuel cell cabable lf converting glucose to electricity at high rate and efficiency. Biotechnology Letters. 2003;**25**: 1531-1535

Chapter 12

Plant Microbial Fuel Cell

M. Azizul Moqsud

Abstract

Living plants can generate electricity with the help of the microbial fuel cell. This is a sustainable way to generate electricity as there is no chance of environmental pollution. In this chapter, plant microbial fuel cells will be discussed thoroughly including their design and mechanism for sustainable power generation through plant microbial fuel cells. This plant microbial fuel cell can provide the necessary bioenergy and the potential food supply at the same time. If we could get green energy and food together, then it would certainly increase the chance of the probable supply of the two important demands of the world, which are shortages of green energy and food for humanity.

Keywords: plant microbial fuel cell, bioelectricity, power density, voltage, bacteria

1. Introduction

We need electricity and new sources of electricity in the world as there are around 1.6 billion people that still live in the dark at night. Many countries of the world can produce only half of their electricity demand. The demand for clean sources of energy has been increasing in recent times [1]. The global warming phenomenon is happening due to the excess amount of greenhouse gas emissions for electricity generation from fossil fuels and nonrenewable energies. Fossil fuel, which is not good for the environment, is used all over the world, and the remaining amount of this fossil fuel is decreasing day by day [2]. So, it is very much important for us to find new sources of green energy for a sustainable world. In this perspective, the plant microbial fuel cell (PMFC) is one of the technologies in which bioelectricity can be generated with the help of living plants and microbial actions. PMFC is an ecological solar cell in which solar energy helps plants in photosynthesis to produce carbohydrates [3]. In the photosynthesis process, living plants can use carbon dioxide from the atmosphere. As this is the main source of electricity generation, PMFC is a carbon-neutral option for clean energy production [4]. The soil microbes that live in the root area are very useful for electricity generation as they break the carbohydrates from the trees and can produce electrons [3]. These geobacteria serve as a biocatalyst resource for PMFCs [4]. Through placing an electrode near the roots, the root exudates can be oxidized by electro-active bacteria, which can transfer their electrons from anode to cathode [5–7]. Recently, the generation of bioelectricity *via* microbial fuel cells (MFCs) has been gaining notice due to its greener nature and simultaneously resolving the environmental problems. MFCs produce bioelectricity *via* the metabolic actions of electrochemically active microbes.

PMFCs are microbial fuel cells consisting of living plants, a supporting matrix, and a conductive anode inserted into the substrate and cathode placed in air or water to convert chemical energy into bioelectricity [8–13]. According to Moqsud et al., two basic designs of PMFCs can be distinguished, namely, aquatic and nonaquatic plant microbial fuel cells [14–18]. Based on these two designs, other models such as tubular and flat-plate PMFCs have been developed and successfully applied [18–25].

2. Advantages of plant microbial fuel cell

Plant microbial fuel cell is a sustainable source of electricity because they will not destroy any food products such as corn or soybeans for bioenergy production [26–30]. Both bioenergy and food can be received at the same time in this innovative system. So, there will be no competition between the green energy source and the food source for humans and animals [31–35]. Secondly, it is a renewable source of energy as only sunlight is used to generate bioelectricity with the help of living plants. There are trees all over the world, and the photosynthesis of these trees is the main source of energy for this system. Finally, unlike fossil fuel, it is a completely environmentally friendly approach to electricity generation. So, there is no further fear of greenhouse gas emissions and consequent global warming in the future. Plant microbial fuel cells can change the future need for green energy and a sustainable future by achieving sustainable development goals.

The other advantages of plant microbial fuel cells are that it is a carbon-neutral or carbon-negative approach for electricity generation, which is the future goal for energy harvesting [36–40].

Plant microbial fuel cells are not only a green way to generate electricity but also a safe way to get electricity. So, there is no danger of getting electricity from the living plants. PMFC can be used for wastewater treatment and bioremediation or phytoremediation of the contaminated soil and environment [41–45].

3. Mechanism of plant microbial fuel cells

The plant microbial fuel cell is a novel technology in which organic matter is converted into electricity using living plants and bacteria in the soil [46–50]. The green leaves of the plants can generate sugar, which is an organic matter due to photosynthesis. The plants use the carbon dioxide from the air and water from the soil to make the organic matter in the green leaves [51–56]. However, about 60% of this organic matter has been deposited in the soil near the root zone. This sugar or organic matter is the food of the bacteria that live in the soil. So, the geobacteria and others gather in the root area and eat the sugar/organic matter and increase the number. While breaking down the organic matter, a significant amount of electrons is released from it. The anode and cathode connected with the external circuit can catch this electron and consequently get the electricity.

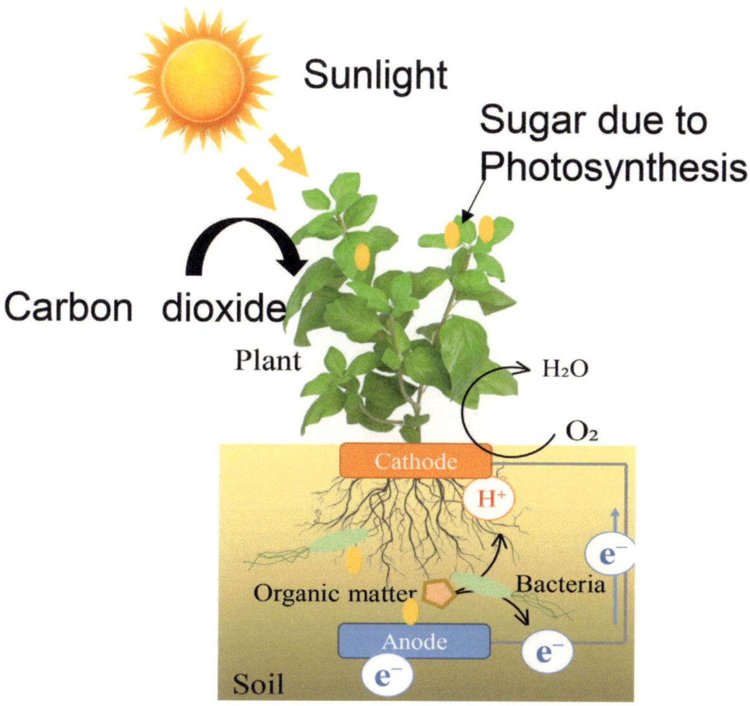

Figure 1.
Mechanism of plant microbial fuel cell.

This is the main mechanism of plant microbial fuel cells. **Figure 1** shows the mechanism of plant microbial fuel cells.

4. Types of plants used in PMFC research

Different types of plants have been used in the research so far in plant microbial fuel cells [51–60]. However, most of the plants are aquatic plants that use a lot of water for their growth such as rice plants [61]. **Table 1** summarizes the amount of power generation by using plant microbial fuel cells using different types of plants. It has been seen that it can be used for electricity generation for a long time while using different types of plants. The reason behind it is that bacteria can get food from living plants continuously. However, many factors can affect the bioelectricity production from the plant microbial fuel cells [62]. Weather factors are one of the main factors that can influence bioelectricity generation in cold regions. The bacterical activities are reduced during the lower temperature, and hence, the bioelectricity amount was also reduced [63].

For aquatic plants, the type of water has an influence on the power generation in PMFCs. It can be seen that seawater plants can generate more bioelectricity than freshwater plants. The reason behind this is that seawater is a more electricity-conductive material compared with fresh water. Soil properties are another important factor for bioelectricity generation. The organic soil shows more bioelectricity than the sandy soil as the organic soil contains more microbes and food for bacteria.

Name of plant species	Duration of operation	Highest Power Density (PD) recorded and reported
Canna indica	140 Days	18 mW/m^2
Canna indica	N/A	320.08 mW/m^3
Typha latifolia	N/A	6.12 mW/m^2
Physcomitrella patens	70 Days	6.7 ± 0.6 mW/m^2
Lolium perenne	70 Days	55 mA/m^2
Oryza sativa	N/A	140 mA/m^2
Sporobolus arabicus	2 Months (60 Days)	120 mW/m^2
Typha latifolia	228 Days	93 mW/m^3
Elodea nuttallii	276 Days	184.8 ± 7.5 mW/m^3
Canna indica	90 Days	5.11 mW/m^3
Phragmites australis	3 months (90 Days)	0.15 mW/m^3
Phragmites australis	160 Days	22 mW/m^2
Spartina anglica	160 Days	82 mW/m^2
Cyperus involucratus R.	5 Days	5.99 mW/m^2
Brassica juncea	30 Days	69.32 mW/m^2
Trigonella foeumgraecum	30 Days	80.26 mW/m^2
Canna stuttgart	30 Days	222.54 mW/m^2
Sedum album	360 Days	0.0024 $\mu W/m^2$
Sedum sexangulare	360 Days	0.0084 $\mu W/m^2$
Sedum rupestre	360 Days	0.0155 $\mu W/m^2$
Sedum hybridum	360 Days	0.092 $\mu W/m^2$
Sedum reflexum	360 Days	>0.001 $\mu W/m^2$
Sedum kamtschaticum	360 Days	>0.001 $\mu W/m^2$
Sedum spurium	360 Days	>0.001 $\mu W/m^2$
Acorus tatarinowii	51 Days	21 mW/m^2
Vetiveria zizanioides Nash	N/A	242 ± 10.5 mA/m^2
Puccinellia distans	114 Days	83.7 mW/m^2
Chlorophytum comosum	100 Days	18 mW/m^2
Epipremnum aureum	60 Days	15.38 mW/m^2
Dracaena braunii	60 Days	12.78 mW/m^2
Chasmanthe floribunda	100 Days	0.21 mW/m^2
Papyrus cyperus	100 Days	1.083 mW/m^2
Chlorophytum comosum	100 Days	18 mW/m2
Spartina anglica	105 Days	1.04 mW/m2

Table 1.
Different plants and their ability to generate electricity.

5. Future challenges in plant microbial fuel cell

Various technological and feasibility challenges are still there in plant microbial fuel cell application. The low voltage generation and the stability of voltage generation are the two main challenges for the future application of plant microbial fuel cell. Some plants have a short lifespan and can die if there are some lacks water and fertilizers. To resolve these problems, more research will be required. High power generation is another serious problem. Most of the small plants cannot generate high power. So, finding out the best plant is another challenge for practical application of plant microbial fuel cell. Most of the plant species used in the plant microbial fuel cells are wild grass type and do not have good economic value. However, using rice plants and wheat plants can change this situation. Some of the researchers are using vegetable plants and can generate electricity successfully. Another most important factor for future challenge is the internal resistance of the plant microbial fuel cell. Biofilm can be generated on the surface of the anode materials and can reduce the conductivity of the electricity. So, in the future research, this problem should be addressed properly. The second challenge for PMFC research is the standard design. It should be standardized the electrode materials, their shapes, sizes and other electrical properties. Most of the researchers are using different types of materials and different shapes for the electrode materials. This causes a serious problem to compare the data and the future development of this method. So, in some point, the standard materials should be used to support the future global development of this power-generating system. The reported plants, electrode materials, and design are different for plant microbial fuel cell that cause a serious problem for the future development of it. So, some experimental protocol should be developed.

6. Future aspects of PMFC

Plant microbial fuel cell is a promising technology for green energy harvesting. Bioelectricity can be generated from various types of plants, which may change the game of energy demand in the world soon. However, a lot of research is required to fulfill the needs of the optimum design and the sustainability of this system in different weather and different places in the practical application. The future of green energy depends on the final output of power generation and the standard design of the total system.

PMFC can be used in mangrove vegetation areas and remote areas for off-grid power supply.

PMFC can be used in the forest and in the slope areas to get more energy for powering the environmental monitoring sensors.

7. Conclusions

Plant microbial fuel cell is one of the new methods for getting green energy for future generations. Plants are available all over the world, and they can be used as a source of green bioelectricity. This system can change the world's energy demand

and the sustainable source of green energy. PMFC can also help to achieve carbon neutrality in the future. However, there is still a long way to go to use it for a practical purpose. The long life span trees and other plants should be introduced rather than grass type of plants only for the better outcome. PMFC can be a source of green energy in which both food and energy can be achieved simultaneously, which will make this new bioenergy system more sustainable and environment friendly. The effect of natural disasters on bioelectricity generation in PMFCs should be studied further for emergency uses as a source of electricity and green energy supply during the most needed time at evacuation shelters and distressed areas. However, most importantly, 25% of the world's population is still living in the dark at night due to the lack of electricity, and PMFC can give a "Light of Hope" to them.

Conflict of interest

The author declares no conflict of interest.

Author details

M. Azizul Moqsud
Department of Civil and Environmental Engineering, Yamaguchi University, Ube City, Japan

*Address all correspondence to: azizul@yamaguchi-u.ac.jp

References

[1] Moqsud MA, Okamoto T. Bioelectricity generation from corn plant microbial fuel cell (PMFC) using natural bamboo charcoal as electrodes and *Shewanella oneidensis*. Bioresource Technology Reports. 2023;**24**:101611

[2] Moqsud MA. Bioelectricity generation and remediation of sulfide-contaminated tidal flat sediment. International Journal of Sediment Research. 2020;**35**(1):91-96

[3] Sekar N, Ramasamy RP. Recent advances in photosynthetic energy conversion. Journal of Photochemistry and Photobiology C. 2015;**22**:19-33

[4] Moqsud MA, Omine K, Yasufuku N, Bushra Q, Hyodo M, Nakata Y. Bioelectricity from kitchen and bamboo waste in a microbial fuel cell. Waste Management & Research. 2014;**32**(2):124-130

[5] Powell RJ, White R, Hill RT. Merging metabolism and power: Development of a novel photobioelectric device driven by photosynthesis and respiration. PLoS One. 2014;**9**(1):e86518

[6] Strik DP, Hamelers HVM, Snel JF, Buisman CJ. Green electricity production with living plants and bacteria in a fuel cell. International Journal of Energy Research. 2008;**32**(9):870-876

[7] Helder M, Strik DPBTB, Hamelers HVM, Kuhn AJ, Blok C, Buisman CJN. Concurrent bio-electricity and biomass production in three plant-microbial fuel cells using Spartina anglica, Arundinella anomala, and Arundo donax. Bioresource Technology. 2010;**101**(10):3541-3547

[8] Wetser K, Sudirjo E, Buisman CJ, Strik DP. Electricity generation by a plant microbial fuel cell with an integrated oxygen-reducing biocathode. Applied Energy. 2015;**137**:151-157

[9] Moqsud MA, Yoshitake J, Bushra QS, Hyodo M, Omine K, Strik D. Compost in plant microbial fuel cell for bioelectricity generation. Waste Management. 2015;**36**:63-69

[10] De Schamphelaire L, Cabezas A, Marzorati M, Friedrich MW, Boon N, Verstraete W. Microbial community analysis of anodes from sediment microbial fuel cells powered by rhizodeposits of living rice plants. Applied and Environmental Microbiology. 2010;**76**(6):2002-2008

[11] Moqsud MA, Omine K, Yasufuku N, Hyodo M, Nakata Y. Microbial fuel cell (MFC) for bioelectricity generation from organic wastes. Waste Management. 2013;**33**:2465-2469

[12] Moqsud MA, Gochi T. Biocementation of slope soil by using native Cytobacillus horneckiae. Bioresource Technology Reports. 2023;**23**:101520

[13] Yadav AK, Dash P, Mohanty A, Abbassi R, Mishra BK. Performance assessment of innovative constructed wetland-microbial fuel cell for electricity production and dye removal. Ecological Engineering. 2012;**47**:126-131

[14] Xu L, Zhao Y, Doherty L, Hu Y, Hao X. The integrated processes for wastewater treatment based on the principle of microbial fuel cells: A review. Critical Reviews in Environmental Science and Technology. 2016;**46**(1):60-91

[15] Bombelli P, Dennis RJ, Felder F, Cooper MB, Iyer DMR, Royles J, et al.

Electrical output of bryophyte microbial fuel cell systems is sufficient to power a radio or an environmental sensor. Royal Society Open Science. 2016;**3**(10):160249

[16] Takanezawa K, Nishio K, Kato S, Hashimoto K, Watanabe K. Factors affecting electric output from rice-paddy microbial fuel cells. Bioscience, Biotechnology, and Biochemistry. 2010;**74**(6):1271-1273

[17] Sudirjo E, Buisman CJN, Strik DPBTB. Electricity Generation from Wetlands with Activated Carbon Bioanode IOP Institute of Physics Publishing Conference Series: Earth and Environmental Science; 2018 March. Bristol, UK: IOP Publishing; 2018. p. 012046

[18] Helder M, Chen WS, Van Der Harst EJ, Strik DP, Hamelers HBV, Buisman CJ, et al. Electricity production with living plants on a green roof: Environmental performance of the plant-microbial fuel cell. Biofuels, Bioproducts and Biorefining. 2013;**7**(1):52-64

[19] Wetser K, Liu J, Buisman C, Strik D. Plant microbial fuel cell applied in wetlands: Spatial, temporal and potential electricity generation of Spartina anglica salt marshes and Phragmites australis peat soils. Biomass and Bioenergy. 2015;**83**:543-550

[20] Helder M, Strik DP, Timmers RA, Raes SM, Hamelers HV, Buisman CJ. Resilience of roof-top plant-microbial fuel cells during Dutch winter. Biomass and Bioenergy. 2013;**51**:1-7

[21] Tapia NF, Rojas C, Bonilla CA, Vargas IT. Evaluation of sedum as driver for plant microbial fuel cells in a semi-arid green roof ecosystem. Ecological Engineering. 2017;**108**:203-210

[22] Deng H, Chen Z, Zhao F. Energy from plants and microorganisms: Progress in plant–microbial fuel cells. ChemSusChem. 2012;**5**(6):1006-1011

[23] Nitisoravut R, Regmi R. Plant microbial fuel cells: A promising biosystems engineering. Renewable and Sustainable Energy Reviews. 2017;**76**:81-89

[24] Strik DP, Timmers RA, Helder M, Steinbusch KJ, Hamelers HV, Buisman CJ. Microbial solar cells: Applying photosynthetic and electrochemically active organisms. Trends in Biotechnology. 2011;**29**(1):41-49

[25] Regmi R, Nitisoravut R, Ketchaimongkol J. A decade of plant-assisted microbial fuel cells: Looking back and moving forward. Biofuels, Bioproducts and Biorefining. 2018;**9**:1-8

[26] Cheng S, Liu W. Microbial fuel cells and other bio-electrochemical conversion devices. In: Kwong-Yu C, Chi-Ying VL, editors. Electrochemically Enabled Sustainability. Boca Raton: CRC Press; 2014. pp. 66-131

[27] Kokabian B, Gude VG. Photosynthetic microbial desalination cells (PMDCs) for clean energy, water and biomass production. Environmental Science. Processes & Impacts. 2013;**15**(12):2178-2185

[28] Rabaey K, Lissens G, Siciliano SD, Verstraete W. A microbial fuel cell capable of converting glucose to electricity at high rate and efficiency. Biotechnology Letters. 2003;**25**(18):1531-1535

[29] Min B, Logan BE. Continuous electricity generation from domestic wastewater and organic substrates in a flat plate microbial fuel cell. Environmental Science & Technology. 2004;**38**(21):5809-5814

[30] Chiranjeevi P, Yeruva DK, Kumar AK, Mohan SV, Varjani S. Plant-microbial fuel cell technology. Microbial Electrochemical Technology. Elsevier; 2019. pp. 549-564

[31] Kwong-Yu C. Chi-Ying Vanessa Li, Editors. Electrochemically Enabled Sustainability: Devices, Materials and Mechanisms for Energy Conversion. first ed. Boca Raton, Florida: CRC Press; 2014

[32] Kouzuma A, Kaku N, Watanabe K. Microbial electricity generation in rice paddy fields: Recent advances and perspectives in rhizosphere microbial fuel cells. Applied Microbiology and Biotechnology. 2014;**98**(23):9521-9526

[33] Wolińska A, Stępniewska Z, Bielecka A, Ciepielski J. Bioelectricity production from soil using microbial fuel cells. Applied Biochemistry and Biotechnology. 2014;**173**(8):2287-2296

[34] Jiang D, Li B, Jia W, Lei Y. Effect of inoculum types on bacterial adhesion and power production in microbial fuel cells. Applied Biochemistry and Biotechnology. 2010;**160**(1):182

[35] Liu H, Cheng S, Logan BE. Production of electricity from acetate or butyrate using a single-chamber microbial fuel cell. Environmental Science & Technology. 2005;**39**(2):658-662

[36] Roesch LF, Fulthorpe RR, Riva A, Casella G, Hadwin AK, Kent AD, et al. Pyrosequencing enumerates and contrasts soil microbial diversity. The ISME Journal. 2007;**1**(4):283

[37] Goto Y, Yoshida N, Umeyama Y, Yamada T, Tero R, Hiraishi A. Enhancement of electricity production by graphene oxide in soil microbial fuel cells and plant microbial fuel cells. Frontiers in Bioengineering and Biotechnology. 2015;**3**:42

[38] Timmers RA, Strik DP, Hamelers HV, Buisman CJ. Long-term performance of a plant microbial fuel cell with Spartina anglica. Applied Microbiology and Biotechnology. 2010;**86**(3):973-981

[39] Mohan SV, Mohanakrishna G, Chiranjeevi P. Sustainable power generation from floating macrophytes based ecological microenvironment through embedded fuel cells along with simultaneous wastewater treatment. Bioresource Technology. 2011;**102**(14):7036-7042

[40] Helder M, Strik DPBTB, Hamelers HVM, Kuijken RCP, Buisman CJN. New plantgrowth medium for increased power output of the plant-microbial fuel cell. Bioresource Technology. 2012;**104**:417-423

[41] Liu S, Song H, Li X, Yang F. Power generation enhancement by utilizing plant photosynthate in microbial fuel cell coupled constructed wetland system. International Journal of Photoenergy. 2013;**172010**:10https. DOI: 10.1155/2013/172010

[42] Habibul N, Hu Y, Wang YK, Chen W, Yu HQ, Sheng GP. Bioelectrochemical chromium (VI) removal in plant-microbial fuel cells. Environmental Science & Technology. 2016;**50**(7):3882-3889

[43] Ali H, Khan E, Sajad MA. Phytoremediation of heavy metals-concepts and applications. Chemosphere. 2013;**91**(7):869-881

[44] Förstner U, Wittmann GT. Metal Pollution in the Aquatic Environment. second ed. Berlin Heidelberg: Springer-Verlag; 2012. p. 475

[45] Kadlec RH, Wallace S. Treatment Wetlands. second ed. Boca Raton, FL: CRC press; 2008. p. 1046

[46] Regmi R. Examining different classes of plants under various operating conditions for bioelectricity production. In: Plant microbial fuel cell [dissertation]. Thailand: Thammasat University; 2017

[47] Hubenova Y, Mitov M. Conversion of solar energy into electricity by using duckweed in direct photosynthetic plant fuel cell. Bioelectrochemistry. 2012;**87**:185-191

[48] Gazali T, Moqsud MA. The effectiveness of animal dungs and leaf Mold for bioelectricity generation using microbial fuel cells with soils. Journal of Sustainable Bioenergy Systems. 2017;**7**(04):165-169

[49] Li H, Qu Y, Tian Y, Feng Y. The plant-enhanced bio-cathode: Root exudates and microbial community for nitrogen removal. Journal of Environmental Sciences. 2019;**77**:97-103

[50] Jethwa KB, Bajpai S. Role of plants in constructed wetlands (CWS): A review. Journal of Chemical and Pharmaceutical Sciences. 2016;**974**:2115

[51] Xu B, Ge Z, He Z. Sediment microbial fuel cells for wastewater treatment: Challenges and opportunities. Environmental Science: Water Research & Technology. 2015;**1**(3):279-284

[52] Timmers RA, Strik DP, Arampatzoglou C, Buisman CJ, Hamelers HV. Rhizosphere anode model explains high oxygen levels during the operation of a Glyceria maxima PMFC. Bioresource Technology. 2012;**108**:60-67

[53] Valipour A, Ahn Y-H. Constructed wetlands as sustainable ecotechnologies in decentralization practices: A review. Environmental Science and Pollution Research International. 2016;**23**(1):180-197

[54] Eynard A, Lal R, Wiebe K. Crop response in salt-affected soils. Journal of Sustainable Agriculture. 2005;**27**(1):5-50

[55] Strik DP, Terlouw H, Hamelers HV, Buisman CJ. Renewable sustainable biocatalyzed electricity production in a photosynthetic algal microbial fuel cell (PAMFC). Applied Microbiology and Biotechnology. 2008;**81**(4):659-668

[56] Moqsud MA. Bioelectricity from organic solid waste. In: Strategies of Sustainable Solid Waste Management. London, UK, London, UK: IntechOpen; 2021. p. 129

[57] Narula D. Generation of Electricity Using Spartina patens with Stainless Steel Current Collectors in a Plant-Microbial Fuel Cell [Dissertation]. Lamar: University Beaumont; 2017

[58] Chen Z, Huang YC, Liang JH, Zhao F, Zhu YG. A novel sediment microbial fuel cell with a biocathode in the rice rhizosphere. Bioresource Technology. 2012;**108**:55-59

[59] Kaku N, Yonezawa N, Kodama Y, Watanabe K. Plant/microbe cooperation for electricity generation in a rice paddy field. Applied Microbiology and Biotechnology. 2008;**79**(1):43-49

[60] Moqsud MA, Bushra QS, Rahman MH. Composting barrel for sustainable organic waste management in Bangladesh. Waste Management & Research. 2011;**29**(12):1286-1293

[61] Moqsud MA, Gazali T. Green electricity by water plants in organic soil and marine sediment through microbial fuel cell. Energy Sources. 2017;**39**:160-165

[62] Moqsud MA, Hayashi S. An evaluation of solid waste management practice in Japan. Daffodil International

University Journal of Science and Technology. 2006;**1**:39-44

[63] Halan B, Tschörtner J, Schmid A. Generating electric current by bioartificial photosynthesis. In: Scheper TH, Belkin S, Bley TH, Bohlmann J, Gu MB, Hu W-S, editors. Advances in Biochemical Engineering/ Biotechnology. Berlin, Heidelberg: Springer; 2018. pp. 1-33